Recreational Fisheries of Coastal New England

Recreational Fisheries of Coastal New England

BY

Michael R. Ross

WITH CONTRIBUTIONS BY

Robert C. Biagi

Drawings by Nancy Haver

University of Massachusetts Press
AMHERST

Printed in the United States of America
LC 90-49253
ISBN 0–87023–742–X (cloth); 743–8 (pbk)

Designed by Jack Harrison
Set in Sabon by Keystone Typesetting, Inc.
Printed and bound by Thomson-Shore, Inc.

Library of Congress Cataloging-in-Publication Data
Ross, Michael R., 1947–
Recreational fisheries of coastal New England / Michael R. Ross,
with contributions by Robert C. Biagi; drawings by Nancy Haver.
p. cm.
Includes bibliographical references and index.
ISBN 0–87023–742–X (alk. paper). — ISBN 0–87023–743–8 (pbk. : alk. paper)
1. Fishery resources—New England. 2. Fishes—New England.
3. Fishing—New England. 4. Cookery (Fish)
I. Biagi, Robert C. II. Title.
SH 221.5.N4R67 1991
799.1'6'0974—dc20 90–49253
CIP
British Library Cataloguing in Publication data are available

Contents

Preface

Books about the fishes of a geographic region have been written by a variety of ichthyologists and fish ecologists. Although many of these books are scholarly in nature, some are written in a manner appealing to the interested lay person as well as the student and professional. Few of these books focus on the fisheries of a region; that is, few present information not only on identification, distribution, and biology of the fishes, but also on the manner and variety of ways that humans interact with them. Those books that do, often address only the lay audience and may not possess substantial material of interest to the student or professional.

Perhaps the "weakest link" in the university training of fisheries professionals is the tendency to graduate students who, although well schooled in biological principles, may not possess broad perspectives on the fisheries management process itself. For example, students do not always have a well-grounded understanding of the major legislation that has shaped the direction and scope of fisheries management. Anglers also would benefit from knowing more about the management process and about how their interests are influenced by it, but such information is not readily available in a format that is easily read and understood.

I have written this book for the students, fisheries managers, anglers, conservationists, and lay naturalists who are fascinated by coastal fishes as integral parts of our natural world and who are interested in how these highly valued natural resources are used and managed by humans. Also, Robert Biagi has contributed information based on his experiences in catching, handling, and cooking fishes, which is specifically designed for the angler or for any other person who simply enjoys the culinary merits of fishes.

Although scientific names for fish species are standard, many of these fishes have been given a variety of common names. Alternative names are presented for many of the species profiled in this book, so that readers can more easily relate the descriptions to names they recognize as familiar. There also is no universally used standard for plural forms of common names of many fishes, nor is there for the word "fish" itself; for example, "hake" may be used as a plural form in some publications, while "hakes" is used in others. The standard followed by the American Fisheries Society, as developed from *Webster's Third New International Dictionary of the English Language Unabridged,* and as updated by the ninth edition of the *Collegiate Dictionary,* is used for all species discussed in this book. Additionally, the term "fish" is used when referring to more than one individual of the same species, whereas "fishes" is used when referring to individuals in more than one species.

The study of fishes and the way in which they are conserved and managed is a highly technical field, but I have tried to present the information in this book in a way that will stimulate understanding by the various people whose vocation or avocation involves pursuing and/or learning more about these aesthetically and economically important natural resources.

I greatly appreciate the assistance given me by many colleagues and friends during development of this book. Steve Murawski and Larry Nielsen provided thoughtful comment on format and content in the initial stages of developing the book's scope. Laura Thorpe and Gary Nelson helped develop some of the graphics and researched some materials used in the species profiles. Various people from the Northeast Fisheries Center (NMFS/NOAA), the Sport Fishing Institute, the Atlantic States Marine Fisheries Commission, the South Atlantic, Middle Atlantic, and New England Fishery Management councils, and the New England state agencies responsible for coastal fisheries management provided information used in the species profiles. In preparing the line drawings of fishes, Nancy Haver consulted the following sources: Henry B. Bigelow and William C. Schroeder, *Fishes of the Gulf of Maine* (Washington, D.C.: GPO, 1953); Office of Biological Services, Fish and Wildlife Service, U.S. Department of the Interior, *Development of Fishes of the Mid-Atlantic Bight,* vols. 2, 4, 5, and 6; A. J. McClane, ed., *McClane's Field Guide to Saltwater Fishes of North America* (New York: Holt, Rinehart and Winston, 1978); Peter Thompson, *The Game Fishes of New England and Southeastern Canada* (Camden, Maine: Down East, 1980); Patricia E. Pope, *A Dictionary of Sharks* (St. Petersburg, Fla.: Great Outdoors Publishing, 1973); W. B. Scott and M. G. Scott, *Atlantic Fishes of Canada,* Canadian Bulletin of Fisheries and Aquatic Sciences No. 219; C. Richard Robins, *A Field Guide to Atlantic Coast Fishes of North America,* Peterson Field Guide Series (Boston: Houghton Mifflin, 1986); and Joy O. I. Spoczynska, *An Age of Fishes* (New York: Charles Scribner's Sons, 1976).

Recreational Fisheries of Coastal New England

Introduction

The Roots of Recreational Fishing

Angling has roots that reach well back into ancient times, when fishing was "a sport of kings and the wealthy" (Royce 1987). The Sumerians (of the Persian Gulf) organized harvest of fisheries resources very early, with hundreds of people fishing in an organized system for the temples (in that society, the government). Such early fishing efforts cannot be classified as recreation, as they were conducted for the benefit, and under the orders, of others. Control over ownership by the temples was strictly enforced; poachers faced severe punishment for their indiscretion (Royce 1987). Fishing became a recreation very early in some cultures. Egyptian drawings of more than 3,400 years ago show leaders engaged in fishing; one can assume they were fishing not for subsistence but rather for pleasure. The Romans were first to propose that fish resources were a common property that could be "owned" only after they were caught. Roman law stated that waterways were public, and no one could be prevented from fishing in them. However, ownership by the aristocracy of waters and the fishes within them continued throughout all but recent history in many Old World societies.

Ownership issues were commonly less rigid in colonial America, but for many people fishing was still a subsistence rather than a recreational activity. Over the last two centuries, however, with the advent of more affordable equipment, ready access to fishing grounds, and increased leisure time, angling has evolved into one of the most popular outdoor recreational activities in the United States. In 1970, one out of every three U.S. men and one out of every nine U.S. women fished. The number of anglers in the United States increased 31% between 1960 and 1970 (Everhart et al. 1975), and continued to grow at a rate of about 3% annually through the 1980s (Royce 1987).

Not only is recreational fishing aesthetically important to an individual's quality of life, it can also have a major impact on local and regional economies. Anglers spent more than $7.2 billion on marine recreational fishing in the United States in 1985. Nearly $500 million was spent to fish in New England coastal waters alone, with about $240 million of additional economic activity occurring due to these expenditures (unpublished report of the Sport Fishing Institute, 1988). Nationally, expenditures by coastal anglers can exceed the value of all commercially harvested marine fisheries resources (Royce 1987).

Why Do We Fish?

Anglers fish for a number of reasons. As in the days when angling was primarily a subsistence activity, fishes are still valued as food by many anglers. However, other experiences associated with angling are often more highly valued than eating, or even catching, fishes. Anglers typically list a variety of reasons for fishing, which may include: (1) temporarily escaping from the daily schedule, (2) relaxing and sharing an experience with family and friends, (3) enjoying time in clean, natural surroundings, (4) experiencing adventure and excitement in the outdoors, (5) meeting the challenge of trying to catch fishes, (6) catching fishes, and (7) eating fishes. Some analyses of angler preferences and motivations indicate that many people who fish consider catching and eating fishes secondary to the other goals listed above (as in Dawson and Wilkins 1981). The order of importance of motivations and goals will generally vary among anglers and according to species of fishes pursued. Note that, to many anglers, meeting the challenge of trying to catch fishes is considered apart from, and perhaps more important than, actually catching them. Indeed, the average angler makes decisions in his or her selection of equipment and gear that may significantly reduce the likelihood of success. "People [often] accept challenges to catch fish in the most difficult way possible: with fragile equipment, with tiny lures, . . . and with lines or leaders having a strength of only a small fraction of the weight of the fish." Fishing provides them with "a personal satisfaction unique to their own [personal] challenges" (Royce 1987). As one writer aptly phrased it,

> Fortunately, I learned long ago that although fish do make a difference—*the difference*—in angling, catching them does not; so that he who is content to not catch fish in the most skillful and refined manner, utilizing the best equipment and technique, will have his time and attention free for the accumulation of a thousand experiences, the memory of which will remain for his enjoyment long after any recollection of fish would have faded. (Miller 1971)

Recreational fishing is a pleasure to many and a passion to some. Because of their enthusiasm for this leisure activity, anglers have generally supported conservation practices that may impinge upon the level of their success on a given day, in order that they and others may enjoy the experience of angling in the future. They have been willing to pay "user taxes" to fund these conservation efforts (see chapter 3). Not surprisingly, this support has varied from time to time and place to place, so that the history of angler support of conservation is fraught with examples of selfishness and personal gain, as well as filled with examples of consideration for protecting the fishing experience for other anglers, both present and future.

What Is a Fishery?

A fishery includes not only a fish population and the complex of interactions occurring between it and its natural environment but also those occurring between it and humans. Thus, the description of a recreational fishery specifically includes the effects that angling has upon a fish population. All New England fish species that are landed by anglers are affected by humans in a variety of other ways. Development, industrialization, urbanization, and intensive agricultural use of coastal lands severely affect a variety of fish species by polluting or changing critical characteristics of their habitat. Species that migrate into coastal rivers and streams as a critical part of their life cycle are often cut off from these habitats entirely by construction of dams and other obstructions. Many New England species that are pursued recreationally are also targets of intensive harvest by commercial fishing interests. Fisheries managers responsible for conservation of fish species that provide recreational pleasure to the angling public often face a multitude of issues related to human interactions with these natural resources; thus, economic, cultural, and sociological considerations are usually entwined within the conservation decision-making process.

This book presents information on New England's coastal recreational fisheries. Part 1 considers coastal habitats, the ecological roles fish species fill in these coastal environments, and the management and conservation processes that have developed to encourage the wise use of these important natural resources. Part 2, the largest part of the book, reviews forty-four specific recreational fisheries, focusing upon the life cycle, general ecology, and management of the fish species associated with these fisheries, as well as useful angling, handling, and cooking information for the angler who is an avid participant in this region's coastal recreational fishing activities.

I

FISHES, THEIR ENVIRONMENT, AND THEIR MANAGEMENT

1 Coastal Habitats

The average depth of the ocean is about 3,800 m (12,467 ft), with nearly half of the ocean floor lying about 3,000 to 6,000 m below the water's surface (Reid and Wood 1976). Particular regions of the ocean's floor are thrust upward into expansive mountain ranges, the peaks of a few breaking above the water as oceanic islands, while most remain submerged in the ocean's depths. Deep furrows, or troughs, which often occur parallel to the ocean's mountains, form the deepest areas. The Puerto Rican trough has a maximum depth of nearly 9,220 m (30,242 ft; Reid and Wood 1976). The continental shelves, submerged extensions of continents that are less than 200 m deep, support the greatest production of fisheries resources worldwide. The continental shelf is more expansive in the Atlantic Ocean than in the Pacific and is particularly broad in the Northwest Atlantic, forming the large fishing banks of New England and the Canadian Maritime Provinces, including Georges Bank off New England. This chapter provides a brief description of the various coastal habitats housing the fishes that contribute to the recreational fisheries of New England.

Estuaries

Estuaries are inshore basins where fresh water mixes with and dilutes salt water (Reid and Wood 1976). Barrier beaches built up by the deposition of sand and gravel typically occur at the estuary's opening into the open coastline. These beach ridges protect the estuary from the full force of tidal currents and waves that strike the coast's shore areas. This protection makes estuaries quiet bodies of water in which the incoming flow of inland streams and tides is slowed.

Many New England estuaries were formed as slowly rising sea levels flooded the mouths and adjacent valleys of coastal rivers after the last major glaciers receded from the region (Reid and Wood 1976). Due to the configuration of the coastal area, estuaries tend to be smaller north of Cape Cod than in southern New England, although several systems in northern New England, such as the Scarboro (ME), Hampton (NH), and Parker River (MA) marshes are large (Teal 1986).

Once an estuary is formed, it slowly fills with sediment, losing its deeper waters and displaying changes in the occurrence and expanse of particular habitat types; thus, estuaries are areas of major sediment deposition, from both incoming tides and outflowing streams. As rising tides carry over barrier beaches, their flow is slowed to the point where large particles such as sand and small gravel can no longer remain suspended in the moving water, thus settling to the substrate. As tidal currents continue into the estuary, their flow is further slowed due to the drag caused by the bottom of the shallow basin and to the counterdirectional flow of fresh water entering the estuary from streams. Due to this further decrease in the rate of flow, smaller silt and clay particles that can no longer be carried by the water's movement also sink to the bottom. Likewise, the velocity of a stream is slowed by the opposing force of tides and the width of the area the stream flows through as it leaves the stream channel and enters the estuary. This process also causes settling of suspended particles. Sediment deposited at the mouth of an estuary is dominated by coarse sand and gravel; sediment farther into the estuary is dominated by silt and clay, forming soft-mud substrates (Nixon 1982).

Tidal intrusions affect other characteristics of an estuary, creating markedly different habitat conditions from site to site. Also, the oscillating nature of tides creates constantly changing conditions at any particular site. Tides are semidiurnal in New England, cycling from one high tide to the next about every 12 hours and 25 minutes. Tidal amplitude (the change in water level between high and low tide) is determined by the shape and bottom contour of the estuary, the size of the estuary's mouth, and other physical factors of the area and region, as well as the phase of the moon (gravitational forces of the moon and sun create greater tidal forces, and thus higher tides, during full and new moon phases than at other times of the lunar cycle). Generally, tidal amplitude is greater north of Cape Cod (averaging about 3 to 4 m, but as extreme as 7 m in northern Maine) than it is in southern New England (about 1 to 1.5 m; Whitlatch 1982). Unless dammed, streams flowing into estuaries do not display major changes in discharge within a day but will do so seasonally. These daily and seasonal changes in the amount of freshwater and saltwater input create extremely variable salinity and temperature conditions in an estuary, not only from site to site but also from time to time.

In many estuaries, lighter fresh water flows seaward over denser, inflowing tidal salt water (salt water has a higher density due to the dissolved mineral salts; Reid and Wood 1976). Thus, differences in salinity occur vertically in the water column. Incoming high tides flow along the estuary's bottom farther inland than do low tides; thus, the salinity above the sediments at particular sites shifts markedly during the tidal cycle. In other estuaries, vertical differences in salinity may be less marked due to mixing of the fresh and salt waters. In such instances, the salinity throughout the water column at many given sites within the estuary is constantly shifting.

Differences in temperature of fresh and salt waters entering estuaries cause similar fluctuations in temperature in the areas of mixing. Due to its much smaller volume, water in coastal streams responds much more rapidly to daily and seasonal changes in air temperature than does water along the coast. Generally, upper waters of an estuary will be cooler in the winter and warmer in the summer than waters near the estuary's mouth (Reid and Wood 1976).

Thus, tidal activity creates a dynamic system that is constantly fluctuating, causing the development of several different habitat types arranged sequentially from the inland edge of the estuary to its mouth. Estuarine habitats are typically classified according to physical and chemical conditions, and to the dominant vegetation that occurs in areas with particular sets of those conditions. Major habitat types, listed from the inland edge of an estuary to its mouth, include the high salt marsh, the low salt marsh, mud flats, and deeper channels and areas that are generally inundated with water. A brief description of each of these habitats follows.

The high salt marsh. This portion of the estuary is unflooded much of the time but is affected by high waters of the tidal cycle. The boundaries of the high salt marsh are most frequently defined by the distribution of vegetation in this habitat rather than by environmental conditions established by tides. Upland from the high marsh may be a narrow band of freshwater wetland vegetation (such as cattail and reeds; Nixon 1982), which grows on substrates unflooded by tides and separates the high salt marsh from typical upland tree and shrub vegetation that is intolerant of salt water. The high salt marsh in New England is dominated by salt-marsh hay (*Spartina patens*) and perhaps to a lesser extent by spike grass (*Distichlis spicata*). The distribution of these plants is generally used by ecologists to describe the area of the high marsh.

Salt-marsh hay was responsible for making coastal marshes an integral part of the colonial New Englander's "life-support system." This plant was a critical source of food for the early settler's livestock. Its harvest was so important to successful livestock rearing that site selection for new townships prior to 1650 was largely influenced by the distribution of high salt marshes (Nixon 1982).

Substrates in high marsh areas consist of fine silt and clay particles, with some organic detritus. This habitat typically has tidal creeks winding and branching sinuously through it, and it is dotted with numerous pond holes or "pannes," shallow depressions of the marsh area lacking salt-marsh hay and often housing small permanent pools of water. Rising tides move through the tidal creeks, finally wetting the land in between and flooding the pannes. Fishes are largely absent from the high marsh except during high tides, although mummichogs and other small-bodied species may be residents of some of the pannes.

Regularly flooded or low salt marshes. The low salt marshes of New England estuaries are flooded by all tides under normal conditions (Teal 1986). These areas are vegetated almost exclusively by *Spartina alterniflora*, a stiff grasslike plant that can tolerate the constantly changing water level and temperature and particularly the high salt concentrations that inhibit the growth of other freshwater wetland plants. Interestingly, although *Spartina* grows luxuriantly in the low salt marsh, it is easily outcompeted by many other plants outside of this salty habitat; thus, its occurrence is generally restricted to the low-salt-marsh system. Low salt marshes house many fishes, including the highly abundant mummichogs (*Fundulus heteroclitus*), killifishes (*Fundulus* spp.), Atlantic silversides (*Menidia menidia*), and other small-bodied species. These residents congregate in tidal creek channels during low tides and spread throughout the *Spartina* as the tide rises (Teal 1986). Other, larger-bodied species move into and out of this habitat according to water levels during rising or dropping tides.

Tidal flats. Tidal flats are sand and mud-bottomed habitats situated between low and high tide limits; they lack emergent vegetation such as *Spartina*. Tidal flats are often bordered by salt marshes landward and by deep channels or eelgrass beds seaward (Whitlatch 1982). These flats can be either a small or an expansive part of the estuary.

Mixtures of sand and clay particles make up the sediments of the tidal flat, with sand dominating where tidal currents are greater, usually toward the mouth of the estuary. Tidal flat sediments contain a relatively high organic detritus content that is brought in and deposited by tides washing over and receding from the salt marsh and tidal creeks (Welsh 1980) and from eelgrass beds of open estuarine waters (Whitlatch 1982). Tidal flats are well known for their highly productive food webs.

Gastropods, such as periwinkles and mud snails, and mussel beds are common on New England mud flats, as are a variety of crabs and shrimp. Polychaete worms, bivalves such as the quahog, soft-shell clam, and hard-shell clam, and amphipods are abundant within the sediments. Seasonal water temperatures affect the distribution of some of these organisms along New England's coast. For example, the hard-shell clam is generally restricted to southern New England, whereas the soft-shell clam is more abundant northward.

Because they are productive feeding areas, tidal flats teem with fishes during high tide. Bait fishes such as silversides, killifishes, and menhaden (*Brevoortia tyrannus*), and bottom fishes such as flatfishes and skates are particularly dependent upon tidal flats for food (Whitlatch 1982); other species such as the scup and the tomcod also actively feed in this habitat. Because of the complex food webs seen in tidal flats, these sites are noted for converting plant production into animal biomass, as opposed to the more landward marshes that are seen largely as exporters of plant materials to flats and other adjacent waters.

Eelgrass beds. The lower (seaward) reaches of estuaries typically have deeper channels than are seen landward and, particularly in larger systems, have extensive areas that are subtidal (substrates remain submerged even during low tides). The substrates of this portion of the estuary may be unvegetated or may possess expanses of a variety of plants called sea grasses. One of the most prominent of these, eelgrass (*Zostera marina*), commonly occurs in New England estuarine systems. Light concentrations in the water column seem to be a limiting factor determining the distribution of eelgrass in estuaries and more open bays. New England estuaries, which tend to be less turbid than those farther to the south along the Atlantic coast (thus allowing light to penetrate more effectively through the water column), may have eelgrass beds occurring in waters as deep as 10 m (Thayer et al. 1984). Eelgrass beds provide several benefits to estuaries in which they occur. "Meadows" of *Zostera* help stabilize subtidal and intertidal habitats, reducing the abrasive and eroding effect of tidal currents and thereby providing better opportunity for growth and expansion of the eelgrass meadow itself (Thayer et al. 1984). Eelgrass beds also are optimal nursery habitat to young fishes, providing them with protection from predators and a plentiful food supply. Because young of some fish species, and all ages of other small-bodied species, utilize eelgrass so extensively, large predators tend to feed around the fringes of eelgrass meadows. Due to the high levels of production of plant tissues accomplished by eelgrass meadows and the algae that live within them, eelgrass habitats have long been considered to have an important role in the food webs not only within the estuary but also within adjacent waters, by both the export of plant materials to the coast via tidal currents and the export of portions of the food web by mobile predators that feed around eelgrass and exit the estuary for open water (Thayer et al. 1984).

The importance of estuaries to fisheries resources. Animals living in estuaries must be capable of tolerating extreme fluctuations in environmental conditions (salinity fluctuations are particularly stressful); otherwise, they must move from site to site within, or into and out of, estuaries to seek conditions best suited to their survival. In spite of what appears to be an extremely harsh environment, estuaries are very productive systems. Although variable from one system to the next, estuaries typically exhibit high levels of primary production (the growth of plant tissues or biomass), which supports active food webs not only within the estuary itself but also within coastal areas seaward from the estuary (Odum 1980; Teal 1986).

The presence of estuaries is critical to many valued coastal fisheries resources. Of the finfish and shellfish species landed by anglers and commercial fishermen along the Atlantic coast, 80% use estuaries during some stage of their life cycles (Arnett 1983). Some New England recreational species, such as the tomcod, white perch, and American eel, live in protected bays and estuaries throughout most of their life cycles, although some tomcod may exit the estuary for more open bays, and the perch and eels are equally well adapted to persist upstream in freshwater habitats. Anadromous fishes such as the Atlantic salmon, American shad, river herring, and rainbow smelt pass through estuaries on their spawning runs, and their offspring spend varying amounts of time in estuaries during migrations to the sea (Deegan and Day 1984). Estuaries serve as nursery grounds for some species, providing habitat conditions that are critical to offspring survival in early life stages. The winter flounder and tautog migrate into estuaries to spawn; other species such as the menhaden, bluefish, black sea bass, and summer flounder spawn in coastal shelf waters, with the young moving into estuaries sometime after hatching. Large juvenile and adult predators such as the striped bass, weakfish, and bluefish also commonly move into estuaries to feed but typically do not remain there for any period of time, residing instead along the coast in more open waters. Smaller fishes generally inhabit tidal creeks and marsh pannes in upper estuarine reaches and eelgrass beds in lower ones; larger juveniles and adults tend to frequent the edges of eelgrass beds and deeper channels, feeding actively on mud flats when flooded by high tides.

Estuaries, critical habitats for coastal fishes as well as other commercially and aesthetically important organisms, are highly affected by human activities. Unfortunately, habitat deterioration and loss occur commonly within these areas (Whitlatch 1982). Upper salt marshes are filled for development or impounded and diked to control water movement (Nixon 1982). More open areas of estuaries have been dredged and channelized for navigational purposes. The discharge of a variety of pollutants, including domestic sewage, agricultural run-off, industrial chemicals, and oils has widely polluted estuarine waters (Whitlatch 1982). Although all of these things are potentially controllable by a variety of legal statutes (see chapter 3 for examples of federal laws), they remain growing problems affecting the health of estuarine biotic communities and the capacity of these systems to support our coastal fisheries resources.

Shoreline Habitats

The coastline of New England is typical of coastal regions that have proceeded through a period of submergence by the ocean. Melting of the polar icecaps since the last great glacial period has created a steady rise in water level (90 m in the past 10,000 years; Reid and Wood 1976). Also, the northeastern edge of North America has slowly dropped, leading to further shoreline submergence. The rocky, steeply sloping shoreline of New England, particularly north of Cape Cod, is characteristic of regions of submergence. Such areas have numerous cliffs rising above narrow beach areas, often composed of gravel rather than sand. Coastal marshes and open estuaries are often small. Most of the shoreline from Maine's northern border southward to Cape Elizabeth is continuously rocky, as are hundreds of islands that dot the inshore waters of this region. The shoreline from Cape Elizabeth southward to Scituate Harbor, just south of Boston, Massachusetts, is characterized by a succession of gravel and sand beaches alternating with rocky shorelines. Rocky shorelines exist on Cape Cod and in southern New England, as in Narragansett Bay and along much of the Connecticut shore, but this region also contains more gently sloping shores with greater expanses of sand beaches and larger marsh areas.

Tidal action upon inshore areas influences the type of biotic communities, and of fisheries resources, that inhabit the shoreline. Sandy beaches are dynamic habitats, the sandy sediments constantly being deposited, eroded, and shifted due to the actions of tides and wind-driven waves. The sediments of the intertidal zone (the area between the low and high tide lines) are particularly unstable, particles constantly being lifted, carried, and deposited by the turbulent water. Due to seasonal changes in the force of shoreline waves (the changes are due to the seasonal frequency and force of storms that drive waves onto the shoreline), the shape and configuration of the submerged portion of the beach area may change greatly throughout and between years. Due to the unstable nature of the sand sediments, few attached plants will be found living on sand beaches, although waves, particularly after storms, will wash in large amounts of algal masses "uprooted" from slightly deeper waters.

Intertidal and subtidal (below the low tide line) areas serve as active feeding sites, because the turbulent waters and shifting sediments stir up small invertebrates upon which many fishes feed. In turn, larger predators pursue the abundant smaller fishes. Anglers find tidal "rips" particularly productive fishing sites. A rip occurs where rising or dropping tides flow essentially unidirectionally through an area, rather than moving directly shoreward in waves that wash progressively higher or lower onto the shoreline. Such currents, moving very much like a rapidly flowing river, occur wherever tides move through or along a physically restrictive area. Such areas may be the constricted opening of an estuary, or they may occur where the tide flows between the shore and a submerged or exposed sand or gravel bar that lies just offshore, parallel to the shoreline. In the latter instance, the tide flows along the shore rather than directly toward it. Tides may flow similarly along certain jetties or other man-made shoreline structures. During rising and receding tides, the forceful movement of water along such areas and the stirring of sediments create a very active feeding site for fishes and fishing site for anglers.

Rocky shorelines provide a more stable substrate than do sand beaches; thus, these habitats contain a broad array of attached algae, including the large green, red, and brown algae that form mats on the rocks and boulders of the intertidal and subtidal zones. The intertidal zone actually possesses a series of algal assemblages (groups of particular species or forms), organized in vertical groupings between the high and low tide lines. The algal forms that live in the intertidal zone are each adapted to survive particular levels of exposure to air during the tidal cycle; thus, they are vertically segregated along the boulders and rocks of this zone (Reid and Wood 1976 provide a description of the intertidal algal assemblages of southern New England). Other algal forms sensitive to exposure to air are abundant below the low tide line in the subtidal zone. A variety of mollusks and other invertebrates inhabit the intertidal and subtidal zones of the rocky shoreline, and fishes such as the cunner, tautog, scup, and sea raven are abundant in subtidal rocky habitats.

Interestingly, many favored fishing sites along the shoreline are features of the human environment. Wharves, piers, small docks, jetties, and other structures not only provide a "setup" site where anglers can deposit themselves and their gear but also provide shelter for fishes. These structures provide some of the basic habitat conditions seen in shoreline reaches with more stable substrates, such as rock and boulder areas.

Open Waters of the New England Coast

The greatest proportion of fisheries resources in the Northwest Atlantic is produced on the region's continental shelf. The shelf, greatly expanded from waters offshore from Cape Cod northeastward through the coastal areas of the Canadian Maritime Provinces, provides great regions of shallow-water habitat that is thoroughly mixed throughout the year. This mixing, caused by tidal currents and storm-driven waves, continually carries nutrients from bottom sediments up into the water column, where planktonic algae utilize them to produce very high levels of primary production. This plant biomass forms the basis of extremely productive food webs, including a variety of fishes widely pursued in recreational and commercial fisheries.

The outer shelf in this region displays a series of offshore elevations, called banks, which are great expanses of shallow water separated from each other and the mainland of the continent by broad, deep channels. The southernmost of these, Georges Bank, extends northeastward from Cape Cod. It is separated from Browns Bank and the adjacent banks to the north, collectively called the Scotian Shelf, by the Northeast Channel and from the Nantucket Shoals and southern New England waters by the Great South Channel (Figure 1). The portion of Georges Bank less than 100 m deep is about 140 miles long and 80 miles wide, thus occupying an area of approximately 11,000 square miles (Hachey et al. 1954). Sediments in waters less than 60 m deep on Georges Bank, and on Nantucket Shoals to the southwest, are

Figure 1. The coastline of New England and adjacent areas. The 100 and 200 m depth lines along the continental shelf are marked.

dominated by coarse sands, as tidal currents and wave turbulence prevent smaller particles from settling to the bottom (Moody and Butman 1980). Sediments in deeper shelf areas south and west of Cape Cod consist of varying percentages of sand, silt, and clay; tidal currents in these waters are not forceful enough to keep these smaller particles fully suspended in the water column.

Just northward along the coast from Cape Cod lies the Gulf of Maine. This deepwater basin, with average and maximum depths of 150 m and 377 m, respectively, is flanked on its seaward side by the Scotian Shelf and Georges Bank (Uchapi and Austin 1987). The Northeast Channel and the Great South Channel connect the Gulf to the deeper continental slope waters. The Gulf of Maine, covering an area of about 35,000 square miles, is topographically complex. Its bottom consists of 21 deep basins that are separated by shallower banks and steeply sloping ridges (Uchapi and Austin 1987). The southern and western sides of the Gulf's basin are relatively smooth, while the greatest variability in bottom contour, including numerous gullies, ridges, and pinnacles, lies in the northern and eastern sides. Sediments range from sand and gravel at the periphery of the basin to silt and clay in the deeper areas; ridges and pinnacles are covered with various percentages of clay, silt, sand, and gravel. Substrates either originated as glacial deposits or were washed into the Gulf from coastal streams.

Current patterns. Currents in the Northwest Atlantic establish two different ecosystems, based upon annual water temperatures, that influence the distribution and occurrence of fisheries resources. Basic patterns of current flow along the Northwest Atlantic coastline are shown in Figure 2. The Gulf Stream directly influences waters offshore from the continental shelf of southern New England and Georges Bank. This current, set in motion by the rotation of the earth, flows northward from the equator along the seaward side of the continental slope of the Atlantic coast to Georges Bank and the Scotian Shelf, where it bends in an easterly and northeasterly direction across the Atlantic to Europe. Shelf waters of southern New England and Georges Bank are influenced by this warm, tropical current, even though currents on Georges itself move in a large clockwise gyre (in a direction counter to the Gulf Stream when adjacent to it), and currents in southern New England originate in part from the Gulf of Maine via the Great South Channel and flow along the coast in a southwesterly direction.

The Gulf of Maine has much less contact with the Gulf Stream. In general, waters flowing southward from the Gulf of St. Lawrence mix with Labrador Current waters flowing southward from the Grand Banks; the Labrador Current does contact the Gulf Stream before it reaches the area of mixing with the outflow from the Gulf of St. Lawrence. Once mixed, these currents proceed south and west along Nova Scotia, folding into the Bay of Fundy and Gulf of Maine proper (Butman and Beardsley 1987). Currents within the Gulf flow counterclockwise, exiting the Gulf through the Great South Channel and the Northeast Channel areas.

Annual temperature regimes. Temperature regimes separate New England coastal waters into two distinctive regions: Georges Bank/southern New England and the Gulf of Maine. Due to the origin and mixing of currents, waters in southern New England and Georges Bank are considerably warmer than those of the Gulf of Maine (Table 1). In part because of these temperature differences, two distinctive fish faunas are seen in New England coastal waters. Georges Bank serves as the northernmost area where species such as the Atlantic bonito, summer flounder, white marlin, and others occur in any abundance. Other species such as the bluefish and weakfish occur sporadically in the Gulf of Maine, usually reaching higher densities during years when they are highly abundant farther south. Cape Cod and Georges Bank are not absolute boundaries for most species, but some such as the tautog are "less regular . . . , less abundant, and more local" in the Gulf of Maine than they are on Georges Bank and southward (Bigelow and Schroeder 1953). Some species, such as the Atlantic wolffish and redfish, are far more abundant north than south of Cape Cod/Georges Bank.

Coastal salinity. The salinity of the

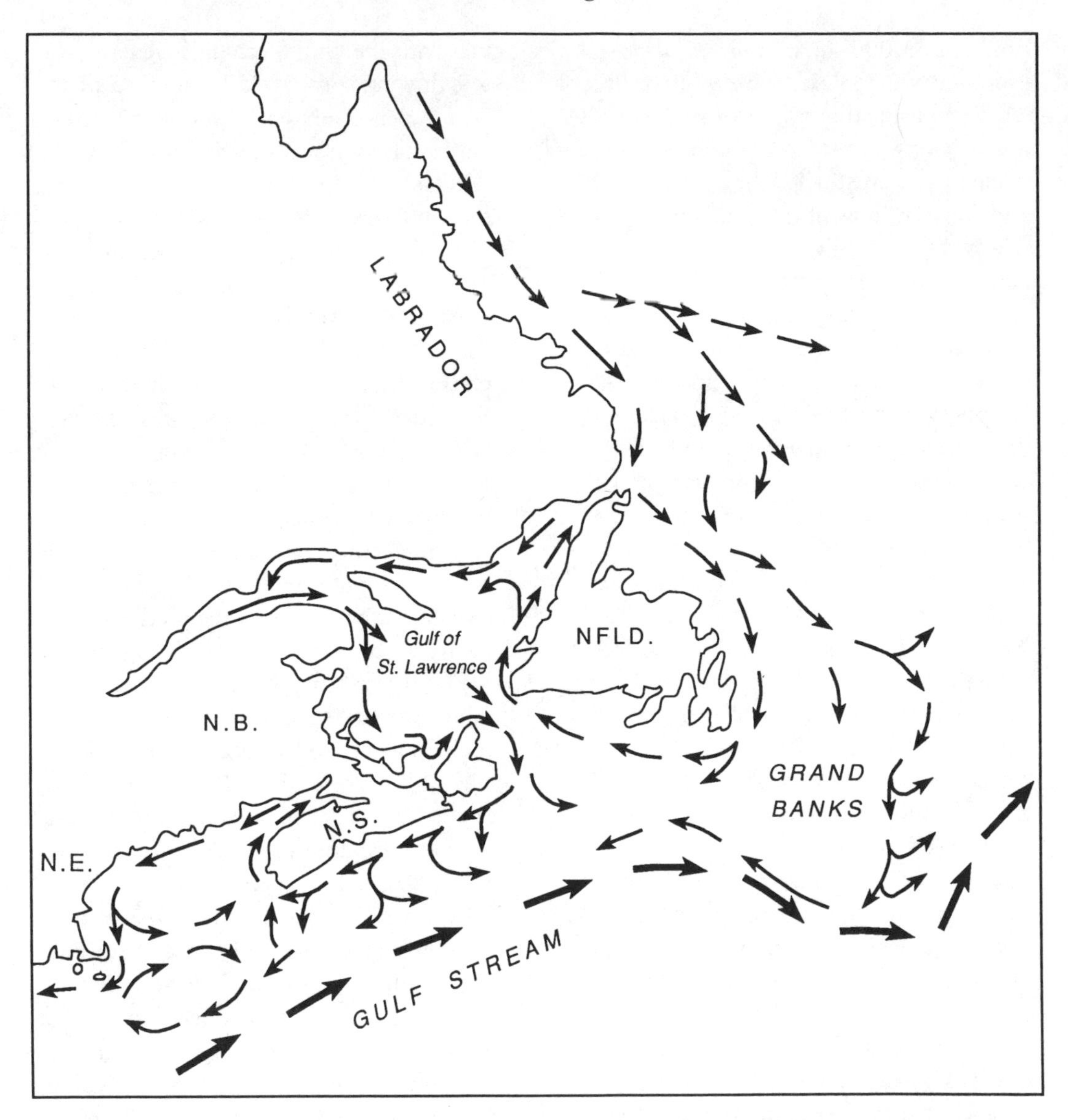

Figure 2. Current patterns along the Northwest Atlantic coastline. Note the counterclockwise current in the Gulf of Maine, which is fed by waters flowing southward from the Labrador Current and the Gulf of St. Lawrence. The clockwise current on Georges Bank is adjacent to the northward-flowing Gulf Stream. Currents along the inner continental shelf of southern New England flow southwesterly, counter to the Gulf Stream off the continental slope. N.E. = New England; N.S. = Nova Scotia; N.B. = New Brunswick; NFLD. = Newfoundland. Current patterns are taken from a more detailed diagram in Scott and Scott 1988.

open ocean seaward from the continental slope varies little from the average level of 35 parts per thousand (symbolized as ‰; this measure reflects the weight in grams of all types of salts that are dissolved in 1 kg of sea water). Interestingly, the flow of fresh water from rivers affects not only the salinity of estuaries and the shoreline but

Table 1. Average seasonal bottom-water temperature (in degrees C) from the Nantucket Shoals, Georges Bank, Gulf of Maine, and Scotian Shelf regions

Location and depth range	*March*	*June*	*September*	*December*
Nantucket Shoals (20–80 m)	3.6	7.5	11.3	9.5
Georges Bank (4–100 m)	4.2	8.6	13.0	8.3
Gulf of Maine (100–377 m)	4.8	5.9	6.7	6.1
Scotian Shelf (40–140 m)	2.2	4.6	7.0	4.6

Source: Schopf and Colton 1966.

also that of waters throughout the continental shelf. Salinity increases steadily from the North American coast seaward until it reaches standard seawater concentrations in the Gulf Stream. Low-salinity waters carried southward by the Labrador Current and from the Gulf of St. Lawrence extend over the outer reaches of the Scotian Shelf and Georges Bank. These currents, and additional freshwater input from New England rivers, similarly reduce the salinity along the continental shelf south of Cape Cod. The salinity throughout New England's shelf region is lower than 32.5 ‰, except for the Northeast Channel and a portion of the Gulf of Maine, where it can approach 33 ‰ (Bugden et al. 1982). Although these salinities are not reduced enough from standard seawater concentrations to limit the distribution and production of most fish stocks (as do the highly fluctuating salinities of an estuary), they do demonstrate the powerful impact that freshwater input has upon the chemical and physical nature of our coastal waters.

Primary production. Georges Bank exhibits the highest level of primary production of all of New England's open shelf waters. Primary production in the shallower reaches of Georges (< 60 m depth) is second only to the inshore shelf off the coast of New Jersey for all coastal areas from Cape Hatteras, North Carolina, to Nova Scotia (O'Reilly et al. 1987). Annual primary production of Georges is three times as great as the average level for all of the world's continental shelves. Primary production levels in the southern New England shelf area are lower than on Georges Bank but significantly higher than those characteristic of much of the Gulf of Maine. Highest levels in the Gulf occur in shallower waters (Yentsch and Garfield 1981).

Primary production on Georges Bank is high from spring through fall, which differs from deeper waters of the Gulf of Maine where production decreases in the summer (O'Reilly et al. 1987). High algal production on Georges and in shallower reaches of the southern New England shelf, and to a lesser level the Gulf of Maine, is due to complete mixing of waters from the surface to the substrate, which continually transports nutrients from sediments into the water column where they are available for uptake by algae. Tidal currents are responsible for the complete surface-to-bottom mixing on Georges Bank. The drop-off of production in deeper waters of the Gulf of Maine during the summer is due to water depth. In deeper areas, neither tides nor currents mix bottom waters into the upper water column, and storm-driven wave action is not forceful enough at this time of the year to force surface-to-substrate mixing in deep waters (this is in part due to the changes in density that occur in water as its temperature increases or decreases; see Reid and Wood 1976 for a detailed explanation of the relationship between water temperature and mixing).

Primary production determines productivity of the entire food web. The high levels of algal biomass on Georges produce high concentrations of zooplankton (Sherman 1981), which in turn serve as a major food source for most larval fishes and for juveniles and adults of some species such as the sea herring and menhaden. Even after the extremely high exploitation levels of the 1960s and 1970s, fish production on Georges Bank was among the highest of shelf ecosystems throughout the northern hemisphere (Cohen and Grosslein 1987). The productivity of fish stocks along the entire New England coast, particularly Georges Bank and Nantucket Shoals southward, makes this region a notably productive fishing area.

2
Life Cycles and Community Interactions

Because they are represented by divergent evolutionary lines with long histories, and because they are the most diverse group of vertebrates, fishes exhibit a wide variety of life cycle patterns. How certain basic components of a fish's life cycle function is of interest both to the ecologist trying to understand why such variety has evolved among fishes and to the fisheries manager, who must predict how harvest and other human activities affect fish populations. This chapter focuses on the features of the fishes' life cycle that are of joint interest to ecologists and managers, namely, (*a*) parameters that determine the reproductive capacity of fishes (collectively referred to as life history characteristics) and (*b*) those interactions among species, such as competition and predation, that determine the role a particular species fills in fish communities.

Adaptations are frequently viewed as aiding populations, or species as a whole. The viewpoint that specialized traits evolved "to help save the species" is a common misconception that does little to clarify the adaptive value of those traits. Species are typically broken down into geographically separated units referred to as populations (also called stocks when referencing fisheries resources). Populations can be defined as groups of animals that reproduce freely among themselves but are reproductively isolated, normally due to geographic separation, from breeding with other individuals of the same species.* Evolution is the result of interaction between a gene pool (the total genetic makeup of a population) and the environment in which the population exists. The genetic makeup varies among individuals within populations, due to mutation and other genetic changes as well as the recombination of genes that occurs when two individuals reproduce sexually (such recombination does not occur when an organism reproduces asexually, such as cell division of single-celled organisms). If genetic differences among individuals influence in any way the relative number of surviving offspring the individuals produce, the frequency of occurrence of such traits within

*For example, the silver hake is separated into two stocks off the northeastern coast of the United States, with the shallow waters of Georges Bank separating a population in the Gulf of Maine from one inhabiting southern Georges Bank to Cape Hatteras, North Carolina (Almeida 1987).

the population will change through time. Traits that tend to increase the average number and rate of surviving offspring an individual produces over the number and rate produced by others not possessing the trait will increase in frequency in the gene pool, whereas traits that lessen the rate of production of surviving offspring will decrease in frequency. Note the importance of the term "surviving offspring": Genetic adaptations occur when a trait is passed from one generation to the next, gradually increasing in frequency because individuals carrying the trait possess some survival and/or reproductive advantage over those that do not. If traits consistently produce such advantages in reproductive success from generation to generation, the genetic material that produced the trait should ultimately become the dominant form in the population.

Thus, adaptations arise in populations because some individuals are genetically predisposed to consistently produce a greater rate of surviving descendants than others breeding in the same gene pool. Evolution is a process that acts upon individuals, and adaptations occur because individuals possessing them are capable of contributing more to the population's future gene pools than are those not possessing them. Evolutionary fitness, a term originally popularized by Herbert Spencer's (1872) phrase "survival of the fittest," is no longer characterized in terms of physical strength and aggressiveness. It is defined solely in terms of an individual's contribution to future gene pools, or it is measured by an individual's success in producing surviving offspring. Thus, adaptations evolve not because of the survival advantages they provide to populations or entire species but because of advantages in reproductive output and survival of individuals. In order to best understand the adaptiveness that specialized traits confer, they will be presented and discussed using this evolutionary perspective.

Life History Characteristics and Reproduction

Fecundity

Certainly, one major factor contributing to reproductive success is the number of offspring produced. The number of offspring produced by a male fish is related to the number of females with whom he mates but is independent of the number of gametes (i.e., sperm) he produces. Male fishes are generally credited with producing nearly unlimited amounts of sperm. However, the number of offspring produced by a female is obviously determined by, and limited to, the number of eggs she produces. Fishes display extreme variations in female fecundity, both among individuals within species and among species.

Fecundity generally increases with the body size of females within a species. Variation in the numbers of eggs produced by small and large females is relative to the absolute fecundity typical of the species. Thus, an Atlantic cod may produce about 200,000 eggs per season when small, and more than 12 million when large; at the other extreme, female blue sharks will produce 4 to no more than 135 offspring (Scott and Scott 1988). The number of eggs that a female produces is related to how much energy she can direct toward gamete production in the face of other critical energy needs.

At any point in their lives, fishes must contribute energy reserves to meet all their current needs (such as the need to avoid predators, to forage for food, to grow, to reproduce, and to retain the proper physiological balance, or homeostasis, that is necessary to maintain health). Energy is allocated to different needs according to the relative importance of those needs to the fitness of the individual. Reproduction may require a substantial portion of the total energy available for use, since it may involve gamete production, migration, spawning, and nest-site and offspring protection. Larger-bodied females are generally believed to have greater available energy reserves; thus, they are capable of producing a greater number of eggs. Further, individuals in many species produce the greatest number of offspring only if they survive to spawn several to many times during their lives. In part due to the high energy demand, reproduction confers a risk of death to an individual; the higher the energy expended in reproduction, the greater the risk of death. Thus, it may be advantageous for young females to produce only a modest number of eggs; greater reproductive effort might decrease the likelihood of surviving to reproduce again. On the other hand, older females that have already completed most of their life-span might allocate a markedly greater proportion of their energy reserves to egg production. In theory, since they have already accomplished most of their lifetime reproductive effort, diverting energy from reproduction to increase the likelihood of survival is less critical than it would be for young adults.

High levels of energy are allocated to reproduction by species that reproduce once in their lives and die. The Atlantic silversides, a common estuarine species in New England, conducts an extended spawning season during which females develop and spawn several separate batches of eggs and die sometime after the spawning season ends (Conover 1985). Such prolonged spawning activity, referred to as serial or fractional spawning, allows markedly greater numbers of eggs to be produced by a female in one season than if all eggs developed at once. In the southern portion of its range, the American shad dies after migrating into rivers and spawning; farther north, a high percentage of adults migrate back to the sea after spawning and survive to spawn in subsequent years. It is not surprising that a southern female shad, which does not allocate energy to survive the rigors of migration, produces several times as many eggs as the average female in more northerly populations (Leggett and Carscadden 1978).

Fecundity is also related to egg size; generally, the greater the fecundity, the smaller the average-size egg. The size of an egg is correlated to the amount of its yolk reserves, larger eggs housing greater energy reserves for embryo development. The most fecund marine species, such as members of the cod and tuna families, the flatfishes, and others, produce very small eggs that are spawned or "broadcast" into the open water column. Extremely small and poorly developed larvae hatch from these pelagic eggs, often after only a few days of development. For example, cod eggs are only about 1.6 mm in diameter, and larvae are 3.3 to 5.7 mm long at hatching (Scott and Scott 1988). Mortality rates of these eggs and larvae are high. Thus, broadcast spawners allocate minimum amounts of energy into each small egg and produce very large numbers of offspring that offset the high rates of mortality the offspring suffer. Extremely high fecundities have been measured among the broadcast spawners; the female ocean sunfish (*Mola mola*) may produce as many as 300 million eggs per year (Hart 1973). Fishes such as the Atlantic salmon, members of the sculpin family, and the Atlantic wolffish display markedly lower fecundities than

Table 2. Maximum fecundity, egg diameter, and length of offspring at hatching or birth of selected species of coastal New England fishes

Species	*Fecundity*	*Egg size (mm)*	*Offspring length (mm)*
Bluefish	1,100,000	1.2	2.4
Bluefin tuna	60,000,000	1.1	3.0
Atlantic mackerel	1,000,000	1.3	3.0
Winter flounder	3,329,000	0.8	3.5
Monkfish	1,320,000	1.8	4.5
Cod	12,000,000	1.6	5.7
Shorthorn sculpin	60,976	2.5	8.6
Acadian redfish	40,000		9.0
Pipefish (*Syngnathus fuscus*)	860	1.0	9.0
American shad (CT R.)	321,000	1.8	10.0
Atlantic wolffish	40,000	6.0	18.0
Spiny dogfish	15	45.0	300.0
Blue shark	135		440.0

Source: Wolffish fecundity from Barsukov 1972; other information from Scott and Scott 1988.

the broadcast spawners but pack greater yolk energy reserves into each egg. Embryos in large eggs typically exhibit longer periods of development and are larger and relatively better-developed larvae at hatching than embryos of broadcast spawners. Once young Atlantic salmon leave the gravel nest in which they were spawned and hatched, they are well-developed juvenile fishes. Thus, by the time they are first exposed to predators and the problems of foraging for food, their advanced stage of development allows them greater opportunity for survival. Ovoviviparous sharks display extremes in egg size and in fecundity. The eggs of these species contain enough yolk reserves to complete exceptionally extensive periods of development while incubated inside the female's reproductive tract. Such sharks give birth to very small numbers of very large juvenile offspring as much as a year or more after the eggs were initially fertilized; exceptionally high rates of offspring survival follow. Thus, egg size not only is inversely related to fecundity but, not surprisingly, is directly related to the rate of offspring survival. Table 2 presents the relationship among fecundity, egg size, and larval size at hatching or birth for a variety of coastal New England fishes.

Fecundity also is clearly related to the level of parental care provided to eggs and larvae. Parental care is often defined as any behavior displayed by adult fishes that directly increases the survival of their offspring. Such behavior may include selection and tending of the site where eggs are laid and protection of eggs and larvae from predators. More broadly, parental care can include any reproductive specialization that increases offspring survival rates (Baylis 1981). This includes not only direct behavioral "care" but also physiological and morphological adaptations that provide offspring with an advantage, such as traits that allow internal fertilization and incubation of eggs and direct nourishment of developing embryos. Generally, the greater the level of parental care provided offspring, the lower the fecundity demonstrated by fish species.

Parental Care

Broadcast spawners. Broadcast spawning represents the least specialized mode of reproduction. Buoyant eggs are broadcast into the water where they are fertilized. They remain in the water column throughout incubation, passively drifting with prevailing currents. Broadcast spawning of buoyant eggs is common among marine fishes but is atypical of freshwater systems. Buoyant eggs in the ocean remain in the surface waters where newly hatched larvae must feed on plankton after yolk reserves are depleted. Although carried great distances from where they were spawned, eggs drifting in ocean currents tend to remain within areas of suitable environmental conditions, since the open ocean is a large, homogeneous, and well-mixed environment (Baylis 1981). Conversely, habitats vary greatly within short distances in fresh waters. Thus, it is possible for fishes to spawn in sites that offer greater opportunity for offspring survival than do other areas close by, because of the variation that occurs in chemical and physical conditions (such as dissolved oxygen concentrations and temperatures), in the availability of shelter from predators, and/or in the quality of food resources for larval fishes. In such ecosystems, buoyant eggs and drifting larvae could easily be carried from favorable to unfavorable habitats by currents and wave patterns. Thus, in freshwater systems natural selection has favored fishes that spawn demersal (sinking) eggs that adhere to substrate or stationary objects in areas offering the best conditions for offspring survival. Most of the marine species that display such reproductive characteristics are coastal or shallow-water species that spawn in habitats exhibiting spatially variable microhabitat conditions (Baylis 1981).

Although broadcast spawners exhibit the least specialized mode of reproduction, some evidence suggests that offspring survival may be enhanced by parent behavior. For example, North Sea plaice migrate to the Flemish Bight to spawn, afterward dispersing over wide areas throughout the North Sea. Eggs and larvae drift from the Flemish Bight with prevailing currents, and the majority become demersal (bottom-dwelling) juveniles by the time they reach the region of the Friesian Islands, which provide optimal nursery conditions for the newly demersal fishes. Thus, adult spawning migrations centralize reproduction in areas that increase the likelihood of offspring settling into optimal habitat when ready to become bottom dwellers (Pitcher and Hart 1982).

A few species of broadcast spawners produce demersal eggs. The anadromous American shad releases eggs that are slightly heavier than water; thus, they tend to sink to the substrate of the rivers and streams in which they are spawned. Unlike most demersal eggs, the American shad's are not adhesive; currents tend to move them far downriver from the sites in which spawning occurs. Some eggs do become lodged in bottom rubble as they drift along the substrate (Scott and Scott 1988). Although this species does not spawn buoyant eggs, it is properly classified as a broadcast spawner because its eggs develop in areas removed from the site of spawning.

A great variety of species exhibit specializations from broadcast spawning that confer survival advantages to offspring. These include selecting the type of habitat and/or substrate on or in which eggs are deposited and protecting or tending offspring. Fish ecologists have developed several schemes for classifying patterns of parental care using these characteristics of reproduction (as in Nikolsky 1963; Balon 1975).

Egg scatterers. Perhaps the simplest specialization from broadcast spawning is the deposition of adhesive eggs that remain in the optimal habitats in which they were spawned. Many fish species release adhesive eggs over specific types of substrate, onto which the eggs stick. The selection of appropriate substrates is critical for optimum offspring survival. Because egg scatterers do not modify the surfaces on which they lay eggs, they are usually separated from those that do, the nest builders.

Egg scattering is widespread among freshwater fishes, is common in anadromous or shallow-water coastal species, but is characteristic of relatively few deep-water marine fishes. Egg scatterers are frequently classified as to the types of surfaces on which they scatter eggs (as in Balon 1975). The winter flounder, the only species of Northwest Atlantic flatfishes that lays demersal, adhesive eggs, typically spawns over sandy substrate in estuaries (Buckley 1989), and the sand lance (*Ammodytes dubius*) scatters eggs over sandy substrates in more open waters (Scott and Scott 1988). The capelin (*Mallotus villosus*) swims onto beaches during rising tides to lay eggs on sand. Following spawning, adults reenter the water with rising waves, which also shelter the egg clusters by rolling a layer of sand over them. The lumpfish (*Cyclopterus lumpus*) lays large egg masses that adhere to rocks and boulders in seaweed beds (Mochek 1973).

Several New England coastal species spawn over aquatic vegetation. The rainbow smelt scatters eggs on a variety of substrate types in fast-flowing coastal streams, including gravel, rocks and boulders, and aquatic vegetation. However, survival rates of eggs are significantly higher on vegetation than on other substrate types (Sutter 1980). The Atlantic silversides spawns on mats of filamentous algae in the intertidal zone (Conover and Kynard 1984). Throughout much of the period of embryo development, eggs are actually out of the water column, due to low shoreline water levels at ebb tide. At such times the eggs are obviously protected from aquatic predators. Adhesion to algal mats may have the double advantage of hiding eggs from predators when the eggs are submerged and preventing them from dessicating when exposed to air during low tides; the mat of algae tends to remain moist throughout such periods. The estuarine mummichog deposits eggs on a variety of substrates, including vascular plants and algal mats (Scott and Scott 1988).

Shelter spawners. Some fish groups seek a specific type of shelter into which eggs are laid. This shelter, which may be living or nonliving, typically serves to reduce the level of predation on eggs and, in some cases, larvae.

Two common New England species, the sea raven and longhorn sculpin (*Myoxocephalus octodecemspinosus*), commonly lay their eggs at the base of, or among the branches of, living sponges. Although no other species of coastal New England lays eggs in or on living organisms, this behavior is displayed by fishes elsewhere. The European common bitterling (*Rhodeus sericeus*) deposits its eggs within the mantle cavities of bivalve mollusks (Lagler et al. 1977), and Pacific snailfishes of the genus *Careproctus* lay theirs in the gill cavities of crabs (Peden and Corbett 1973). Species of damselfishes (Pomacentridae) lay their eggs among the tentacles of sea anemones (Balon 1975). Juveniles of one New England species, the red hake, find refuge inside the mantle cavity of larger bivalves such as the sea scallop and surf clam. However, the eggs and larval stages of this species are pelagic. Thus, such shelter seeking is a juvenile behavior, not a result of parental care.

Some Northwest Atlantic species lay

their eggs in shelter sites in the substrate. The longhorn sculpin and shorthorn sculpin lay eggs in V-shaped crevices and cavities of hard substrates. The Atlantic wolffish and the ocean pout (*Macrozoarces americanus*) typically lay eggs in large, protected crevices or in holes under large boulders.

Nest builders. Numerous species not only select specific substrate types but modify the substrate in some manner. This nest-building behavior may serve to cover and hide the eggs from predators and/or improve the microhabitat conditions that might affect offspring survival during embryo development.

Construction of gravel nests is common in freshwater stream fishes and is typical of some anadromous species that spawn in the same flowing-water habitats. The male sea lamprey (*Petromyzon marinus*) constructs a shallow pit of gravel and small rocks by pushing materials around with his mouth (Scott and Scott 1988). The action of moving gravel and rocks dislodges fine silt particles into the current, which carries such materials away from the nest. Thus, the substrate of the nest consists solely of coarse, clean gravel. Removal of silt particles allows excellent percolation of water through the gravel and across the batches of eggs, which provides optimum aeration and removal of waste products from the egg clusters. Eggs that settle into gravel are also protected somewhat from egg predators. The female Atlantic salmon builds an elongate pit (called a redd) of coarse gravel cleaned of silt particles. She accomplishes this construction by turning on her side against the substrate, then vigorously and continuously flexing her body. This action flushes small silt particles out of the gravel and forms the pit (Chapman 1988). After eggs are spawned and fertilized in the pit, the female covers them with gravel, using the same finning motion. After spawning is completed, the clusters of large eggs are buried 12.7 to 25.4 cm (6 to 10 in) deep in cleaned gravel. Such redds are as much as 1 m (3 ft) wide and nearly 6 m (about 20 ft) long. The coastal fourhorn sculpin (*Myoxocephalus quadricornis*) sweeps shallow depressions in soft-bottom muds or flat algal mats for spawning (Scott and Scott 1988).

Other groups of fishes construct simple to elaborate nests from vegetative materials. Perhaps the extreme instance of persistence in nest construction is that displayed by the West Coast garibaldi (*Hypsypops rubicunda*). The male garibaldi tends a nest, consisting of a patch of a particular form of red algae, throughout the year, even during nonspawning periods. The male "farms" his algal patch by clipping off sections, gradually shaping it into a particular form. This constant tending is apparently necessary to attract females, who lay eggs on the algal patch (Clarke 1970).

Some of the most elaborate nests made from plant materials are seen in the stickleback family Gasterosteidae. Using fragments of aquatic vegetation, the male ninespine stickleback (*Pungitius pungitius*) constructs a tunnel-shaped nest with an opening at each end. The nest is usually built on weeds above mud-bottom substrates. The male uses a secretion of the kidney to glue the vegetation fragments together (Scott and Scott 1988). Males of other species of sticklebacks build nests of several different shapes (tunnel, spherical, cuplike), either on the substrate or attached to projections above it (Balon 1975).

Other types of nest builders, such as froth nesters (which construct nests of mucous-covered bubbles) and hole or burrow diggers, are not represented by New England coastal species.

Guarding. Shelter spawners and nest

builders can also be classified according to the presence or absence of guarding behavior by parent fishes at the nest site. Guarding is a common behavior of fishes that lay eggs in protected sites. Males are the guarders in most species (Baylis 1981), although females are in some taxa, such as the ocean pout (Keats et al. 1985). Fish ecologists have presented several hypotheses concerning the preponderance of nest-site guarding by males, and Baylis (1981) provides a logical premise explaining why evolution has selected for such male behavior. Although nest-site guarding may continue throughout egg incubation and after hatching, males in many guarding species defend the nest site only until spawning ceases, whereupon they abandon the site and the eggs they have parented. Thus, defense of a nest site serves some basic function beyond the protection of offspring housed in the nest. The evolution of reproductive territoriality and defense of the nest site is characteristic of species that require specific habitat conditions for successful survival of eggs and, in some instances, larvae. Habitats that provide optimum conditions for egg incubation and offspring survival are probably limited in availability for most species. Thus, natural selection should strongly favor individuals that are able to monopolize such spawning sites, that is, to defend the sites from other members of the same sex and parent all the offspring that are spawned on the site. The number of offspring that a male produces is limited only by the number of females with which he is able to mate, since males are generally considered to produce enough sperm for multiple mating acts. Thus, monopolizing an optimal spawning site to which multiple mates are attracted can significantly increase the number of offspring a male produces. Alternatively, much less reproductive advantage would accrue to a female that spends time and energy to defend a nest site, since the number of offspring she produces would still be limited to the number of eggs she develops. In other words, defending an optimal spawning site will increase the number of offspring a male produces as well as maximize their rate of survival, whereas a female benefits only in terms of maximizing the rate of survival. Thus, it is not surprising that evolution has led to male nest-site guarding in the majority of species that exhibit this behavior.

It is likely that the defense of the spawning site often evolved as a social behavior (i.e., a behavior that is directed toward other members of the same species), with the advantage of increased spawning frequency by males that successfully monopolized optimal sites. If offspring survival would be enhanced by constructing a nest or guarding eggs and larvae from predators, selection would strongly favor individuals displaying such behavioral tendencies. Thus, although males of some species only defend spawning sites from other males, males in other species have evolved behavioral patterns that improve the suitability of the site and/or reduce offspring mortality caused by predation (Baylis 1981).

The ability to successfully exclude other males from a spawning site requires a male to establish his social superiority over others. Social dominance, a prerequisite to successful spawning-site defense, is frequently related to the size of a male, with the largest males in a population generally gaining dominant status (as in Ross 1977; Ross and Reed 1978). Most interactions between males vying for a spawning site involve not actual fighting but merely the display of behaviors (referred to as "threat behaviors") that demonstrate a male's superiority were a fight to occur. Males "matched" against physically superior opponents normally retreat from the site, since the risk of serious injury and possibly death would jeopardize their likelihood of future reproduction. When fighting does occur, it happens most often between similar-sized males that are unable to establish dominant/subordinate status by threat displays.

Male nest-site defense should favor the evolution of a female's ability to "select" the best mating opportunity (this process, of course, is not a conscious choice made through a decision-making process but rather a simple response to appropriate stimuli). Since a female's reproductive potential is limited to her egg production, evolution should favor a female that will reproduce in situations that provide the greatest opportunity for her offspring to survive. Several advantages may be gained by females that mate with males capable of defending an optimal spawning site. First, if the site offers advantageous environmental conditions, the survival of the female's offspring will be enhanced. Second, a male that protects a prime site successfully from other males may also be most effective in protecting eggs and larvae, if these behavioral patterns are characteristic of the species; size, a critical criterion for successfully defending a nest site from other males, might also be important to successful defense against egg predators. Finally, a genetic advantage might be gained by the offspring of a female that spawns with males on optimal sites. Males capable of defending an optimal site are typically superior, at least in size and aggressiveness, to other males. Even if size is their major advantage, males have to survive substantial portions of the typical life-span to gain that advantage, since size is related to age. If the genetic makeup of a male contributes in any meaningful way to his gaining superior social status, such traits would be passed on to offspring, enhancing their chances for survival and/or future reproductive success.

Brooding. Brooding, or bearing (the carrying of offspring throughout at least a portion of early life stages after fertilization), is the most specialized mode of parental care. Egg brooders not only display behavioral specialization but may possess morphological adaptations that allow internal fertilization, egg development within specialized pouches or the female's reproductive tract, and in some instances the nourishment of embryos from the female's circulatory system. Brooders, or bearers, may be divided into two major groups: (1) external brooders, where offspring are carried outside of the female's reproductive tract, and (2) internal brooders, which carry developing embryos within ovaries, oviducts, or uteri.

External bearers have evolved a variety of ways to carry young. Some species in the freshwater family Cichlidae (the cichlids) release and fertilize eggs on the substrate. After fertilization, one of the parents gathers the eggs in its mouth, where they are incubated until hatching. Other families of fishes have external-bearing species that carry externally fertilized eggs in the mouth, attached to the body, or even in the intestinal tract (Lagler et al. 1977).

Male sea horses (*Hippocampus erectus*) and northern pipefish carry eggs in specialized abdominal pouches. Females lay eggs directly into the pouch, where the male fertilizes them and carries them until after hatching. As in nonbearing fishes, yolk reserves of the eggs are the only source of energy for developing offspring.

All internal brooders fertilize eggs internally by passing sperm into the reproductive tract of the female. Male fishes have specialized copulatory organs that are inserted into the female for sperm passage. These organs have evolved from several different structures, depending upon the fish group. Male sharks and rays have specialized pelvic fins, called claspers, that serve this function, and males of the freshwater family Poeciliidae (including common aquarium fishes like the guppy, black mollie, and others) have a specialized anal fin that serves the same purpose.

Although some shark relatives such as the little and winter skates fertilize eggs internally and subsequently release egg cases into the water column where they complete development and hatch, most bearers incubate eggs at least until hatching. Many of these species are ovoviviparous; that is, even though eggs are incubated within the reproductive tract, the eggs' yolk reserves provide all nutrition supporting development. The redfish or ocean perch produces markedly greater numbers of offspring than most other live-bearers, with an annual fecundity of about 40,000 eggs. The redfish provides much less yolk material to eggs than most ovoviviparous species, so its pelagic larvae are small and much less well developed at hatching and birth. Live-bearing in this instance serves to protect eggs from predation, but the pelagic larvae suffer levels of mortality that are probably more comparable to larvae of broadcast spawners than to other marine live-bearers.

Ovoviviparous shark species (such as the spiny dogfish) display markedly lower fecundities than the redfish. However, such sharks deposit such great reserves of yolk within eggs that offspring accomplish an unusual amount of development before birth. These offspring are extremely large before ever being exposed to predation or competition; thus, their survival rates are very high.

Some live-bearers such as the blue shark are viviparous, the embryos gaining nutrition directly from the female. Females may commit much less energy to production of yolk reserves, instead nourishing their young through a placenta. These sharks have a uterus, with the capillary attachment occurring between the uterine wall and the offspring's circulatory system (as in Pratt 1979). The young of some sharks gain nourishment while in the uterus by feeding on unfertilized eggs or uterine fluids (Scott and Scott 1988).

All species of fishes expend tremendous energy reserves when reproducing. The use of these reserves is ultimately necessary to address the evolutionary pressure caused by high rates of mortality in early life stages. Some species have evolved extremely high fecundities (such as the broadcast spawners), committing minimal energy to each individual offspring and succeeding in reproduction mainly because the enormous number of offspring produced outweighs the high rate of mortality the offspring suffer. Other species have evolved a variety of levels of parental care, committing greater amounts of energy to each offspring and in doing so improving the probability that a particular offspring will survive. Energy reserves necessary to accomplish such a strategy result in fewer total offspring. Thus, in basic terms of energy allocated to offspring and the rate of offspring survival, one should not consider any of the strategies described above as being superior; they are different approaches that evolved due to the pressures of offspring survival and the success of an individual's reproductive effort.

Egg and Larval Mortality

Fish eggs and larvae suffer the greatest rates of death of any stage of a fish's life cycle. Although juvenile and adult death rates of many large-bodied species may average about 5% to 10% annually (Woodhead 1979), and mortality of species that are heavily preyed upon may be much higher, eggs and larvae of these same species may suffer similar rates of mortality on a daily or weekly basis. Between 90% and 99% of smelt eggs die before hatching (Sutter 1980); 70% to 85% of larval American shad in the Connecticut River die from 4 to 9 days after hatching (Crecco et al. 1983). Winter flounder and striped bass larvae suffer about a 1.2% and 1.5% daily death rate, respectively (Pearcy 1962; Dey 1981). Pitcher and Hart (1982) suggest that a 2% to 10% daily death rate may be typical of many species that do not exhibit parental care (i.e., the broadcast spawners).

Larval mortality can be distinguished by its cause: either physical and chemical conditions or biological circumstances such as starvation and predation. Larval mortality rates generally are not caused by one factor but are a result of a multitude of factors affecting larval life stages (Pitcher and Hart 1982).

Environmental causes of mortality. Physical conditions, particularly water temperature and rate of current flow, can cause extreme fluctuations in larval survival from year to year. Survival of North Sea plaice larvae is greatly affected by the water temperatures and current speeds to which they are exposed while drifting from the Flemish Bight to the Friesian Islands (see section on Parental Care). In years when water temperatures are lower and/or current speeds are higher than the normal range, eggs and larvae may be swept far past optimum nursery grounds before they are ready to become demersal. Conversely, if water temperatures are higher and/or current speeds are lower than normal, larvae settle to the bottom before reaching the optimum nursery grounds. In either case, increased mortality is believed to result (Pitcher and Hart 1982). Spawning in areas where currents carry larvae to nursery grounds with high densities of food is common among broadcast spawners (Fortier and Leggett 1982). Drifting of larval Pacific hakes (*Merluccius productus*) affects their survival rate in a manner similar to that of the North Sea plaice (Bailey 1981). Ocean current patterns are believed responsible for fluctuations in offspring production of the bluefish (Norcross et al. 1977) and a variety of other open-ocean spawning species (Iselin 1955).

The average river discharge (volume of water passing downstream within a given amount of time), water temperature, and total monthly precipitation are all significantly correlated to larval survival of American shad (Crecco and Savoy 1984). These factors do not act independently to affect survival rates. During years of high June rainfall, the high river flows and resulting low water temperatures are believed not only to affect egg and larval development rates but also to reduce densities of zooplankton. Since zooplankton serve as the major food of American shad larvae, unfavorable feeding conditions result in high larval mortality.

The level and frequency of onshore winds affect the survival rates of larval capelins (Frank and Leggett 1981; Leggett et al. 1984). Wind-induced wave action is required to free capelin larvae from the sand and pebble substrates of the beaches where they were spawned. If time intervals between onshore winds exceed several days, larvae remain buried in beach substrates too long, resulting in total yolk resorption and rapid deterioration of their physical condition. Similarly, onshore wind-driven currents that transport larval Atlantic menhaden from offshore spawning to onshore nursery grounds are critical for their survival (Nelson et al. 1977). Although physical parameters of the environment are clearly correlated to mortality, in many instances these factors merely place larvae into unsuitable conditions or affect them in some manner that weakens their condition; in such situations, starvation or predation is the ultimate cause of death.

Biological causes of mortality. Starva tion and predation can cause high levels of mortality in early stages of the life cycle. Invertebrate plankters such as copepods and chaetognaths (Pitcher and Hart 1982) represent the greatest proportion of fish larvae predators, although small fishes may also eat substantial numbers of smaller larvae. Predation has been identified as the cause of major reductions in larval densities (as in Moller 1984). However, predation has its greatest effect when it acts in combination with other factors. Low water temperature or poor feeding conditions that slow growth rates can increase the proportion of larval mortality that is caused by predation. Many predators, particularly invertebrates, feed only upon smaller fish larvae; as the larvae grow, they may become a desirable size for other fish predators, but they become too large to be eaten effectively by invertebrates. The longer it takes larvae to grow through such critical size ranges, the longer they will be vulnerable to invertebrate predation. Thus, poor feeding and/or growth conditions affect both starvation and predation rates.

Field collections and laboratory experiments provide evidence that poor feeding conditions directly cause mortality of larval fishes. Food availability during the period after yolk resorption is felt to be of great importance to survival (Pitcher and Hart 1982). This time is frequently referred to as the critical period, because most fishes that cannot accomplish some minimal level of food gathering at this time die, even if they ultimately encounter suitable densities of food before they starve. Haddock typically begin feeding about 2 days after hatching, but yolk reserves are not totally depleted until 4 to 5 days later. In laboratory experiments Laurence (1974) starved newly hatched larvae for varying numbers of days, after which they were provided with high densities of zooplankton food. Over 50% of larvae deprived of food for 6 days or less after hatching survived. However, larvae starved for 8 to 14 days all died, even though some larvae had actively fed after being provided with food. Similar results have been shown for the anchovy *Engraulis mordax* (Lasker et al. 1970).

The larvae of many species require minimum densities of food. A general threshold is considered to be 100 plankters per liter of water, since larval growth rates decline and survival rates become highly variable at densities below this level (Pitcher and Hart 1982). At densities above 100 plankters per liter, larval growth rates level off, indicating they are approaching a maximum rate of growth, and survival rates become more consistent. Such thresholds hold great intrigue for fisheries scientists, since the typical food densities measured in the laboratory that are necessary to allow high larval survival are several times greater than the average plankton densities found in the ocean (Pitcher and Hart 1982). Hunter (1972) and others have invoked the "plankton patch hypothesis" to explain how fish larvae find suitable densities of food to survive. Prevailing oceanic currents and surface flow patterns created by wind-driven waves tend to concentrate plankton into localized areas of high densities. These "patches" are separated by oceanic reaches of very low plankton concentrations. Thus, localized patches have markedly higher densities than the average density of a large oceanic region. Concentrations of plankton in these patches provide optimum feeding conditions for larvae that encounter them. Although some biologists (such as May 1974) do not believe that such high densities are necessary for survival of at least some larval fishes, other studies suggest that fish larvae might be able to survive only if they encounter such plankton patches (Werner and Blaxter 1980).

In theory, encountering these patches might not be difficult for larvae. There is a strong relationship between the time of spawning of many fish species and the time of peak plankton production (Pitcher and Hart 1982). Since larvae are planktonic, ocean currents could concentrate them with their planktonic food. If optimal growth is attained within a patch, then even if survival were possible outside of patches, the larvae within them receive greatest advantage. Drifting in the same currents that concentrate zooplankton should keep larvae associated with their food when they most critically need it. However, yearly variation in currents and wind-driven surface waves might not predictably drift larvae into suitable plankton patches during the critical period, or may not produce large enough patches to provide food necessary for large numbers of newly feeding larvae.

Thus, early life stages of fishes encounter multiple hazards of mortality. If optimal food resources are not available, larvae may die due to starvation, or their slow growth and poor physical condition will increase their vulnerability to predation or other mortality factors. These factors result not only in extremely high but also in extremely variable and unpredictable death rates. Reproductive cycles of fishes have evolved in response to the variable and unpredictable nature of the severity of these causes of larval mortality, as will be described in the Reproduction Schedule section below.

Growth

Unlike warm-blooded vertebrates, which grow at a somewhat predictable rate and stop growing at some specific average age, fish growth is indeterminate and highly plastic. That is, potentially, a fish can grow throughout its life cycle; the rate at which it grows is highly variable and dependent upon life stage and conditions to which the fish is exposed.

Growth according to life stage. The rate of growth is greatest in early life stages. As a fish reaches sexual maturity, some portion of energy previously available for body growth is diverted to reproductive activities such as gamete production, migration, and spawning. The amount of energy contributed to reproduction can be great; thus, the growth rate displayed by the fish slows markedly due to the reduction in energy reserves available for growth. Fishes that approach maximum age typically exhibit additional decreases in growth due to the combined needs of reproduction and maintenance activities (i.e., foraging, shelter seeking, maintaining internal physiological balance). Energy allocated to maintenance activities is felt to increase with age and size, due to the increased energy needs required to maintain a larger, older body.

Although the relative increase in size (the weight or length gained during a given time period compared to the size of the fish at the beginning of that period) is greatest in larval and early juvenile life stages, the actual length or weight gained is small, since the fish is small. Thus, if the size of an individual from a long-lived species is compared to its age, the relationship

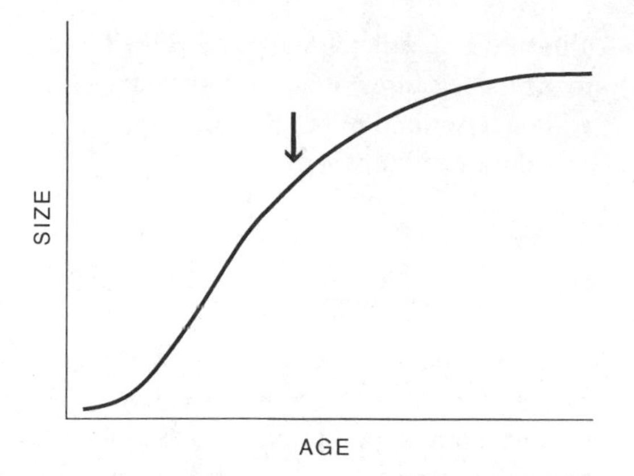

Figure 3. The size of an individual at age. Age at sexual maturity is marked by an arrow.

forms a sigmoid (S-shaped) curve, as in Figure 3. Sexual maturity occurs at the point where the increase in size from one age to the next begins to decrease, and the ultimate flattening of the curve is due to the combined energy costs of reproduction and maintenance.

Growth variation among individuals. Rates of growth typically vary among individuals of a given age within populations. Differing capacities to grow are likely carried throughout life; thus, although 1-year-old fishes of a long-lived species may not vary much in size, differences in size between fast- and slow-growing individuals increase with age. Ultimately, fast-growing individuals may be larger than older but slower-growing ones.

Growth rates may also vary significantly between males and females within a species. Typically, the sex that displays larger average sizes at a given age achieves greater growth at least in part by deferring maturity (that is, by reaching sexual maturity at a later age than the other sex). The sex for which maturity is deferred spends a prolonged period of life as fast-growing juveniles, thus reaching larger sizes before growth is slowed because of the energy demands of reproduction. That sex that matures at an older age may also realize a longer average life-span than members of the other sex. In such species, all of the largest individuals in a population are usually older members of the same sex. Thus, all large striped bass are commonly called cows, since it is well known by anglers and scientists alike that they are inevitably females. The reproductive advantages of these growth patterns are discussed in the section on Age at Sexual Maturity.

Environmental factors and growth. Early experimental work clearly demonstrated that water temperature affects the metabolic rate of fishes, thus influencing the efficiency and rate at which food is converted into energy and tissue for growth. Generally, the warmer the average water temperature, the faster the growth, as long as the temperature falls within a range that is suitable (i.e., nonstressful) for that species. In temperate regions of the world, many fishes tend to grow more rapidly during warmer seasons than colder. Differences in growth rates between cold and warm seasons are so great that hard tissues of fishes, such as bones and scales, exhibit yearly growth rings, reflecting the effect that temperature-controlled metabolic rates have upon the density, and composition, of bone and scale tissues produced during cold and warm seasons. Such marks provide the fisheries scientist with unique opportunities to age fishes accurately and thus to determine the rate of growth an individual has experienced during its life-span.

The same relationship between temperature and growth can produce measurably different average growth rates among populations of the same species residing in different geographic regions, if the regions differ consistently in annual temperature regimes. Likewise, changes in average annual temperatures can affect growth rates of fishes within populations. Growth is positively related to mean annual water temperatures to which Atlantic herring (*Clupea harengus*) are exposed, individuals growing more rapidly during years when average water temperatures are highest in the Northwest Atlantic than in years when temperatures are lower (Anthony and Fogarty 1985).

Salinity has been recognized as another environmental factor affecting rates of growth, due to the energetic costs of adjusting internal physiology to withstand the stresses of fluctuating salinity levels (Pitcher and Hart 1982). Also, in freshwater systems, water-level fluctuations have clearly affected the growth rate of the walleye, *Stizostedion vitreum* (Carlander and Payne 1977).

Density-dependent growth. Competition for food resources has been supported as a major cause of changes in individual growth rates in populations with fluctuating abundances. Theoretically, if the food resources that are harvested by a particular fish population vary around some average level of availability, then an increase in population abundance should result in a decrease in growth rate, since the food ration available to each fish would be smaller. Conversely, if population abundance declines, individual growth rates should increase due to a higher available food ration.

Early experimental work (such as Swingle and Smith 1942) established that this relationship between population abundance and growth clearly exists, if all other variables that might affect growth, such as fluctuations in water temperature, are held constant. Indeed, natural populations that suffer a major, rapid decline in abundance have displayed subsequent increases in individual growth (as reviewed in Ross and Almeida 1986). Although some studies have found that fluctuations in abundance produce consistently appropriate increases or decreases in growth, others have not (Ross and Almeida 1986). Fluctuations of environmental factors such as water temperature and water level occasionally are found to have much greater impact upon growth rates than changes in population density (as in Carlander and Whitney 1961; Carlander and Payne 1977). Accumulated information about the relationship between population density and growth has been so inconsistent that fisheries scientists have independently presented all the following hypotheses: (1) abundance is clearly demonstrated to be a major factor influencing growth (Beverton and Holt 1957); (2) abundance consistently affects growth in natural populations but often is not discovered due to the difficulty of gathering accurate, precise measures of abundance (Backiel and Le Cren 1967); and (3) abundance is *not* "generally and systematically related to growth" (Weatherley 1972). Such conclusions hardly establish a consensus concerning the importance of population density as related to growth in natural populations.

Analyses of the Atlantic herring (Anthony and Fogarty 1985) and the silver hake (Ross and Almeida 1986) strongly support the importance of the effects that abundance has upon growth rates of individuals in populations during series of years when abundance is relatively high; these same biologists concluded that abundance has little effect upon growth rates in populations during prolonged periods of low densities. Anthony and Fogarty found that water temperature determined growth rates of herring during periods of low abundance. When population density is low over the long term, especially during periods of intensive human exploitation,

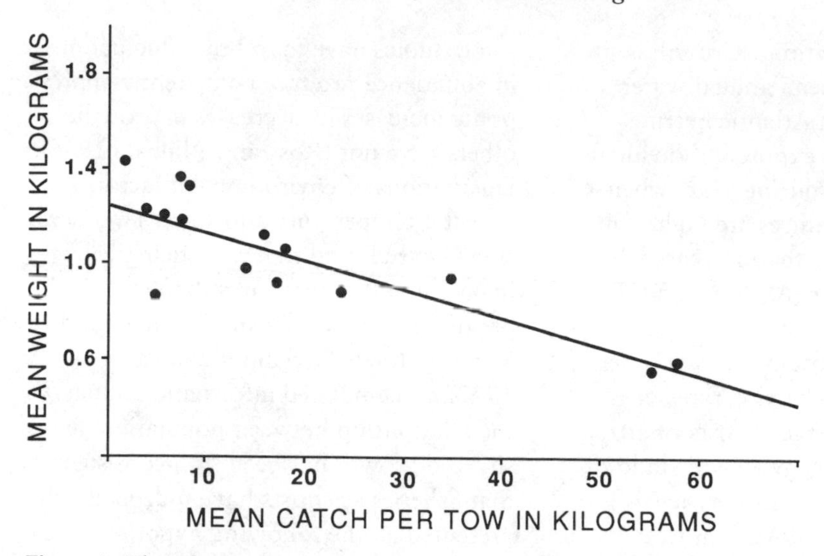

Figure 4. The relationship between weight of age-2 haddock from Georges Bank and population density. Density is an index measured as the average weight of all haddock captured per net tow during bottom-trawl research cruises of the National Marine Fisheries Service. Regression analyses compare the average weight of age-2 fishes captured during each year's cruise to the average biomass (density index) of the stock. Since weight at age is a result of growth conditions fish have been exposed to in the past, average weight of age-2 haddock is compared to the average biomass for the population 12 to 24 months prior to the time that individuals were captured and weighed. Regression line $r^2 = 0.73$.

food resources are not limited in availability to the relatively small number of fishes feeding upon them. Theoretically, annual fluctuations in population abundance will not change the food ration available to each individual, since the ration is not limited. However, when population density and competition for food are high, yearly fluctuations in the population's abundance will change the average ration available to each fish, thus affecting each fish's annual growth.

This hypothesis further suggests why studies of natural populations frequently do not show a correlation between fluctuations in population size and responses in growth rate. Many studies are focused upon economically and/or recreationally important species of fishes that are suffering high rates of fishing mortality and thus are being maintained over long periods of time at low population densities. One should not expect to find density-dependent growth in these conditions; if annual fluctuations in growth occur in these instances, it is likely that factors other than population density are stimulating the changes. Ross and G. A. Nelson (manuscript in preparation) have recently found strong correlations between population density and individual growth rates for two Northwestern Atlantic groundfish species, the haddock and yellowtail flounder. Figure 4 illustrates the correlation between weights of 2-year-old Georges Bank haddock, and an index of population abundance. Note that the data points that diverge farthest from the regression line are from time periods when population abundance was low. Ross and Nelson also found that changes in temperature as well as population density were correlated to changes in growth rates of haddock and yellowtail flounder during periods of low population abundance.

Age at Sexual Maturity

Fishes of coastal New England display a broad array of ages at which individuals reach sexual maturity. Individuals in species such as the Atlantic silversides spawn at age 1; at the other extreme, the Atlantic sturgeon (*Acipenser oxyrhynchus*) may not spawn until 28 years of age (Bigelow and Schroeder 1953; Scott and Scott 1988).

Although small-bodied species generally mature at younger ages than do large-bodied ones, this trend is most precise when comparing species of greatly different body sizes. The relationship is less clear when comparisons are made between species with only moderate differences in body size.

Age at sexual maturity is generally believed to be correlated to growth rate. An acceleration in growth rates and a subsequent lower age at first reproduction are typical life history responses during a decline in population abundance (as reviewed for herring in Ware 1985). North Sea Atlantic herring increased their growth rates by 25% and lowered their age at sexual maturity by 2 years during a period when the stock was undergoing a decline in abundance due to a growing fishery (Murphy 1977). Although this species displayed a significantly younger age at maturity, body size at first reproduction remained constant due to accelerated growth. Such observations have led many to conclude that maturity is size-dependent (as in Alm 1959) and is related to the body size that is required to provide sufficient energy to reproduce successfully and, for many species, to survive the spawning season. Roff (1982) suggests that maturity is most size-dependent in large-bodied, older-maturing species; constraints such as the energy and time necessary to develop eggs fully may require small-bodied, early-maturing species to reach a certain age before spawning regardless of their juvenile growth rates. Thus, for many species, age at sexual maturity may vary greatly between geographic areas, and among years within geographic areas, based at least in part upon the prevailing growth conditions that individuals are experiencing.

Mortality suffered by a species influences the evolution of age at maturity. Species that have consistently high juvenile death rates generally mature at relatively young ages. Thus the scup, which displays annual mortality rates for juvenile and adult fishes of up to 80%, spawns at 2 years of age (Finkelstein 1969, 1971). Although many Atlantic silversides survive their first reproductive season, nearly all adults die during the subsequent winter's offshore migration (Conover 1985); thus, nearly all individuals spawn at age 1 and die before age 2. The spiny dogfish, noted by Scott and Scott (1988) to have few natural enemies, may not spawn until 12 years of age. The Atlantic sturgeon, which may not mature until the age of 28 years, represents the other extreme from fishes such as the scup and silversides in the Northwest Atlantic. Some of the largest-bodied and probably longest-lived shark species have not been analyzed for age at maturity.

Characteristics such as age at maturity should evolve in a way that provides the greatest likelihood that an individual will achieve the maximum rate of offspring production of which it is capable. If juvenile mortality is consistently high, an individual has little likelihood of surviving an extended period of life as a juvenile before maturing at a relatively old age; the risk of death is far greater than any benefits associated with spawning for the first time at older ages. Thus, early age at sexual maturity should predictably evolve (Stearns 1976).

Many species suffer only moderate rates of juvenile mortality but do display relatively high adult death rates due to the intensive stresses caused by reproduction. If death induced by the stresses of reproduction is the greatest risk to future production of offspring, then older ages at maturity may evolve. By extending the period of life before sexual maturity, fishes can reach markedly greater sizes before they reproduce. Larger fishes should have greater energy stores available for activities associated with reproduction, since the surplus energy available for reproduction is clearly related to body weight, particularly for females (Ware 1985). For example, Ware has shown that the ovary weight of Pacific herring from British Columbia is a function of body size and does not change due to fluctuations in environmental conditions or population density. Thus, greater energy stores should allow an individual to produce a greater number of offspring. Therefore, the fish is risking spawning-induced death at a time when it is capable of producing markedly greater numbers of offspring than it might had it matured at a younger age and smaller size. Later age at maturity might also allow greater opportunity to survive the stresses connected with reproduction (production of gametes, migration, spawning, nest construction, offspring protection), since larger, older fishes presumably have greater stored energy reserves. This would allow a greater opportunity to live to spawn during some subsequent spawning season (Roff 1982).

If reaching sexual maturity at a relatively old age carries the joint advantages of increased fecundity and postspawning survival, is there an advantage to spawning when young, other than for those species that are suffering high rates of juvenile mortality? Theoretically, the decrease in fecundity, and thus in the number of offspring produced, caused by spawning at a young age and relatively small size may be somewhat offset by the reduction in time that is required to produce a new generation of fishes. By increasing the number of new generations occurring during a given number of years, an individual may produce more "descendants" during that time period than a more fecund individual that matures at an older age. Figure 5 illustrates the advantage of decreasing age at maturity, even at the expense of a substantial decrease in fecundity at maturity. Of course, such a simplistic representation does not consider such factors as surviving to reproduce in subsequent seasons, rates of mortality of both offspring and parent, the effects that growth after reproduction might have upon future fecundity, and other conditions that might influence the probable offspring production rate of an individual. However, it does demonstrate that increasing fecundity and lifetime production of offspring by increasing the age at maturity is not inevitably advantageous.

Fast-growing Atlantic salmon tend to remain in the ocean longer as juveniles before maturing and migrating into fresh water to spawn than do slow-growing individuals (Schaffer and Elson 1975). This seems counter to the general hypothesis that fishes tend to become sexually mature upon reaching a specific size, those growing faster thus reaching size at sexual maturity at an earlier age. However, Schaffer and Elson argued that the risk to the slow-growing Atlantic salmon of remaining at sea for long periods before maturing was

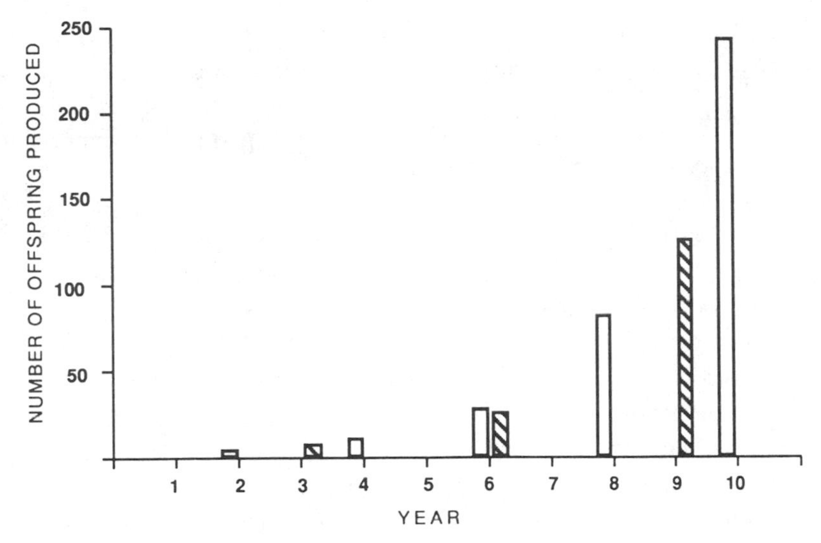

Figure 5. The hypothetical number of descendants produced within a 10-year time span by a female that matures at 2 years of age and produces three female offspring, and a female that matures at 3 years of age and produces five female offspring. This simplistic graph assumes that all females die after reproducing and all offspring live to reproduce once. The female that matures at 2 years of age is represented by the solid bars, the female that matures at 3 years of age by the hatched bars.

greater than the benefits gained from extending the juvenile life stage. Since such fishes are slow-growing, extra time spent as juveniles would not substantially increase the size at maturity; if fecundity is related to size, there would be little gain in numbers of offspring produced by the extended period of time spent as juveniles. However, this action would increase the time period of vulnerability to predation before such fishes ever spawn. Alternatively, the gain in fecundity accomplished by fast-growing fishes with prolonged juvenile life stage may be a greater benefit than the extended period of prespawning vulnerability to predation is a risk. Thus, slow-growing salmon may mature after as little as 1 year at sea, whereas fast-growing ones may remain at sea for as long as an average of 2.6 years before maturing. Schaffer and Elson (1975) found a highly significant correlation between age at first spawning and juvenile growth rates for salmon from 14 river systems in the Atlantic Canadian provinces.

Deferred maturity. Age at maturity may also consistently differ between sexes within populations of fishes. Females of many coastal species reach sexual maturity at a significantly older age than males. Such instances of female "deferred" maturity are common among the flatfishes (Roff 1982) and many other broadcast-spawning or egg-scattering fishes. Female deferred maturity is felt to be adaptive due to the increased body size gained by remaining a juvenile for a longer period of time. Larger body size allows greater egg production and may confer an advantage in surviving the spawning season. The production of suitable numbers of sperm by males is not constrained by body size. Thus, males that mature at a younger age have the advantage of reduced generation time but experience no disadvantage in gamete production.

Males mature at significantly older ages than females in some species. Male deferred maturity has evolved in species that possess intensive male social hierarchies as a means of competing for mates. In such species, body size of a male is a critical element in determining his capacity to defend a nesting site and/or to mate as often as possible. Small males in such species typically fertilize a limited number of eggs (as in Gross and Charnov 1980; Ross 1983), yet still face the risk of mortality associated with spawning activity. Since body size is of utmost importance for males in such species, maturity comes at an older age for males than for females, due to the additional growth that males must achieve while juveniles.

Reproductive Schedule

Another factor influencing the reproductive fitness of an individual is its expected reproductive schedule, which may be defined as the number of spawning seasons a fish participates in, modified by the probability that it will survive to participate in each successive spawning season.

Semelparity and iteroparity. Some fishes, such as the American eel and Atlantic silversides in New England and the salmon genus *Oncorhynchus* of the Pacific Ocean, have evolved a semelparous life cycle (that is, all adults die after their first spawning season). Some species that are iteroparous (having the potential to survive to spawn in two or more seasons), such as the scup, suffer such high mortality rates as adults that few individuals spawn in more than one year; individuals in many other iteroparous species, however, may spawn during multiple seasons before dying. Thus, when grouped, iteroparous fishes form a continuum of species, from those having only a moderate chance of spawning more than once to those in which the average life-span includes multiple spawning years.

Intuitively, longer life-spans that include multiple spawning seasons before death would appear to have an advantage over onetime spawning, due to the greater number of offspring a fish might produce in a prolonged adult life. However, many short-lived species displaying semelparity or limited iteroparity do not fit this pattern. Several theories have been advanced to address the evolution and advantages of various reproductive schedules in fishes (theories addressing the evolution of life histories in general are nicely reviewed in Stearns 1976).

First, semelparity should not necessarily be considered disadvantageous. Reproduction is generally highly stressful to fishes, requiring such great energy expenditure that many individuals from iteroparous species die from the physiological stresses or succumb to infection and disease or predation while in postreproductive weakened health. Energy reserves needed to resist stress and possible death obviously cannot be put into offspring production. In theory, long-lived species should maintain relatively modest levels of offspring production in order to maintain energy reserves necessary to survival and future reproduction. Alternatively, semelparous fishes may display markedly greater rates of offspring production since energy need not be reserved for survival. Thus, the inability of semelparous fishes to produce offspring over a series of years may be countered by the production of extremely high numbers of offspring in the one season in which they do reproduce.

There are several ways in which fishes increase the chance of surviving through multiple spawning seasons by conservation of energy use.

1. Gamete production by individuals may be lower than in fishes that spawn and die.
2. Maturity might occur at an older age, since older, larger individuals might have greater energy reserves available for spawning and postreproductive survival.
3. Individuals may not spawn in every year after reaching maturity, in order to grow and build up the required energy reserves.

Geographically separated spawning populations of the American shad exhibit both semelparity and prolonged iteroparity as reproductive schedules (Leggett and Carscadden 1978). The shad is a homing species, individuals typically returning to the river in which they hatched once they are ready to spawn. These authors have shown that shad spawning in river systems in the southeastern United States are semelparous, whereas increasing percentages of adults spawn at least twice in their lives as one proceeds northward along the Atlantic coast to Canada. Female shad spawning in Florida mature at a younger age, and smaller size, yet produce about three times as many eggs as do first-time spawners in iteroparous populations in the Miramichi watershed of New Brunswick (Leggett and Carscadden 1978). Indeed, even though spawning only once, the average St. Johns River (FL) female produces nearly twice as many eggs in her lifetime as does the average iteroparous female of the Miramichi River. Regardless of where they spawn, American shad are widely migratory along coastal North America during the oceanic, growth phase of their lives, generally experiencing the same array of environmental factors that may affect their growth rates. Yet southern shad grow significantly more slowly than do northern ones. Glebe and Leggett (1981) found that female shad in the St. Johns River use between 70% and 80% of available energy reserves to migrate into fresh water and spawn; females in the Connecticut River use from 40% to 60% of available energy reserves to migrate, spawn, and return to the sea. These researchers estimated that Florida females allocate about 1.6 times more energy to egg production than Connecticut River females. Thus, southern shad contribute greater energy reserves to reproduction and less to growth and postreproductive survival than do northern ones.

The semelparous life cycle produces a greater number of offspring for the female shad than does iteroparity. Additionally, the time it takes for a semelparous female to produce her next generation is shorter, further increasing her rate of production of descendants over that possible for an iteroparous female. (As discussed earlier, the number of possible descendants a female contributes to future gene pools depends not only on how many offspring she produces but on how rapidly she and her descendants produce each new generation. Animals that produce relatively modest numbers of offspring can still out-reproduce others that are more fecund but mature at older ages.)

Semelparity seems so advantageous for the American shad that one must question why this way of life does not dominate all shad populations. Leggett and Carscadden (1978) suggest that differences in the severity of egg and larval mortality account for the reproductive schedules that have evolved for different shad populations. They reason that rivers in the southeastern United States are relatively stable environments; thus conditions that might affect survival of young shad, particularly water temperature, are reasonably stable from year to year. Alternatively, northern rivers display greater year-to-year variability in environmental conditions, with annual fluctuations in temperature occurring in a random, unpredictable manner. Because southern shad spawn in conditions that are favorable for consistent yearly larval survival, maximum rates of offspring production have evolved, even though the adult does not survive to reproduce more than once. However, shad genetically "tuned" to spawn once and die in northern rivers would often contribute little or nothing to future gene pools, since the mortality of all (or nearly all) offspring in the years when conditions are unsuitable for survival would cause failure of an individual's entire lifetime reproductive effort. Under such circumstances, a life history should evolve that allows more than one opportunity to spawn, even if iteroparity requires that the rate of offspring production be reduced from that possible for individuals that spawn once and die. Spreading the reproductive effort over at least 2 years would increase the likelihood that some offspring would be spawned under conditions suitable for survival.

Several theorists (Murphy 1968; Charnov and Schaffer 1973; Schaffer 1974) have hypothesized that iteroparity and prolonged adult life-spans should evolve when the success of reproduction in a given season is highly unpredictable. The term "bet hedging" has been applied to this type of reproductive schedule, since the animals spread their reproductive effort through time in order ultimately to achieve some level of reproductive success, rather than putting all of their reproductive output into one effort that may be spectacularly successful but is more likely to fail totally. The differences in life history among populations of the American shad provide convincing support for this hypothesis.

Lambert (1984) suggested that environmental variability within seasons, which leads to differing levels of larval mortality, depending on the timing of hatching and first feeding, can lead to the evolution of prolonged periods of reproduction within spawning seasons. He felt that environmental unpredictability has led to the evolution of batch spawning in herring and capelins. Batch spawning, or producing offspring persistently throughout a prolonged spawning season, increases the likelihood that some larvae will hatch and initiate feeding during periods of optimal conditions and high food densities.

Nonannual spawning. Schaffer and Elson (1975) found variability in the reproductive schedules of Atlantic salmon spawning in different coastal watersheds. Salmon that migrate relatively short distances into fresh water to spawn mature at a younger age than do salmon that must make markedly longer freshwater excursions to reach spawning grounds. Additionally, salmon conducting longer freshwater migrations remain at sea at least 2 years between their first and second spawning seasons, whereas most or all that spawn in shorter river systems and survive will reproduce the following year. Schaffer and Elson related these differences to the energy reserves required to survive freshwater migrations. Surviving long river migrations requires greater energy expenditure than does surviving shorter ones; thus, older age at maturity and nonannual spawning (iteroparous species that do not reproduce in consecutive years after reaching maturity) evolved in populations traversing longer watersheds during their spawning migrations.

Although long-lived fishes are generally believed to spawn annually after reaching sexual maturity, a few species are known to be nonannual in their spawning cycles, as seen above for the Atlantic salmon; even though some portion of the adult population reproduces yearly, most individuals spawn on a less frequent than annual basis. The striped bass (Jackson and Tiller 1952) and the shortnose sturgeon *Acipenser brevirostrum* (Dadswell 1979; Taubert 1980) are two other nonannual anadromous species found in New England waters; the sturgeon family in general has a nonannual spawning cycle (Roussou 1957; Scott and Scott 1988). Individual female shortnose sturgeons may spawn no more frequently than every 8 years after maturity, although spawning frequency appears variable among populations. Although male shortnose sturgeons are also nonannual, they spawn more frequently than females; the spawning frequency of males is generally greater than females in nonannual species.

Bull and Shine (1979) suggested that lifetime offspring production might be increased by spawning nonannually, if reproduction itself significantly increases adult mortality or if survival requires storage of energy that would otherwise be channeled into gamete production. This might be particularly applicable if stresses of spawning cause a higher probability of adult mortality than does predation. If present reproductive activity serves as the greatest risk to future offspring production of an individual, then the most effective reproductive schedule might be one that includes prolonged periods between consecutive spawning seasons. These periods could allow fishes to grow substantially, since little or no energy would be diverted into reproduction during the nonspawning periods. Growth not only allows for greater storage of energy reserves to aid in surviving reproduction but, in the case of females, allows for greater egg production, since fecundity is related to body size. Thus, by spawning on a nonannual basis, a female markedly increases her offspring production each time she spawns, while increasing the probability that she will survive to spawn again in the future. Many nonannual spawners, such as the striped bass and shortnose sturgeon, can live for extremely long periods as adults. Therefore, even though they spawn nonannually, individuals can still participate in multiple spawning seasons. Since females benefit more from growing, in terms of gamete production, than do males, it is not surprising that they display longer time periods between consecutive spawning seasons.

Reproductive effort should increase with age, since an older individual risks less potential future offspring production if it dies due to present spawning activity than does a younger one (Williams 1966; Gadgil and Bossert 1970; Schaffer 1974; see section on Fecundity). Thus, nonannual spawning fishes should decrease the time between spawning seasons as they age. This has been shown for the freshwater nonannual white sucker *Catostomus commersoni* (Quinn and Ross 1985).

It should be noted that the rate of offspring production potentially suffers greatly when reproduction is less than yearly (see section on Age at Sexual Maturity). Thus, the benefits of increasing egg production and the number of reproductive seasons must outweigh the slowed reproductive schedule in order for nonannual spawning to have evolved. Based upon the relatively small number of species that are known to display this type of life history, these conditions may not often be met.

Iteroparity in the presence of high offspring survival. Although many species' reproductive schedules follow predictions established by the bet-hedging hypothesis, not all fishes do so. Sharks, and even the egg-laying skates and rays, are generally long-lived species, their life cycle typically including the potential to survive multiple reproductive seasons before death. They produce unusually low numbers of offspring during a reproductive season (see section on Fecundity); however, the energy required for a female to produce offspring in a given season cannot be assumed to be low, since live-bearing is an energy-intensive approach to reproduction. The life history of sharks and shark relatives suggests that offspring mortality should be high and should vary greatly from year to year based upon prevailing conditions, if the bet-hedging hypothesis best explains the adaptive nature of their reproductive cycle. However, offspring mortality of sharks is generally believed to be markedly lower than for most other fishes; further, offspring survival probably varies only modestly from year to year. What might explain the evolution of shark life histories?

One of the earliest hypotheses developed to explain the evolution of particular life history patterns was the "*r* and *K*" selection hypothesis (Stearns 1976; *r* represents the rate of population increase through reproduction, and *K* represents standing crop, or current population abundance levels). In this hypothesis, *r*-selected organisms live in a highly variable environment in which mortality rates are high. Such organisms mature at a young age, display maximum rates of offspring production, and have short adult life-spans. The constant, high rates of mortality keep population abundance low, providing ample opportunity for new individuals to enter the population and compete for limited resources such as food. Alternatively, *K*-selected organisms live in stable environments, in which mortality rates are low and adult populations (or standing crops) are maintained at consistently high levels of abundance. High population densities "saturate" the available habitats, making it unlikely that more than a few new individuals (i.e., offspring) can successfully enter the population and compete for limited resources in a given year. Thus, *K*-selected organisms are long-lived and produce relatively few offspring at a given time but reproduce over many seasons. The difficulty in explaining life histories of many animals using the *r* and *K* selection hypothesis (for example, American shad display iteroparity and long life-spans when offspring mortality is high, and shorter life-spans and semelparity when offspring mortality is lower) led ecologists to develop alternative hypotheses, including bet hedging. However, some life histories may be explained quite reasonably by the general tenet of *r–K* selection.

Juveniles and adults of many shark species are eaten by few predators and thus suffer relatively low levels of mortality. Further, although the marine environment certainly imposes very high rates of mortality on early life stages of many species, shark young-of-the-year apparently suffer low mortality rates. Prolonged periods of incubation, and live-bearing, have essentially provided shark offspring with a stable "environment" (inside the adult female), thus generating low mortality rates. In the absence of human exploitation (a mortality factor for which no species has a truly effective life history), sharks probably come as close as any fishes to representing animals that maintain saturated population levels. Stable environments, specifically the female's uterus, and low mortality rates are in this case associated with a long life-span and the seasonal production of a very small number of offspring throughout an extended period of years spent as adults.

Hermaphroditism. One final life history pattern displayed by the New England coastal fauna is worthy of note, due to its extremely unusual character. The black sea bass, a member of the marine family Serranidae, is a protogynous hermaphrodite; that is, at sexual maturity nearly all individuals reproduce as females, and at some time later in the life cycle the same fishes "switch" sexes and reproduce as males. Thus, at adulthood, black sea bass have fully functional, egg-producing ovaries; at the time of sex reversal, ovarian tissues become nonfunctional and previously inactive testes mature and initiate sperm production.

Although single individuals of many fish species have been found that possess both ovaries and testes, the vast majority of these are nonfunctional hermaphrodites, incapable of any reproductive activity due to an "accident" of embryological development. True hermaphroditism, in which individuals function fully as both sexes at some time during their life cycle as the normal mode of reproduction for that species, is an unusual phenomenon among all vertebrate groups and occurs in only a limited number of evolutionary lines of fishes, including the sea bass family.

Although the black sea bass is a protogynous (female-first) hermaphrodite, other species may be protandrous (male-first at maturity) or synchronous (producing eggs and sperm at the same time) hermaphrodites. Individual synchronous hermaphrodites do not self-fertilize as a rule but mix their eggs or sperm with gametes produced by other individuals. There are many interesting questions concerning this unique life-style; for example, how and under what conditions is sex reversal accomplished in sequential hermaphrodites (see Fishelson 1970 and Robertson 1972 for factors controlling sex reversal)? The rest of this section will focus on the adaptive value of hermaphroditism to the reproductive output of such individuals.

Early descriptions of hermaphroditism typically presented analyses of individual species, or small taxonomic groupings, of hermaphroditic fishes and did not present hypotheses that address the general adaptive significance of this reproductive system. Smith (1967) proposed that the adaptive significance of synchronous hermaphroditism lies in the increased number of offspring that can be produced by a population in which all adults are egg producers. He reasoned that total egg production in a nonhermaphroditic species is the product of the average number of eggs produced by females in a population and the number of adults in the population, divided by 2. The division is necessary because approximately one half of the adult population is male. He further reasoned that synchronous hermaphroditic populations will produce roughly twice as many offspring as nonhermaphroditic populations with similar average fecundities, since all, not just half, of the adults would be producing eggs. The total offspring production would likely be less than twice as large, since synchronous hermaphrodites must use some energy to make sperm that otherwise might have been used in egg production. Although this reproductive system indeed should produce greater numbers of offspring by the population, it is useful to discuss its evolution in the framework of reproductive advantages gained by individuals, since this is the means by which adaptations arise in populations. Offspring production of an individual will be the sum of all its eggs that are successfully fertilized by other individuals, plus all the eggs produced by other individuals that are fertilized by its sperm (Warner 1984). This number is obviously higher than if the individual were a female in a standard two-sex species. Also, this total offspring production is arguably higher than what many males might also accomplish in nonhermaphroditic species.

Warner (1975, 1984) has provided an elegant explanation of the adaptive advantage associated with sequential hermaphroditism. Protogynous hermaphroditism should have evolved in concert with a male-social-dominance hierarchical mating system. As described earlier (see section on Parental Care), male-dominance hierarchies provide greatest advantage to the larger, older males due to the size advantage associated with age. However, small young males are largely disassociated from reproduction, typically producing very limited numbers of offspring. Warner hypothesizes that young females produce more offspring than do young males in such mating systems. Further, large males produce markedly more offspring than do large females, since a dominant male may mate multiple times with numerous females, whereas offspring production by a female is still limited to the number of eggs she produces. Thus, an individual will maximize the number of offspring at each stage of its adult life if it reproduces as a female when younger and reproduces as a male starting at an age and size when it is capable of dominating the male spawning activity within the social system typical of that species.

Second, Warner proposed that when protandrous hermaphroditism evolves, it should do so in association with species that display random mating systems in which most males have essentially equal opportunity to spawn. Under such conditions, the average number of offspring a small male can produce will be essentially equal to the average fecundity of all sizes of females in the population, since even small males may have opportunity to mate with any size female in the population. The number of offspring a small female may produce will be somewhat less than the average fecundity of all females; since fecundity is related to size, small young females will produce smaller numbers of eggs (and offspring) than the average number produced by all sizes of females. Thus, young males will produce more offspring than young females in a random mating system. Conversely, large males will parent about the same number of offspring as small males (in other words, the average number for all females in the population), whereas the largest females will produce more than the average, since they are producing the greatest number of eggs. Thus, in random mating systems, an individual can maximize offspring production throughout its life by reproducing as a male when young and small and as a female when older and larger than the average-sized individual in the population. Warner presents a reasonable correlation between protogynous and protandrous hermaphroditism and the types of mating systems that are associated with them, as predicted by his hypothesis.

If maximizing production of offspring is a strong force in evolution, one might be tempted to ask why more, or even all, species of fishes are not hermaphroditic. However, this question is essentially irrelevant. Evolution is limited to natural selection acting upon the available genetic variability housed in gene pools, which will not always contain enough genetic potential to allow the evolution of what may appear best theoretically. We cannot question why something has not evolved; it simply has not. However, as Warner has shown, it is exciting to attempt to understand the adaptiveness of life history patterns that have evolved.

The Manager's Use of Life History Information

Knowledge of life history characteristics is centrally important to the development of management strategies based upon controlling harvest. Obviously, if an exploited population is to be maintained at some desired level, reproductive activity and growth must compensate for the portion of the population lost to harvest. Managing harvest cannot be successful without understanding the capacity of a stock to replace lost numbers and biomass. We have been far from perfect in managing harvest, as shown in chapter 3 and in the Species Profiles. Even though the unpredictable nature of successful reproduction certainly has an impact on the fisheries scientist's precision in predicting the effects of fishing, sociological and economic conflict has often influenced management's failures far more than any inability to predict the outcome of specific levels of harvest properly.

The history of fisheries management is fraught with an inability to match harvest levels with the potential of populations to withstand the depleting effects of fishing. It is abundantly clear that species have markedly different capacities to resist the effects of harvest. One of the more interesting future management activities in the New England region may be associated with the very abundant spiny dogfish and skate stocks off our coastal waters. These stocks reached historic highs in the 1980s (see section on Competition later in this chapter), when many other stocks had declined or collapsed. Federal programs aimed at aiding the fishing industry supported the development of domestic marketing opportunities for these species in order to take advantage of their tremendous stock sizes. Simultaneously, concern grew among fisheries scientists that these stocks would be unable to withstand substantial fishing pressure. Their late age at maturity and extremely low rate of offspring production make dogfish and skates extremely vulnerable to overexploitation.

When harvest management has protected stocks, it is because exploitation was successfully matched with rates of offspring and biomass production. The Chesapeake Bay striped bass stock offers an example of this interaction (see Striped Bass in Species Profiles). Although the decline of this stock from the 1960s to the mid 1980s was affected by a myriad of factors, ineffective control of harvest levels certainly influenced the rate of decline. The recovery of that stock that occurred in the late 1980s came about largely due to the absolute protection of fishes hatched in 1982, the first year of successful reproduction by that population in over a decade. Harvest of this age group of striped bass was prohibited throughout the rest of the 1980s, until they had matured and produced a strong year class of offspring. Complete protection from harvest allowed the 1982 year class to be markedly better represented in the spawning stock of the late 1980s than it would have been had harvest not been curtailed.

Interactions within Fish Communities

Along with their concern relating to the dynamics of fish production and the effect that exploitation has upon the capacity of populations to maintain particular biomass and abundance levels, fisheries scientists and managers are also interested in the nature of the effects that species have upon each other within fish communities.

In theory, removal of a portion of a fish population through fishing may change the predator–prey and competitive relationships of a species within the community in which it lives. The ramifications of depleting a particular fish stock may be far-reaching, affecting a variety of other fish species with which it normally interacts.

Predator–Prey Interactions

Early fisheries biologists felt that predators were capable of decimating populations upon which they preyed. Baird (1873) credited the decline of a variety of coastal New England fish species in the mid 1800s to predation caused by one species, the bluefish. Although this species has a reputation as a voracious predator (see Bluefish in Species Profiles), it hardly was capable of causing simultaneous declines of numerous fish stocks with which it had coexisted previous to that time.

This early philosophy of the detrimental effects that predators have upon fish communities persisted well into the 20th century. Interestingly, this thinking, essentially consistent with general ecological theory of the time, troubled the fisheries manager only when the prey species in a predator–prey relationship was a prized fishery; fish predators were taking individuals that humans might otherwise catch. However, when the predator was the prized member of a predator–prey interaction, its value as a game fish was enhanced by its reputation as a predator. Predator control programs conducted by fisheries management agencies paralleled similar programs conducted by wildlife biologists during the same period (i.e., "vermin" control programs aimed at eliminating terrestrial predators such as wolves, coyotes, cougars, foxes, owls, hawks, and other birds of prey in order to "protect" game wildlife species). Such fisheries and wildlife efforts shared a common thread: Humans were unwilling to recognize that their uncontrolled harvest made them the "predators" most responsible for the decline of the species they proposed to protect.

The most energetic predator control program conducted by fisheries biologists was directed toward the sea lamprey in the upper Great Lakes (Lake Huron, Lake Michigan, and Lake Superior) of the central United States. After its arrival in these watersheds, this introduced species was blamed for the collapse of native lake trout stocks, even though these stocks had suffered through well-documented prolonged population declines before the lamprey had become established (see figures in Berst and Spangler 1972). These declines were the result of fishing. No one can determine what effect this newly introduced predator might have had upon lake trout stocks had they not been so severely overfished.

By the mid 20th century, fisheries scientists were developing a better understanding of the manner in which predator and prey interact. Efficiency of food gathering by predators correlated closely with availability of prey, with prey abundance serving as a major factor in determining availability. Predator–prey interactions between the predaceous walleye and the yellow perch (*Perca flavescens*) provide an excellent example of interactions between a single predator and a single prey species. When walleye populations increase in abundance, their feeding activity may lead to a population decline of the perch. However, as this prey species declines in abundance, the efficiency with which the walleye can harvest it also declines. As a result, walleye switch to other prey (including younger, smaller walleye), which leads to gradual recovery of the yellow perch population (Forney 1977). Such cyclic changes in population abundances of predator–prey complexes have been identified elsewhere in fish communities. Recruitment rates (the addition of new individuals to a stock due to successful reproduction and survival of offspring) of the Pacific cod are positively correlated to the abundance of a major food, the herring, and recruitment rates of herring are inversely related to the intensity of cod predation. Thus, population densities of these two species cycle up and down through time, probably due at least in part to the nature of their predatory–prey interaction (Walters et al. 1986).

In theory, switching from one prey to another occurs when the cost of energy expended to capture prey begins to have a negative impact on the energy gained from eating it. Such switching does not always provide fully suitable feeding conditions for the predator. Magnusson and Palsson (1989) demonstrated that Icelandic cod consume their major prey item, the capelin, at a rate that is positively correlated to the capelin's abundance. When the capelin stock declines, cod consumption of capelin decreases accordingly. Additionally, total consumption by cod of all foods declines; thus, this predator cannot fully compensate for the loss of a major prey by feeding upon other species. Although predators certainly influence the status of prey populations, the availability and abundance of food determine the foraging success and ultimately affect the population status of the predator species.

In this sense, humans are not functioning as just another predator in the system, as fishing is sometimes characterized. As the amount of energy needed to catch a certain amount of food goes up for the fish predator, it changes its foraging habits and pursues other, more abundant prey. Indeed, when fish stocks decline, the commercial fisherman may find that his or her costs rise in relation to profits earned in the harvest, since more time and money are spent to catch "prey." However, as the biomass of fishes harvested per given amount of fishing effort goes down, the fisherman continues to fish even harder, as long as the increasingly less available fishes earn more dollars per pound at the market; typically, as the abundance of a highly desired food species goes down, its value will increase due to its rarity. Thus, fewer fishes may still bring a profit if the value per pound rises suitably to counteract their increasing rarity. Humans also possess a trait that predatory fishes do not; they are stubborn. Fishermen tend to stick to a particular "foraging" behavior, even after the efficiency of that behavior begins to decline. This will be particularly true if more abundant fisheries resources that are available for capture have little market value. Thus, though fish predator–prey interactions may tend to achieve some general stability in time, human predator–prey interactions have much less tendency to do so.

Competition

The thinking of fisheries scientists concerning the effects of competitive interactions in fish communities has proceeded through an evolutionary process similar to that seen for theories concerning predator–prey systems. Early researchers investigated competition in fish communities by conducting stomach analyses of species in freshwater communities. These studies led researchers to conclude that all species are generalist feeders, eating a wide variety of food types resulting in a wide overlap of diets among species. Overlap was felt to constitute competition among species, which was credited with shaping the makeup of the fish communities that were studied (Hartley 1948; Larkin 1956). Keast (1965, 1966) was one of the first fish ecologists to question this hypothesis. General ecological thinking supported the view that animals are able to coexist in a community only if they fill separated, largely nonoverlapping niches. Further, many morphological adaptations exhibited by particular species, such as body form, mouth size, and size and shape of teeth, evolved as a means of allowing animal species to become niche specialists (Keast and Webb 1966), gathering particular food resources in an efficient manner while reducing interspecific competition to a level that results in coexistence with other species. Keast demonstrated that a large portion of the feeding overlap found by earlier researchers was an artifact of somewhat crude analytic techniques. Identification of food items found in fishes' stomachs was usually very general, with foods being categorized no more precisely than "unidentified fish remains, insects, microcrustaceans," and others. A group of fish species can easily feed primarily within one or more of these broad categories without eating the same species of prey. Keast (1965, 1966) further showed that species that do eat similar types of food may gather them from different habitat or microhabitat types; thus, they would not realistically be in competition for the same food resources. Since individuals within the same species are most similar in morphology and behavior, intraspecific (within species) competition should be the major force, excluding human interference, shaping the status of fish populations. Subsequent work, such as that by Zaret and Rand (1971), clarified the difference between feeding overlap that truly might exist and actual competition. These researchers found broad feeding overlap among species only during times of the year when food resources were rapidly increasing in abundance; during times when this was not occurring, food resources were more limited, and individual fish species exhibited specialized foraging activity that markedly reduced the variety of food types they were gathering. This specialized foraging significantly reduced the level of feeding overlap between and among fish species. Zaret and Rand concluded that, in the fish community they studied, feeding overlap was indicative of the absence rather than the presence of competition, since overlap occurred during periods of high food abundances and did not occur when food abundances were low. (By definition, competition only occurs when food resources are limited in relation to the needs of the consumers).

Werner and Hall (1976) conducted controlled experiments to determine the effects of interspecific competition in a pond community. They found differences in habitat use and foraging behavior of sunfish species, depending on whether other closely related species were present or absent. They felt this change in foraging was the result of direct competition for food and shelter resources. As a result, interacting species exhibited a reduction in growth rate and reproductive rate and an increase in mortality. Thus, these researchers demonstrated that, although species may be morphologically and behaviorally specialized enough to allow coexistence, interspecific competition can affect population characteristics of the interacting species.

Interactions in the Northwest Atlantic

The accelerated rates of harvest occurring in the Northwest Atlantic in the last several decades offer an opportunity to see the results that interspecific interactions may have in our coastal ecosystems. The Atlantic herring and Atlantic mackerel are major components of the Northwest Atlantic pelagic fish community. Analyses of long-term data sets have indicated that these two species have alternated in dominating the biomass of that fish community. Skud (1982) determined that population size of whichever species happened to be dominant (that is, most abundant) during a particular period of time would increase during years of higher water temperature and that the abundance of the nondominant species would simultaneously decline. When dominance (in terms of total biomass) changed from one species to the other, the manner in which the biomass of each species responded to temperature would change accordingly. The periodic changes between species in dominance occurred in part due to changes in patterns of fishing. Skud concluded that these species experience some level of competition for food and that each acts as a predator upon the young of the other. Shifts in dominance were stimulated by the fishing industry, creating a competitive edge for the species that was fished less heavily. Increased water temperature enhanced growth and survival, and thus biomass and abundance, of the dominant species; the resulting increased competitive pressures between the species would cause a decline in abundance of the nondominant one.

Interestingly, Skud noted that the larvae of a third species, the sand lance, increased about 20-fold in the late 1970s when both the herring and mackerel stocks were simultaneously experiencing periods of low abundance and poor recruitment. Sherman (1981) explained this and a similar sand lance population increase that occurred in the North Sea as being the result of the replacement of midsized predators by smaller, faster-growing species when the predators were being depleted by fishing. The diets of the sand lance and mackerel are very similar, and small sand lances are fed upon by adult mackerel. Bowman et al. (1984) felt that the decline of both herring and mackerel in the late 1970s left the sand lance with abundant food supplies and less predation pressure, resulting in the rapid increase in lance abundance. The increase in abundance evidently stimulated other responses in the pelagic fish community of the Northwest Atlantic. By the early 1980s the humpback whale, a major predator of the sand lance, was densely concentrated in areas that had the greatest abundance of this prey species (Payne et al. 1986). The shift in whale distribution to the Stellwagen Bank and other areas of the Gulf of Maine that are close to the New England coastline stimulated an immediate growth of the whale-watching industry, a multimillion-dollar industry based upon people's extreme interest in seeing these giants of the ocean. By the mid 1980s, this industry became increasingly concerned about a redistribution of whales to more offshore areas, out of reach of many half-day and even whole-day whale-watch tours. This redistribution occurred simultaneously with a decline in abundance of the sand lance population. The sand lance decline coincided with a recovery of the mackerel stock, an interaction that had been predicted by Bowman and his colleagues in 1984. Thus, changes in abundance of closely interacting species stimulate changes throughout the biotic community in which they live.

The spiny dogfish and a number of skate species, particularly the little and winter skates, are central figures in another instance of species replacement that has occurred in the Northwest Atlantic. As late as 1980, Grosslein et al. published detailed analyses that indicated that the very high rates of exploitation of Georges Bank bottom species, such as haddock, cod, and yellowtail and other flounders, had led to significant declines in abundance of these species but had neither stimulated a shift in the species composition of that community nor created a large increase in abundance of any largely unexploited species. However, elasmobranchs (principally the spiny dogfish and skates), which had constituted about 21% of the total weight of research vessel catches on Georges Bank in the 1960s, made up about 75% of the total weight of catches by 1989 (Murawski and Idoine, in press). Murawski and Idoine also indicate that, although the species makeup of the survey catches had changed markedly through the years, the size composition of the catches had not. Apparently, not only will the available food resources of the Georges Bank region support an average total biomass of fishes, but that biomass will possess a certain distribution of sizes of fishes, regardless of the abundance of particular species. The highly selective fishing of the U.S. fleet may be replacing larger-bodied marketable species such as haddock, cod, and flounders with similar-sized unmarketable species such as spiny dogfish and skates. Murawski and Idoine suggest that this replacement has occurred because the elasmobranchs are taking advantage of food resources that had previously been "shared" with gadids and flounders; dietary overlaps between flounders-skates and cod-dogfish are typically broad (Grosslein et al. 1980). The fact that Grosslein and colleagues had not yet recorded this replacement in their 1980 publication is not surprising. The reproductive rates of spiny dogfish and skates are low enough that population expansion would not occur rapidly, even if food resources and other needs became markedly more available.

Fisheries scientists are concerned about this replacement in the groundfish community of Georges Bank. If elasmobranch abundance has stabilized at the markedly higher level of the 1980s, recovery of the traditionally important groundfish species may be quite difficult, even if fishing pressure upon them is sufficiently controlled. Ironically, one possible alternative that might stimulate recovery of traditional species would be the development of a market for elasmobranchs that would be large enough to lead to their overexploitation and subsequent stock decline.

3

Managing Our Fisheries Resources

In order to accomplish government-initiated programs in fisheries resource conservation, there must be an organizational structure (for example, an agency) that has the authority to conserve, and programs that the agency carries out. The structure or framework of management is created by the passage of laws. These laws may establish agencies or other entities, or they may specifically define the responsibilities those agencies, or government in general, have in fisheries conservation. The framework of the fisheries management system in the United States has undergone a prolonged process of change, due in part to the ineffectiveness of early approaches and in part to the changing needs of conservation in the face of ever-expanding human use of fisheries resources.

Once the organizational framework is created and responsibilities are defined, programs must be developed to help achieve conservation goals. It should be pointed out that conservation is often not synonymous with preservation. By definition, fisheries resources consist of those marine populations that interact in some way with humans. Note that "fisheries" resources are not limited to fish populations but reference other aquatic organisms as well that are used in some way by humans. Uses are most frequently consumptive; in other words, populations are harvested for use as human foods, animal feeds, fertilizers, or in a variety of other products. However, some resources interact with humans in a nonconsumptive way. The whale-watching industry depends upon such a nonconsumptive, or nonharvesting, use (I am defining whales as marine fisheries resources, since federal legislation places the responsibility of overseeing their conservation with the National Marine Fisheries Service).

State and federal efforts at fisheries conservation are usually multifaceted in design and may include public education, habitat protection, provision of public access to the resources, and regulation of harvests. Controlling harvest often brings the manager closest to direct control of the interaction between humans and the resource. Also, regulations are usually the most visible actions taken by resource managers in the eyes of many recreational and commercial users of fisheries resources.

In this chapter, a historical look at the

evolution of the management system will be presented, focusing on how major federal legislative action has changed the approach we use to manage our coastal fisheries. State agency functions will be mentioned, but the major focus will be toward federal initiatives, since most coastal fisheries resources are cosmopolitan enough to span the borders of numerous states, thus requiring federal assistance to accomplish conservation needs. After this review, fishing regulations typically used to control recreational and commercial harvest will be discussed along with an overview of the role that particular types of regulations are intended to play.

The Evolution of the Management Process

The history of fisheries management in New England coastal waters dates back to the period of European settlement of the region. Exportation of salted fish to Europe was a major component of the economy of colonial New England. Coastal fisheries resources, and the waters in which they lived, were felt to be public (or "common") property; thus, fishes were available to all and could be owned only when caught (Royce 1989). However, it was recognized that common property required public responsibility. Thus, during the colonial period the first regulations to address conservation of these resources were established. For example, in 1639 the General Court of the Massachusetts Bay Colony prohibited the further use of cod and striped bass as fertilizer for farming. Revenues from the sale of striped bass were taxed to support the first public school in the New World and to provide aid to widows and orphans of men who had formerly served the Massachusetts Bay Colony (Alperin 1987). Plymouth Colony restricted early-season mackerel fishing in the 1670s and prohibited the use of seines in the fishery by 1684 (Hennemuth and Rockwell 1987).

National concern over fisheries was displayed soon after the American Revolution. The Massachusetts fishing industry, at that time focusing upon the salting of cod and the rendering of whale for oil, was encountering stiff import duties on its products in Europe and at the same time facing competition with European fishing industries that were receiving generous subsidies from their governments. These factors reduced the availability of a market for New England fish exports and had an impact on domestic sales of these products. This severely burdened the economy of Massachusetts, since at that time about 4,000 people were directly engaged in each of the fisheries, the total more than twice as high as the number of commercial fishermen active in Massachusetts in 1980 (Royce 1989). In 1791, Secretary of State Thomas Jefferson asked Congress to support the Massachusetts fishing fleet so that it could compete favorably with subsidized foreign fleets. In 1792, Congress granted subsidies to licensed cod vessels, a program that lasted until 1838 (Hennemuth and Rockwell 1987). Thus, the first national concern addressed the health of the fishing industry, not of the fisheries resources that supported it. This pattern persisted well into the 1800s, with most federal actions consisting of international treaties concerned with jurisdictional disputes over landing rights of fishing fleets rather than disputes over fishing rights or restriction of harvests for conservation purposes (Graham 1970).

Although New Englanders first fished Georges Bank in the mid 1700s (Hennemuth and Rockwell 1987), exploitation of the vast offshore fisheries resources was limited until the mid 1800s, when technological advances such as the invention of steam power and the concurrent development of the otter trawl vastly improved the efficiency of offshore fishing. Cod, mackerel, and halibut were the targets of these fishing efforts; the halibut stock showed the first signs of decline by about 1850. However, inshore and anadromous fish stocks had already exhibited distinct signs of stress prior to this period. Settlement of New England coastal watersheds led to very early decreases in abundance and, in many cases, loss of local populations of the Atlantic salmon, American shad, river herring, rainbow smelt, and striped bass. The combined effects of overfishing, habitat degradation caused by deforestation and subsequent agricultural and industrial land-use activities, and the complete elimination of migratory pathways due to dam construction severely reduced the abundance of anadromous fish spawning runs in numerous rivers and streams. For example, overfishing had eliminated the striped bass and sturgeon from the Exeter River in New Hampshire by 1762, and construction of dams had destroyed an alewife run in the same watershed prior to 1790 (Bowen 1970). Atlantic salmon runs were eliminated, and American shad runs severely depleted, in the Merrimack River shortly after construction of the first dam on the river's main stem in Lawrence, Massachusetts, in 1847 (Stolte 1981). Such losses of anadromous fisheries were being noted throughout New England coastal regions, and localized stocks of near-coastal species were declining due to the growing harvests imposed upon them by the mid 1800s.

The Establishment of Fisheries Management Agencies

The depletion of New England's coastal fish stocks was a major impetus in the development of government agencies whose responsibilities focused upon fisheries conservation. New England states were among the first to establish agencies to study the condition of fish stocks and to recommend conservation measures to protect them. In 1856, the Massachusetts legislature appointed the state's first fish commission to study the status of the state's migratory fish stocks (Bowen 1970). A New Hampshire Fish and Game Commission report in 1857 decried the loss of shad and salmon in the Merrimack and Connecticut rivers and recommended actions to restore them to their historical range (Stolte 1981). In 1865, the New Hampshire Fish Commission initiated a program to restore migratory fisheries to these watersheds. This became an interstate effort a year later, when the Massachusetts Fish Commission joined the inquiry (Stolte 1981). All six New England states had fish commissions by 1871, when the U.S. Congress established the Office of the Commissioner of Fish and Fisheries, appointing Spencer F. Baird as first commissioner. Baird, a renowned naturalist who had served for years as assistant secretary of the Smithsonian Institution, first became aware of the growing problem of fish stock depletions, and the decline of the fishing industry, during collecting trips to coastal New England. Prior to the establishment of the commission, he had already influenced the Smithsonian to expend funds to study these declines, with surveys being conducted out of Woods Hole, Massachusetts, in 1870 (Bowen 1970). Baird was the major figure lobbying Congress for a federal commissioner's office, and the initial mission of the agency as written in the legislation was a reflection of his thoughts: The commissioner was to "ascertain whether any and what diminution in the number of foodfishes of the coast and inland lakes had occurred." Baird had not originally proposed the study of inland lakes, but these waters were included in the legislation to gain critical congressional support from midwestern states. It was no accident that Baird was the first fish commissioner. The legislation requiring that the commissioner be a federal employee "of proved scientific and practical acquaintance with the fishes of the coast" was written in order to guarantee his appointment (Bowen 1970). In less than 2 months, Baird had successfully lobbied for the creation and funding of a new federal agency and had managed to become its first commissioner.

The first surveys of the Commissioner's Office were conducted in the summers of 1871 and 1872 out of Woods Hole. These surveys focused upon fisheries resources of the southern coast of New England. Interestingly, in his first report to Congress (Baird 1873), Baird laid blame for the decreased abundance of fish stocks on the voracious appetite of the bluefish (see Bluefish in Species Profiles for a colorfully anthropomorphic quote), although he did allow that fishing and other human activities had some influence on the status of the fisheries resources. Based upon this latter conclusion, he recommended that the states (and if they did not act, the federal government) regulate the use of pound and trap nets in coastal waters. These recommendations were not enacted.

Development of the federal hatchery system. After its initial years of surveys, the commission embarked upon an ambitious rearing and stocking program, to increase the abundance of depleted stocks and to establish new species in areas where they did not naturally occur. Various states had instituted hatchery programs prior to start-up of the federal effort; for example, over 92 million shad fry had been stocked in the Connecticut River before 1873 (Baird 1874). In the late 1800s, regulating harvest by using scientific analyses of the condition of fish stocks had not yet developed as a management tool, due to a lack of understanding of the dynamic nature of reproduction, growth, and mortality in fish populations. However, knowledge about rearing fishes in captivity was expanding rapidly. Thus, fisheries management programs were expressions of what managers of that time knew best, the operation of hatcheries and stocking of fishes. Coastal waters were a primary target of the federal hatchery activities throughout the period between 1873 and 1940. By 1940, more than 98% of the eggs and 75% of the fry released were marine species, mostly flounders and members of the cod family (McHugh 1970). Throughout this time, New England coastal waters were a focal point of the coastal stocking activity.

The federal program reached astounding proportions shortly after its inception. In 1900, over 265 million cod fry, 87 million flounder fry, and 81 million lobsters were released along the New England coast. The growth in hatchery activity in New England was stimulated by the ills of the fishing industry; annual commercial landings had dropped by 260 million pounds (nearly 40%) between 1889 and 1900 (Bowers 1901). Stocking efforts continued to increase steadily up to 1940, with the focus on cod, flatfishes, and lobsters, although other species such as the Atlantic mackerel, pollock, haddock, and sea bass were also stocked in large numbers (Bowers 1911; Smith 1921; Bell 1940). In 1938, nearly 0.5 million winter flounder fry and over 5 billion flounder, haddock, pollock, and cod eggs were stocked out of a federal facility in Gloucester, Massachusetts. In that same year, a hatchery at South Boothbay, Maine, stocked over 1.2 billion cod, flounder, and haddock eggs and fry, and Woods Hole contributed another 500 million flounders. Not all activity was conducted out of hatchery facilities. Federal agents also stripped eggs and sperm from fishes captured for market by commercial boats and returned the fertilized eggs to the water while still at sea. This program was intended to put fertilized eggs back into the ocean that otherwise would have been lost since the adult fishes were headed for market. A curious twist on this practice was the apparent use of the U.S. Fish Commission steamer *Fish Hawk* to strip gametes from fishes, then hatch the eggs on board, returning the larvae to sea immediately after hatching (Royce 1989); essentially, other than protecting eggs from natural mortality while on board the vessel (while risking mortality for eggs and adults by handling them), the *Fish Hawk* was merely doing what the fishes would have accomplished had they been left alone!

During the period of expansion of the federal hatchery system, the center of federal activity for anadromous species was in the Middle Atlantic states, even though depletion of New England anadromous stocks was an initial impetus in establishing the U.S. Fish Commission. New England waters did receive a portion of this hatchery production. The U.S. Commission stocked more than 8 million shad fry in the Connecticut River between 1872 and 1880; however, the Massachusetts and Connecticut fish commissions stocked 90 million into the same watershed in 1872 alone. States generally exceeded federal hatchery production of anadromous species all along the Atlantic coast (Bowen 1970).

Although hatchery programs have recorded some success in stocking freshwater and anadromous species, the massive stocking of marine species was terminated by the mid 1900s because there was no evidence that fish stocks were even minimally influenced by the effort. Some simple calculations provide a probable reason for this failure. The species receiving greatest attention all exhibit extremely high levels of egg production, and their eggs and larvae suffer extremely high natural mortality rates (see individual species Profiles and the section on Egg and Larval Mortality in chapter 2). The 5 billion fertilized flounder, cod, haddock, and pollock eggs stocked out of Gloucester in 1938 represent roughly the same number of eggs produced by 5,000 females, if one assumes the average female carries about 1 million eggs (a conservative number for these species). Even if stocked eggs and larvae suffer no more than normal mortality rates after their release, few adult fishes would result from such an effort. Considering the vastness of the Northwest Atlantic, and the great number of adult fishes present even in depleted stocks, replacing the spawning effort of a few thousand fishes would not have any effect upon the abundance of populations. Thus, the massive human effort of the federal hatchery system was minuscule compared to the level of offspring production necessary to enhance the size of coastal fish stocks in the face of uncontrolled exploitation.

Trend toward management of harvest. The condition of coastal fisheries resources continuously declined from the latter half of the 19th century onward, in large part due to technological advances, including the invention of the steam vessel, the improvement of gear such as the otter trawl, and the development of freezing systems that allowed better on-ship storage and subsequent shipping of fishes to interior regions of the continent that historically had been too distant to serve as a market for many New England fish products. By the late 1800s it had become increasingly clear that managing harvest was necessary to retaining the quality of the nation's coastal fisheries resources.

Regulation of fishing of inland, freshwater fish stocks for conservation purposes was common by the turn of the century; states restricted catch and even prohibited sale of some species. Thus, the states had gradually established the first fisheries programs to attempt conservation by limiting harvest. During that time, states also placed increasing emphasis on managing fisheries resources for recreational rather than commercial uses (Royce 1989). This priority was largely due to the growing political power of the angling public in inland regions of the country.

Regulating harvest of coastal fisheries was not as easily accomplished. State jurisdiction in coastal waters extended but 3 miles out from the shoreline. Nearly all heavily exploited marine fish stocks of New England waters extended along the coast across multiple state boundaries and, in many instances, far oceanward to waters well beyond state and U.S. territorial boundaries. Thus, managing harvest required interstate and/or international agreements. Such agreements are often ineffective, since participants in such fisheries are most concerned with garnering their "fair share" of the benefits available from harvesting fisheries resources; too many fair shares often result in poorly managed, depleted fish stocks.

Although recreational fishing interests were more influential than commercial fishing ones in most inland regions of the country, initially they were not influential in coastal regions, where the fishing industry had a long history of economic and cultural importance. It is much easier for recreational anglers than it is for commercial fishermen to identify the value of conservation through harvest management. Fishing for many anglers is in part an aesthetic experience. Even though catching fish is important, anglers generally place a higher value on other aspects of the angling experience, such as enjoying nature, getting away from pressures, and being with family and friends (Dawson and Wilkins 1981). Limiting present harvest to provide the opportunity for future angling experiences should be in the self-interest of this user group. On the other hand, commercial fishermen, whose living is based upon the level of current fishing success, have often had great difficulty parting from the concept referred to by Royce (1989) as "the public common of piscivory"; that is, fisheries resources are available for unlimited use by all. Regulations are viewed in terms of the negative impact they might have upon today's profit with little regard for the value a stable resource might have in the future. Harvest restrictions are not easily supported when future value must be compared to earning a living in the present. Thus, it seems inevitable that advances in fishing technology, which were spurred by the increased profits they afforded the commercial fishing industry through more efficient harvest, led to increasing overharvest and depletion of fisheries resources.

Change in structure of the federal management system. The natural conflict between philosophies held by recreational and commercial interests has been reflected in the frequent reorganization of the federal agency responsible for managing fisheries resources (the history of the reorganization is shown in Table 3). In 1903, the Office of the U.S. Fish Commission was renamed the Bureau of Fisheries and was placed in the newly created U.S. Department of Commerce and Labor. In 1913, the Department of Commerce, with its Bureau of Fisheries, was separated from

Table 3. Placement of the responsibility for managing coastal fisheries resources within the federal government system

1871	Office of the Commissioner of Fish and Fisheries established by Congress
1903	Office of the Commissioner placed in the newly created Department of Commerce and Labor; renamed the Bureau of Fisheries
1913	Department of Commerce, and its Bureau of Fisheries, separated from the Department of Labor
1939	Bureau of Fisheries and Bureau of the Biological Survey transferred into the Department of the Interior
1940	Bureau of Fisheries and Bureau of the Biological Survey joined as the new Fish and Wildlife Service
1956	Fish and Wildlife Service reorganized into the U.S. Fish and Wildlife Service, consisting of the Bureau of Commercial Fisheries and the Bureau of Sport Fisheries and Wildlife
1970	Bureau of Commercial Fisheries abolished; responsibility for overseeing marine fisheries returned to the Department of Commerce in the newly created National Marine Fisheries Service within the National Oceanic and Atmospheric Administration

Labor. Thus, the emphasis on commercial fishing and its economic well-being (which played a major role in the original establishment of a federal fisheries resources agency) was clearly reflected in the placement of the agency within the federal system. In 1939, under President Franklin D. Roosevelt's government reorganization plan, the Bureau of Fisheries was transferred to the U.S. Department of the Interior, where it was combined in 1940 with the Bureau of the Biological Survey to form the Fish and Wildlife Service. It is no surprise that this reorganization occurred during a period when fishing as recreation and support for conservation and protection of fisheries resources in the interest of the public as a whole were gaining greater national influence. The conflicting needs of commercial and recreational interests ultimately led to the creation of the Bureau of Commercial Fisheries and the Bureau of Sport Fisheries and Wildlife, both within the Fish and Wildlife Service. The emphasis on recreational use of inland fisheries as opposed to the emphasis on the commercial importance of coastal ones ultimately led to the termination of the Bureau of Commercial Fisheries; the responsibility of managing marine fisheries resources was transferred back to the Department of Commerce, within its National Oceanic and Atmospheric Administration (NOAA); the fisheries bureau was named the National Marine Fisheries Service (NMFS). This placement reaffirmed the priority of the economic health of the fishing industry in the plans for management of the country's coastal fisheries resources. This organizational framework remained through the 1980s, although there was occasional pressure, particularly from national recreational fishing interests, to return the NMFS to the Department of Interior. The clear intent of such pressure was to place the management of coastal fisheries resources with a parent agency clearly identified with overseeing the management of such resources for the benefit of the public as a whole, both now and in the future.

Managing Offshore Fisheries Resources

International Convention for the Northwest Atlantic Fisheries. Collapse of the New England haddock fishery in the early 1900s was perhaps the single greatest impetus leading to the first efforts to manage exploitation in waters offshore from state jurisdiction (Graham 1970). Haddock landings in New England increased from 50,000 mt in 1925 (40% higher than cod landings that year from the same region) to over 125,000 mt (275 million lbs) in 1929. Landings decreased to half that level by 1932, due to declining abundance of haddock. Voluntary efforts by New England fishermen to reduce catches of small haddock by using trawl nets with larger mesh sizes were short-lived. Since the fishery was being conducted in international waters where many vessels continued to fish using unrestricted gear, those using less efficient large-mesh trawls were in essence "volunteering" themselves into an economic disadvantage.

The haddock fishery progressively worsened, leading nations fishing the offshore banks of the Northwest Atlantic from the Canadian Maritime Provinces to Rhode Island to sign the International Convention for the Northwest Atlantic Fisheries (ICNAF) in 1949. This convention established a commission whose responsibility was to conserve fisheries in a manner that would allow "maximum sustained catches" from fisheries of that region. ICNAF and several other conventions negotiated in the same period (including the International Whaling Convention of 1946, the Inter-American Tropical Tuna Convention of 1950, and the International North Pacific Fisheries Convention of 1953) required that the federal government assume a much greater role in fisheries management. These conventions led Congress to establish the Bureau of Commercial Fisheries within the Fish and Wildlife Service in 1956 (Royce 1987; see Table 3).

Mesh-size regulations were established for the Georges Bank haddock fishery within 2 years after ICNAF initiated activities. These regulations were intended to allow small, preadult fishes to escape through the net, thus preventing harvest of the prereproductive segment of the population. (Otter trawls typically cause high mortality of fishes landed from deeper waters, even if returned overboard immediately after capture.) Mesh regulations essentially reduce the efficiency of collecting gear (see Regulating Gear later in this chapter). The intent of such regulations is to control harvest without restricting vessels from fishing where or as often as they wish. Individual fishermen are given free choice to continue to fish, but they are required to work harder to overcome the mandated inefficiency of the gear. Thus, when using mesh regulations alone, managers are attempting to protect the resource while restricting the efficiency, but not the activity, of the fishermen. However, such a management strategy may simply stimulate fishermen to fish longer and harder to make up for the loss of catch created by the gear restriction (Everhart et al. 1975). At the time, this was not considered a problem by ICNAF. The U.S. fleet, the only one fishing for Georges Bank haddock, was not large enough to increase effort to the level where the purpose of the gear restriction could be overcome. The fact that the U.S. fleet was the only one involved in the harvest also made passage of this regulation easy for ICNAF; no other country had a vested interest at that time in the decision (Graham 1970).

However, within a decade distant-water fishing fleets, particularly from the Soviet Union and other Eastern European countries, started harvesting fisheries resources of the Northwest Atlantic. With the arrival of these fleets came a period of unprecedented growth in levels of exploitation. The fishing capacity of these fleets was astounding. Due to their distance from their home countries, distant-water fleets were composed of extremely large harvesting vessels that pulled trawl nets with mouths (the open end at the front of the trawl into which fishes are gathered) up to several hundred feet wide. These vessels processed, froze, and stored large volumes of fishes onboard. In some instances, fishing vessels delivered their catches directly to even larger "mother ships," huge floating processing and freezing plants. The Spanish distant-water fleet employed a pair-trawling system, in which two ships cooperatively hauled a trawl between them. The nets they pulled were up to twice as large as standard single-vessel trawls used by foreign fleets, which themselves were markedly larger than trawls pulled by the relatively small domestic vessels. Pair-vessel trawls, some with mouths more than 500 ft wide, could capture as much as 180 mt of fishes in one haul under optimal conditions. In 1971 Spanish pair-trawlers made up only 4.5% of the total tonnage of foreign vessels fishing in the Northwest Atlantic, yet they harvested 28% of the cod catch (Warner 1983). For an interesting account of the vessels and fishermen participating in the foreign fleet's fishing activities during the ICNAF years, and the domestic fleet's opinion of them, refer to Warner's book, *Distant Water* (1983).

Such fleets harvested tremendous biomasses of fishes before returning to their distant home ports. On Georges Bank, this marked increase in effort led to increased catches of traditional species such as haddock, cod, and flounders, as well as fishes that had not previously been heavily exploited when that region was fished solely by the U.S. domestic fleet. Mesh regulations established to protect the haddock fishery proved wholly inadequate in the face of uncontrolled increases in fishing effort. The intensifying fishing activity, in concert with unusually successful reproduction in 1963, which had temporarily increased haddock abundance, led to a record haddock catch of 155,000 mt (341 million lbs) in 1965. This was more than three times the average catch of the previous 30 years (Graham 1970). Such exploitation inevitably led to a collapse of the haddock stock and its associated fishery.

ICNAF established total catch quotas for haddock and yellowtail flounder in 1970 and 1971, respectively. In December of 1971, member nations approved modification of the ICNAF wording, which allowed the international commission to assign allocations within total quotas for each member nation participating in the fisheries. By 1973, the program of national allocation of catch levels had expanded to include nearly 20 fish stocks (Anthony and Murawski 1986). Meanwhile, discontent with ICNAF was rapidly growing within the New England fishing industry.

Throughout the 1960s and early 1970s, while the world's total harvest of fisheries resources consistently increased, the U.S. catch actually declined. Concurrently, while many fisheries resources off its coastline were being extensively utilized by other nations, the United States was importing more fish products than any other country in the world (Royce 1987). The New England fishing industry became increasingly convinced that the domestic fleet was severely disadvantaged when competing with the distant-water fleets of foreign nations. In order to control operating costs, fishing fleets of coastal nations typically develop only a short-range harvesting capacity, which is closely dependent upon shoreline processing facilities and which cannot easily be shifted to other fishing areas (Graham 1970). This is particularly true of New England, where the long-standing tradition of the family-owned boat, dating back to colonial times, essentially restricted the size and range of vessels due to cost constraints. On the other hand, distant-water fleets must be mobile and capable of harvesting over extended periods before returning to distant home ports. Such fleets can accomplish unusually high levels of harvest over brief periods of time. When mesh-size restrictions were the only major form of regulation enacted by ICNAF, the fishing power of the distant-water fleets simply overwhelmed the regulation, leaving the U.S. domestic fleet feeling outcompeted along its own coast. The subsequent system of allocation of catches to member nations was intended to divide the catch among member nations in a manner that seemed reasonably equitable to all participants while preventing overharvest.

Even with quota and allocation systems in place, stocks continued to decline in the early 1970s; total catches for heavily exploited species were simply exceeding calculated allocations due to high levels of by-catch. By-catch consists of fishes captured while nets are being fished for other species. Thus, if quotas for a particular species, such as the haddock, are reached in a given year by directed catch levels (the catch of vessels fishing specifically for that species), any vessels harvesting it incidentally as by-catch while fishing for other species would cause the total catch of haddock to exceed the quota for the year.

Due to by-catch problems, allocation systems were modified in the mid 1970s so that the total directed catch accumulated as the year progressed would be added to predicted by-catch levels for that same species. Once the sum of directed catch and predicted by-catch equaled the fishery quota, the directed harvest of that species would be terminated for the remainder of the year. In this way, any subsequent by-catch of that species would have already been accounted for in the original calculations of allocations, with the result that by-catch would not cause the total harvest of a species to greatly exceed the annual quota.

Dividing the catch equitably among numerous nations in an allocation system that allows all countries to be satisfied with their "share" often becomes a greater political problem than a biological one. Deriving an allocation system that is agreeable to all participants can be extremely difficult, since it is in each country's self-interest to receive maximum immediate landings in order to gain the greatest economic benefit from its investment in its fishing industry. Since no country owns a "common resource," interest in reducing present catches to gain the long-term benefits provided by conservation of fish stocks can be quite limited. It is commonly held that ICNAF failed to manage fish stocks effectively due to the lack of interest of most member nations in conservation of fisheries resources offshore from New England. However, there is evidence that, once by-catch levels were incorporated into the total allocations of nations, at least some severely depleted fish stocks were being provided protection from further depletion. After a decade of steady decline, the abundance of principal groundfish species of Georges Bank and the Gulf of Maine increased annually from 1974 to 1976, the last full year that the United States participated in multinational management of those resources (Anthony and Murawski 1986).

In spite of the improvement in conservation practices of ICNAF in the mid 1970s, the U.S. fishing industry's discontent with foreign fishing had grown so great that the dissolution of this management system had become inevitable. The view that foreign nations were benefiting from the fisheries resources of its coastline at its expense had led the United States to join a United Nations resolution calling for the preferential rights of coastal nations to manage exploitation of fisheries resources off their coastlines. In 1971 Congress passed the Pelly Amendment to the Fishermen's Protection Act. This amendment required the secretary of commerce to notify the president whenever a foreign fleet failed to abide by international fisheries conservation programs. The president was authorized to prohibit the importation of all fishery products from the offending country. Finally, dissatisfaction with the inability of ICNAF to develop effective conservation measures led the United States to join the consensus reached at the United Nations Third Conference on the Law of the Sea in 1974, which proposed that coastal states have sovereign rights over living and nonliving resources of the seas within 200 miles of their coastlines; adherence to this resolution eventually nationalized nearly 95% of the exploited fisheries resources of the oceans, since nearly all prime fishing areas occur within 200 miles of some coastline (Royce 1987). Ultimately, the United States signed into law the Fishery Conservation and Management Act (FCMA, also called the Magnuson Act or the "200-mile-limit law") in 1976. This legislation established a 200-mile exclusive economic zone (EEZ; also called the fisheries conservation and management zone, FCZ or FMZ), within which the United States would have sole responsibility for conservation and management of fisheries resources. This action took effect on March 1, 1977, at which time the United States permanently withdrew from ICNAF.

Thus, the passage of FCMA in 1976 was intended to identify ownership of the coastal resources; ownership brought with it the responsibility to apply conservation measures that would insure future health of the coastal fish stocks.

The Fishery Conservation and Management Act of 1976. The findings (or justification for taking legislative action) listed in the FCMA identified the economic importance of commercial and recreational fishing, and the seriously depleted condition of the fisheries resources that provided opportunities for these activities. Further, the findings clearly relegated blame for this problem to the "massive foreign fleets" and to the inadequacy of international agreements in preventing serious and ongoing overfishing. The findings concluded that a national program "is necessary to prevent overfishing, to rebuild stocks, to ensure conservation, and realize the full potential of the Nation's fishery resources." Although development of underutilized fisheries (that is, fish stocks historically not actively fished due to lack of marketing opportunities) was recognized as a goal of the legislation, the focus of the act was clearly toward issues of conservation, not economic development.

The United States assumed exclusive authority for managing all fishes within the fisheries conservation zones, "except [for] highly migratory species," such as tunas and billfishes, whose range typically extends well beyond 200 miles from the coastline and in some instances throughout the entire North Atlantic. Management goals were to be accomplished through fishery management plans (FMPs), to be developed by regional management councils and approved by the secretary of the Department of Commerce. The focus of management policy was toward optimum sustainable yield (OSY), or that harvest that provides "the greatest over-all benefit to the Nation." OSY differs from the maximum sustainable yield (MSY) of ICNAF, in that relevant social or economic factors as well as the biological condition of the resource may be considered as criteria when establishing harvest regimes; in other words, FMPs may recommend harvest levels that consider not only the health and condition of fish populations but also the needs of the human user groups that exploit them (basically the fishing industry and, generally to a lesser extent, recreational anglers).

Under the act, any foreign fishing is to be conducted only by specific agreement of the U.S. government. Foreign fishing is restricted to that portion of the OSY that is not harvested by U.S. vessels; this largely limits foreign fishing to those species that U.S. fleets cannot market. The secretary of state and the secretary of commerce determine the allocation of such excess resources to the foreign nations that have requested permission to fish. Including the Department of State in the fisheries management process had actually been initiated prior to the Magnuson Act when the Bureau of Oceans and International Scientific Affairs was created in the State Department in 1974. Our coastal fisheries resources were viewed as a potentially important tool in maintaining international relations. Thus, foreign fishing was politicized, with permission for any foreign nation to fish subsequent to 1976 based not only upon such factors as historical tradition of the applying country in the identified fishery, its cooperation in conservation matters, and its willingness to help collect important scientific information to be used by regional councils but also upon its relationship with the Department of Sate.

Since the FCMA is a national program, the act specifically states that management measures cannot discriminate among fishermen of different states, and it must be fair to all potential domestic participants in the fisheries.

Regional councils develop the management plans, which typically address conservation and harvest of particular fish stocks. The voting membership of each regional council is composed of the principal marine fisheries management official within each coastal state of that region (Maine, New Hampshire, Massachusetts, Rhode Island, and Connecticut for the New England Fishery Management Council, the NEFMC), the regional director of the National Marine Fisheries Service (NMFS), and appointees selected by the secretary of commerce from a list of names submitted by the governors of each state. There are no restrictions as to the makeup of the appointed portion of the council; the majority of the individuals in this portion of the membership come from the fishing industry or related interests. Appointed members make up a majority of the voting membership.

Councils may recommend a broad spectrum of conservation measures in their management plans, including limiting types of harvesting equipment and vessels, restricting levels of harvest, limiting the times and places where harvest may occur, and limiting the number of vessels that may participate in the fishery (normally referred to as restricted or "limited" entry).

After over a decade of management since passage of the Magnuson Act, attempts at conservation of New England fisheries resources are for the most part a failure. By the end of the 1980s, almost all New England fish stocks with high levels of public interest and marketing potential either had shown little or no recovery or were in a more depleted condition than they had been before passage of the Magnuson Act. The Georges Bank haddock stock declined 76%, and the yellowtail flounder stock 85% between 1977 and 1989 (Campbell 1989). Even the resilient cod stock, the focus of a fishery since colonial times, reached its lowest recorded abundance in 1989. In 1988, even though New England commercial fishermen spent nearly 65% more time fishing than they had in 1978, they marketed 25% less catch. Those species that showed distinctive recoveries, such as the silver hake and the Atlantic mackerel, were ones that had previously been fished only by foreign fleets, and which had limited or no value in U.S. markets.

Why has the management process been so unsuccessful? One factor is related to the management options chosen to regulate fishing. Catch quotas can provide firm control over the level of exploitation allowed in a fishery. However, quotas can create political, sociological, or biological problems when established (see Regulating Harvest later in this chapter). In its early years, the New England Council used quotas to protect cod, haddock, and yellowtail flounder stocks. However, when quotas were reached, concern over hardship within the fishing industry led the council to increase quotas to support continued fishing for the rest of the year (Campbell 1989). By the late 1980s, the New England Fishery Management Council was attempting to manage the biomass of fish assemblages, rather than single species, in a way that would allow that biomass to reach some optimal level, even if the recovery of some species was not achieved. Quotas were not included in these management schemes.

Harvest can also be controlled by limiting the number of vessels that are allowed to fish (limited entry). If the numbers of participants in the fishery are suitably limited, this system can provide opportunity for success to each of those who participate in the harvest (see Regulating Effort in this chapter), even if quotas are used concurrently. Although a few fisheries in other regions of the United States have been managed by limiting entry, this process has not been supported by the NEFMC and has received vigorous opposition from many New England fishing interests. Restricting any fishermen from fishing for a "common" resource has just not been accepted in New England.

Thus, in the late 1980s, the New England Council opted to exclude two of the most powerful controls on harvest, quotas and limited entry, from its management strategy. Instead, it chose to control exploitation by establishing minimum size limits and closing fishing in particular areas and seasons in order to protect the reproductive activity of specific stocks, and it regulated the efficiency of gear, such as by restricting mesh sizes. Why did these management schemes not work?

For the most part, the failure of the Magnuson Act has been due to the philosophy with which the management of fisheries has been carried out. Although the stated intent of the Magnuson Act was to practice sound conservation to achieve long-term resource stability and human benefit, many who supported its passage viewed it simply as a means of keeping foreign fleets from fishing off the U.S. coast (Royce 1987). Prior to 1976, the New England fishing fleet was not large enough to harvest fisheries resources at the same levels as those of all the ICNAF fleets collectively. However, within a short period of time after 1976, the harvesting capacity of the domestic fleet had expanded to the point where fish stocks were more depleted than they had been during activity of distant-water fleets. The New England domestic fleet grew by 42% within the first 3 years under the Magnuson Act; groundfish vessels alone increased from 570 in 1976 to 921 by 1981 (Campbell 1989). Along with this growth, the New England Council developed management strategies that minimized regulatory obligations on fishermen (Campbell 1989). This philosophy was supported by the view that broad harvest restrictions would effectively drive marginally successful vessels into bankruptcy. Further, enforcement at sea of even simple regulations such as minimum mesh-size restrictions on trawls had always proven to be a costly nightmare, which became progressively worse as the resources became harder to catch (Campbell 1989). Vessels that abided by regulations spent money to improve their capture efficiency, purchasing the newest technology in electrical fish-finding equipment and upgrading the power of their vessels to meet the growing pressures of harvesting depleted stocks. As had happened with the first mesh regulations of ICNAF, the intensity of the fishing effort of the domestic fleet had simply overwhelmed controls on harvest. How had the New England fishing fleet, which was still largely made up of single, family-owned vessels, grown to this level?

Passage of the Magnuson Act unrealistically raised the hopes of commercial fishermen. Studies suggested that developing new domestic fisheries could create as many as 43,000 new jobs while substantially improving the U.S. trade balance (Royce 1989). In 1980, Congress passed the American Fisheries Promotion Act, which among other things provided for increased grants to industry, directed more funding into fishery development programs of NOAA, and provided loan funds that would allow boat and facilities owners to avoid defaulting on private loans. In a very short time period the New England fishing fleet, with the support of federal development programs, doubled its fishing power. Thus though the Magnuson Act promoted elimination of foreign fishing and conservation action to aid recovery of depleted stocks, the Fisheries Promotion Act supported marked growth of the domestic fleet that continued to harvest those same stocks. Essentially, domestic fishermen were simply working harder at a greater cost to themselves to harvest the same biomass of fishes. Between 1979 and 1984, total fish landings in U.S. waters remained constant, whereas the value of fisheries products imported from other countries increased by nearly 50% (Royce 1987).

The fishing industry resisted those regulations that the New England Council used to control harvest. Angered by what were viewed as ever-changing regulations that threatened their livelihood, and suspicious that analyses of the condition of fish stocks were inaccurate, some New England fishermen violated regulations, misreported catch data, and dumped thousands of tons of illegal fishes overboard in order to land smaller quantities of other fishes that were not as strictly regulated (Royce 1989). The growth of the domestic fleet, coupled with an inability to manage the catch by regulating gear efficiency, resulted in all-time recorded low abundances for a growing number of fish stocks.

Ultimately, the major failure of the management process in U.S. territorial waters has been the inability to reconcile the economic interests and needs of the local community and region with the broader national interest, which includes conservation practices that protect the health of the fisheries resources. By the end of the 1980s, emphasis in at least some offshore fisheries seemed to be shifting toward the national interest, due to the severely depleted condition of fish stocks, the strengthening voice of the recreational angler demanding protection of those stocks, and the relatively minor importance of the fishing industry to the health of the nation's economy as a whole. As had happened much earlier in inland freshwater regions of the United States, recreational interests gained political strength because of numbers (recreational anglers outnumber commercial fishermen by nearly 300 to 1) and because marine recreational fishing generated economic activity roughly equal to that of the commercial fishing industry (Royce 1989). Elements of the commercial industry, long recalcitrant in accepting any regulations, also were beginning to support firmer conservation efforts as the only means to retain their livelihood in the future (Campbell 1989).

Management within State Territorial Waters

The idea that fisheries resources are "common property" has created some of the same conflicts in inshore waters as were seen during ICNAF's management of offshore fisheries. Early attempts to manage harvest often failed due to lack of agreement between adjoining states on strategy. A law passed by the Connecticut legislature in 1868 restricting the use of pound and trap nets in shad and salmon fisheries was short-lived, in part because the Massachusetts and Rhode Island legislatures failed to take similar action. In the first report of the activities of the Office of the U.S. Fish Commissioner, Spencer Baird recommended that New England states enact legislation requiring broad limitations on the use of pound nets in order to aid recovery of fish stocks; no state followed that recommendation. Even though New England states were among the first to create fish commissions to study the decline of fisheries resources, harvest restrictions were not easily accomplished; in the early years of state activity, management processes were usually directed elsewhere. For example, New Hampshire and Massachusetts, with support from Maine's salmon hatchery operations, cooperated in the 1860s and 1870s to restore the Atlantic salmon to the Merrimack River. The restoration program included requiring dam operators to construct fish-passage facilities to allow upstream migration and stocking salmon fry raised in hatcheries (Stolte 1981).

For the latter half of the 19th and early 20th centuries, state fish commissions placed most of their emphasis on hatchery operations and stocking programs, particularly focusing upon anadromous species. Even as it became apparent that managing harvest and restricting other human activities related to development and pollution were necessary to accomplish conservation goals, establishment of fishing regulations met consistent resistance, for the following reasons:

1. Uneasiness among commercial fishermen and anglers about regulation of a "common" resource
2. Reluctance within individual states to restrict the harvest by their citizens unilaterally, as long as fishing for the same stocks continued unregulated in adjoining states
3. Hesitation by states to enter into multistate agreements that might restrict their own management options

A marked, prolonged reduction in abundance of striped bass stocks, followed by their rapid recovery in the late 1930s, stimulated interest among Atlantic seaboard states in cooperative management of this valuable resource. In much the same way that a decline in haddock stocks led to the creation of the ICNAF, concern over striped bass led Congress to pass the Atlantic States Marine Fisheries Compact in 1942 (Alperin 1987). This compact created the Atlantic States Marine Fisheries Commission (ASMFC) in order to accomplish coordinated multistate management of Atlantic coast fisheries resources. All 15 states of the Atlantic seaboard signed as members of this commission, each state appointing three representatives: the executive officer of the state agency responsible for managing coastal fisheries resources; a member of the state legislature; and a citizen appointed by the governor. During drafting of the legislation, the majority of legislatures of the involved states demanded that they retain their statutory authority for managing resources within state territorial waters. Thus, the commission did not administer management programs directly; it advised the governors and the legislatures of member states on the condition of fish and shellfish stocks and recommended management actions that addressed specific conservation goals. The commission could only recommend those programs supported by a majority of members on the commission who had an interest in the species in question. The original compact was amended in 1950 to allow two or more member states to designate the ASMFC as a joint interstate regulatory agency that could manage particular species identified by those states.

The ASMFC facilitated the development of interstate management programs by developing fishery management plans (FMPs). These plans focused upon entire stocks, not just the portion of those stocks within particular state borders. Even with the 1950 amendment, the commission could not dictate policy; adoption of programs required the consent of member states. Thus, the commission depended upon cooperation of its members and their willingness to support programs that would consider regional and national interests rather than confer advantages on some states to the disadvantage of others. Similar agreements (see Managing Offshore Fisheries Resources) on the international level had failed to accomplish proper conservation of fish stocks. However, in spite of a tendency for government units (in this instance, state governments) to protect their own priorities to the detriment of others, this compact was a milestone in the evolution of the management process in coastal inshore waters. It provided a formal, organized, and ongoing means for cooperative conservation and placed the responsibility for unselfish action firmly "on the shoulders" of its member states.

The commission remained an advisory body for more than three decades until passage of the Atlantic Striped Bass Conservation Act of 1984. The striped bass fishery, arguably the most important of all coastal recreational fisheries, had progressed through a period of steady decline, during which fishing took greater numbers of fishes while reproduction steadily failed from 1971 onward (see Striped Bass in Species Profiles). The Chesapeake Bay striped bass stock, the most important to the coastwide fishery, displayed its first year of successful spawning in over a decade in 1982. However, this heartening development did not prevent growing concern that the adult stock was continuing to collapse in an unabated manner. That some coastal states were not complying with recommendations of the Striped Bass Fishery Management Plan became a hotly argued issue that was carried to the U.S. Congress. Several bills detailing action were considered by Congress in 1984, the most controversial one requiring an absolute moratorium on striped bass fishing for a 3-year period. The legislation that was ultimately voted, commonly referred to as the Studds Bill, broadened the decision-making power of the federal government in managing this species within state territorial waters. The Striped Bass Act mandated compliance by all states with regulations listed by the Striped Bass Fishery Management Plan of the ASMFC. If any state failed to follow guidelines of the plan, the commission was required to report such violation to the secretaries of commerce and interior, who would initiate closing the striped bass fishery within that state.

For the first time, the federal government had chosen to manage an inshore fishery directly through the ASMFC, rather than facilitate interstate cooperation through an advisory system. Thus, this legislation suggested that, at least for this highly visible fishery, management by the federal government was preferable to allowing multiple states any further time to develop and fully support a coastwide management strategy. The inability of the multiple states to agree upon an overall strategy without guarding "personal" interests (the activity of the fishery within particular state boundaries) was similar to the difficulty that multiple nations had in effectively managing offshore resources under ICNAF.

The Anadromous Fish Conservation Act. Anadromous fisheries, which had created the original impetus toward creation of state fisheries management agencies in New England, retained the attention of both state and federal agencies throughout the 1900s. National concern over the decline of salmon stocks on both coasts, as well as over other species such as the striped bass and American shad, led to the passage of the Anadromous Fish Conservation Act of 1965. This legislation authorized the secretaries of interior and commerce to enter into cooperative projects with states to study the condition of anadromous fish stocks and the watersheds into which they migrate and to initiate actions to improve migration, feeding, and spawning conditions of those river systems. It also provided that the federal government would share the cost of addressing these issues with the states involved.

The Wallop–Breaux Amendment. Perhaps the single most important piece of federal legislation supporting the management of recreational fisheries resources within state territorial waters is the Wallop–Breaux Amendment to the Federal Aid in Fish Restoration Act of 1950 (also called the Dingell–Johnson or D–J Act). The 1950 legislation established a means to fund the management and conservation of recreational fisheries resources on nonfederal lands. It created a federal excise tax on sport fishing tackle, with the proceeds going to the states to manage fisheries having "material value in connection with sport or recreation in the marine and/or freshwaters of the United States." Revenues from the excise tax are apportioned to each state by the secretary of the Department of the Interior using a formula based upon (1) the area of each state in proportion to the area of all states, and (2) the number of license holders in each state in proportion to the total number in all states. Thus, apportionment to each state is based upon fishing activity, measured by the amount of water requiring management (area of the state) and the fishing pressure applied to those waters (the number of license holders). No state receives less than 1% or more than 5% of the total money available in any given year. The Department of the Interior may keep no more than 8% of the available money to operate the federal aid program, but it may expend any funds unused by the states.

Moneys are returned to states on a reimbursement basis. States fund projects with receipts from the sale of fishing licenses or other revenues; at the end of the fiscal year the states are reimbursed 75% of the total approved costs of their projects. Thus, the revenues brought in by sale of fishing licenses can be increased up to fourfold by this program.

In 1986, the D–J Act underwent extensive expansion under the Wallop–Breaux Amendment. The 1986 legislation

- expanded the excise tax to include all fishing equipment
- imposed a tax on electric trolling motors and fish-finders
- directed a portion of marine fuel tax receipts into the program
- diverted duties on imported fish tackle, boats, and pleasure craft into the program.

In addition, the legislation requires states to use a portion of the available funds for recreational boating programs. Most importantly, it requires all coastal states to fund marine as well as freshwater projects, in proportion to the relative importance of saltwater and freshwater angling within the state.

The mandate requiring coastal states to fund marine projects is very significant. As of 1989, no New England state required anglers to purchase a general marine recreational fishing license.* Because of this lack of saltwater fishing licenses, states had no means to measure fishing pressure in coastal waters, a mandatory factor used to determine the amount of money apportioned to each state under the D–J formula. The lack of proof of activity (fishing license counts) limited the access of coastal agencies to D–J funds. Further, since each state is allowed only one designated agency to oversee expenditure of D–J funds, in many coastal states marine fisheries were overlooked because they were managed by an agency separate from that administering freshwater fisheries activities and the D–J program. Thus, the Wallop–Breaux Amendment greatly expanded the source of money available to develop state coastal fisheries management programs and "breathed" new vitality into agencies that previously might have been long on philosophy and ideas but short on funding.

In the 1980s, the executive branch of the federal government made several attempts to divert Wallop–Breaux funds from the use outlined in the legislation. Rather than wait for possible reversion of unexpended funds by the states, the executive Office of Management and Budget argued that substantial Wallop–Breaux funds should support programs of the De-

*The lack of a saltwater fishing license results from a curious dichotomy in the thinking of the angling public. In general, anglers have favored the conservation of fisheries resources and have accepted freshwater fishing licenses as a necessary cost of enjoying recreational opportunities made available by management programs. National and local angling associations were leaders in the lobbying efforts that led to design and passage of the D–J Act and Wallop–Breaux Amendment. This history of willingness to pay for the benefits one derives from angling, and in doing so provide others similar opportunities, should be a matter of pride to the angler. However, the relationship between users and fees has not been easily extended to saltwater angling. Historically, states attempting to establish license fees for saltwater fishing have met with strong lobbying against such action. The tourism industry opposes such fees as a barrier to attracting tourists. Legislators are often unwilling to support legislation establishing mandatory licenses because of the "emotional objection of some anglers to 'paying' for a resource that traditionally has been thought to be 'free'" (Scogin 1983).

partment of the Interior, justifying this stance by noting that some states were not fully using the formula funds available to them. This executive branch "budget-balancing" approach to changing the manner in which Wallop–Breaux funds were to be expended was defeated each time by Congress, with the strong support of national recreational fishing and conservation organizations.

The Dingell–Johnson Act and Wallop–Breaux Amendment were landmarks in the evolution of fisheries management and conservation policy in that they established a user-fee concept supported by the users who gained personal satisfaction from angling. The angler's role in supporting this user tax (via lobbying of organized angling associations) served to demonstrate the willingness of this interest group not only to pay for the personal enjoyment it receives but to support conservation programs that protect natural resources to the benefit of the nonangling and angling public alike.

The outcome of inshore management efforts. The success of management programs directed toward New England's inshore fisheries has at best been uneven. Though programs initiated by individual states, or cooperatively by several states to restore lost or rebuild depleted anadromous fish stocks have been successful in some watersheds, stocks of the same species remain seriously depleted in others. Anadromous fisheries have presented the manager with a multitude of problems. Water-quality problems have prevented the restoration of anadromous stocks in many watersheds. Managing harvest has also proven difficult. For example, northern New England Atlantic salmon stocks are heavily fished commercially in the open ocean outside of state and U.S. territorial waters. Such stocks are susceptible to decline and collapse even if coastal and riverine harvest is strictly regulated. Managing Atlantic salmon fisheries ultimately requires international agreement to address such harvest (see Atlantic Salmon in Species Profiles).

Some anadromous fisheries housed largely within U.S. waters have fared little better. In spite of recent indications that striped bass stocks are rebuilding (see Striped Bass in Species Profiles), this species has suffered a long-term downward spiral in abundance. Until the federal government mandated compliance with a coastwide management plan, individual states had great difficulty in agreeing to participate in a coordinated management approach directed at this fishery.

Management of some inshore fisheries is doubly complicated because of intensive harvesting efforts by both commercial and recreational interests in both inshore state-managed waters and offshore federally managed waters. Thus, conflicts may arise between the interests of user groups as well as among governmental units with different perspectives.

Even stocks that are harvested largely by recreational fishing almost solely within state territorial boundaries may not be managed successfully. In the 1980s the bluefish, fished largely by recreational anglers in state inshore waters, exhibited population declines significant enough to stimulate development of an ASMFC coastwide management plan. Unmanaged recreational harvest and waste were felt to be the major contributing factors to the bluefish's decline. Enormous numbers of bluefish, particularly prereproductive juveniles, are wasted after landing by many anglers, who neither return them live to the water nor eat them. Whatever proportion of the angling public behaves so is hardly exhibiting a well-developed sense of the conservation ethic. It is easy to place the blame for decline in fisheries resources with the commercial industry and its interest in immediate economic gain. Although this is the case in many fisheries, evidence does suggest that, if the angling public is left to its own resourcefulness to harvest a highly desired species, the fisheries resource may suffer similar problems.

Habitat Protection

Proper conservation of fisheries resources obviously includes providing and protecting suitable habitats in which the fishes will live. Recognition that habitat deterioration and loss were stressing fisheries of New England appeared early in the development of the fisheries management process. Some of the earliest actions taken by New England fish commissions, such as requiring operators of dams to construct passage facilities for migrating fishes, were a clear indication that managers understood the necessity of providing habitat. In his first report to Congress, U.S. Fish Commissioner Spencer Baird considered the importance of pollution as a factor in the decline of coastal fisheries.

Several states assumed the lead in habitat protection by mandating the removal of barriers to riverine migration in the 1860s and 1870s. Unfortunately, lack of technical knowledge necessary to construct fish-passage facilities generally led to failure in facilitating upstream migration well into the 20th century.

The federal government clearly entered the habitat protection arena in 1899 with the passage of the Rivers and Harbors Act. This act, as amended, required a federal permit for all work conducted in navigable waters of the United States. It also established a mandatory permit system for the discharge of refuse into waterways. The permit program was overseen by the U.S. Army Corps of Engineers.

Although this act did not specifically address issues of fisheries conservation, it laid the groundwork for future legislation that included fisheries resource issues in the approval process for projects in navigable waterways. The Fish and Wildlife Coordination Act of 1934, as amended, required consultation with the U.S. Fish and Wildlife Service and relevant state wildlife and fisheries agencies before federal or federally approved impoundment, diversion, or channelization projects could be started. This law also authorized the secretary of the interior to study the effects of pollution on fish and wildlife. Thus, although most major legislation dealing with habitat conservation lay in the future, the process of habitat management for fisheries resources had begun.

Rapidly expanding public awareness concerning environmental deterioration and the value of protecting our natural resources led to an "explosion" of federal legislation addressing these issues in the 1960s and seventies. Some of this formed the framework within which regional and national programs for conservation of fisheries habitat are presently conducted.

The Water Quality Act of 1965 established the Water Pollution Control Administration and required that all states develop water-quality standards for interstate waterways. The passage of the National Environmental Policy Act (NEPA) of 1969 was a clear statement by the federal government of its intent to address issues related to the importance of the nation's natural resources and the environment. The act itself was much less powerful in its capacity to control environmental degradation than many realize. NEPA required that a detailed public report outlining the potential environmental impact be written and approved through a rather lengthy process before projects involving federal agencies or funding could be started.

While broad legislation dealing with the quality of our environment was being passed in Washington, focus was also placed directly upon those habitats and ecosystems that were critical to the conservation of coastal fisheries resources. Severe deterioration of coastal estuaries, which are critical to the production of many coastal fish stocks, was rapidly increasing by the mid 20th century due to accelerating rates of pollution and development. The Estuary Protection Act of 1968 demonstrated Congress's interest in the value of estuaries to the nation. The act was intended to provide a balance between the national interest in protecting estuarine natural resources and the local and regional interests, which focused upon economic development of these habitats. The secretary of the interior was instructed to inventory the nation's estuaries, considering their recreational and aesthetic importance and their importance to the health of coastal fisheries resources. After the inventory was reported to Congress in 1970, the secretary was to enter into agreements with states on the conservation and development of state public areas on a cost-sharing basis. This legislation also specifically required federal agencies to consider the value of estuaries when planning their development and to consider a review and recommendation by the secretary of the interior concerning development plans. This act was followed by the Coastal Zone Management Act of 1972, which established a mechanism for funding programs intended to "preserve, protect, develop, and . . . restore or enhance, the resources of the Nation's coastal zone." The act authorized the secretary of commerce to conduct a grant-in-aid program, which funds state management plans for the coastal zone that are consistent with federal guidelines and objectives.

Congress also cast its interest in habitat protection seaward in 1972 with the passage of the Marine Protection, Research, and Sanctuaries Act. This law created a program within the Environmental Protection Agency to regulate ocean dumping. It also required analyses of the long-range effects of pollution on ocean ecosystems and authorized the secretary of commerce to designate certain areas as marine sanctuaries in order to protect them from deterioration caused by human activities.

In 1982 Congress enacted the Coastal Barriers Resources Act. Coastal barriers, including inshore islands, elongate spits, and sandbars, are formed by wave action and tide-generated erosion and deposition. These natural structures form the seaward margins of coastal estuaries and marshes, thus protecting these wetland areas from destruction by storm-driven winds and waves. Concern over expansive development, which had caused notable destruction of these coastal barriers, led to the act's passage. Its intent was to discourage overdevelopment and to protect the coastal barrier system. One of the goals of this protection was reduction of damage to wildlife and fish habitats that are dependent upon the barrier system.

Thus, during the last 3 decades, the management of fisheries resources has been broadly expanded to include more thorough protection of the environments in which they live. In spite of legislation establishing the means to accomplish habitat protection, some coastal habitats remain in a seriously deteriorated condition. As the interest in using our coastal fisheries resources for recreational purposes continues to grow, critical habitats continue to diminish in quality and extent (Harville 1983). Restoring spawning migrations of anadromous fishes in many New England rivers remains a problem due to suboptimal habitat quality and lack of habitat availability. Many inshore shellfisheries are closed to harvest due to unacceptably high levels of domestic pollution in the waters where they grow. Some near-coast areas, such as those around Boston Harbor, have become infamous for the contaminated condition of the fishes found within them. Pressure to develop estuaries and coastlines is a great and growing problem in many reaches of the New England coast. Development can cause habitat degradation or loss. One of the most severe problems facing the striped bass fishery involves habitat issues outside of the jurisdiction of New England states. Habitat deterioration in Chesapeake Bay has been identified as a major cause of the decline of many inshore fisheries of that region. Chesapeake Bay is the spawning and nursery area for striped bass that historically has contributed the most adults to the New England fishery. Whereas overfishing has been a major cause of the decline in adult abundance, reproduction has certainly been impeded by the loss of habitat quality. Similarly, other species, such as the summer and winter flounder, use the coastal estuaries as spawning and/or nursery grounds. Still others, such as the bluefish, rely heavily upon forage fishes occurring in those near-shore ecosystems. Thus, ongoing habitat disturbances seriously affect the health of these fisheries, even if other factors such as overfishing were to be regulated.

Regulations

Regulations are a major means of carrying out resource conservation objectives. Most regulations are intended to allow a fish population to maintain its abundance and biomass at desired levels in the face of harvest. Fisheries managers applied angling regulations to an array of freshwater and marine fisheries in the early part of the 20th century, often simply because it seemed wise to do so. Analytical methods that allow managers to clearly assess the condition of, and the effects of fishing upon, fish stocks had not been developed; thus, managers often established regulations by guesswork, rather than by calculating how specific regulations would benefit a fish population's condition.

By the 1940s, regulations had been deemphasized in many fisheries management programs, in part due to studies that suggested that restricting angler harvest had no beneficial effect on reservoir and lake fish stocks that had previously been unregulated (Jenkins 1970). The general deemphasis of regulations continued into the 1960s, with many fisheries scientists concluding that regulations functioned mostly as a sociological tool, satisfying the angling public's desire to see some type of conservation action taken. Even in the 1970s, respected books and review publications still frowned upon the usefulness of many management regulations (Jenkins 1970; Everhart et al. 1975). Statements such as "Fisheries have been regulated on the basis of politics, social pressure, gear competition, predjudice, whim, and sometimes for biological reasons" or "a few laws have been helpful, but of many it can only be said that the fishery survived in spite of the regulation, not because of it" or "Fishing regulations are usually aimed at control of the fishermen with little concern given to the biological health of the fish" (Everhart et al. 1975) probably did reflect the frequent lack of connection between the early use of regulations to manage fisheries resources and effective resource conservation.

New analytical techniques that allowed fisheries scientists greater precision in predicting the need for, and the effect of, regulations stimulated a broadening of the use of this basic technique of fisheries management once more beginning in the 1960s. It is now generally assumed that regulations that restrict the behavior of anglers and commercial ventures not only are a useful means of resource management but may be the only effective approach to protecting the viability of many of our fisheries resources.

Regulations may be grouped into three basic categories, according to the control they are intended to impose upon a fishery. They may control harvest, fishing effort, or the nature and efficiency of that effort. The regulations most commonly applied to recreational fisheries in New England control harvest, by establishing either size limits, catch limits, or fishery closures. In this section, an overview of the impact that specific types of regulations have upon the conservation process will be presented.

Regulating Harvest

Minimum legal size limits. Minimum legal size limits are the most common type of regulation applied to recreational fisheries by the New England states. These regulations are generally considered to be an effective means of controlling harvest, when characteristics of the fishery, and the condition of the fish population itself, are well understood. Size limits often receive greater support from anglers than some other types of regulations. Many anglers readily accept the concept that returning small fishes to the water will allow them to grow, thus making them more desirable for harvest in the future when they are larger (Everhart et al. 1975). Since fisheries resource management is most effective when supported by the resource's users, regulations that are most agreeable to anglers will likely be chosen over equally effective (or, in some cases, theoretically more effective) alternatives.

Minimum legal size limits are intended to provide absolute protection to all fishes smaller than the designated size. Such restrictions may be intended to provide "trophy" fishes to anglers by preventing harvest until individuals have grown to a larger size than they might have reached under unregulated intensive fishing pressure (Dawson and Wilkins 1981). Size limits that produce trophy fishes address anglers' interests but are not necessarily intended to protect the reproductive capacity of a stock or improve the stock's status in any measurable way over that achieved without the regulation. Thus, the fisheries manager is addressing goals of the user, not goals of conservation of the resource. However, minimum size limits established to produce trophy fishes typically do not harm the condition of the stock. Since fisheries resource managers attempt to manage the interaction between a resource and its human users, addressing the interests of anglers without negatively affecting the viability of the fishery resource can be a logical objective of a fisheries management program.

Most minimum legal size limits are established to protect the spawning potential of a population or to increase biomass production of the population by providing fish with longer periods of growth before they are harvested. Minimum-size-limit regulations are often intended to prevent fish from being legally harvested until they have matured and spawned at least once. Since, for any species, most individuals reach sexual maturity at some general average length, it is relatively simple to determine what size limit would allow the vast majority of fish the opportunity to spawn once before any are harvested.

Such regulations, although assuring some level of reproduction, may not be adequate to counteract otherwise unregulated harvest. The youngest, smallest females in the population generally produce fewer eggs than older, larger ones. If most of the reproductive effort of a population comes from young, newly matured fish, the total egg production of the population may be fairly limited due to the low fecundity of these small individuals. Further, many species of long-lived fishes exhibit marked fluctuations from year to year in reproductive success; high offspring survival may occur no more frequently than every several years at best (see chapter 2). Long life-spans and the potential to spawn in multiple years have evolved in many species that face unpredictable offspring survival rates from year to year. If nearly all fish in such species spawn once and are harvested, several consecutive years of relative reproductive failure would leave very few individuals alive for future spawning. Thus, when establishing minimum size limits that allow fish to reproduce once prior to being available for legal harvest, the manager must consider whether the reproductive cycle of the fish and the intensity of harvest of legal-size individuals are suited to this strategy. If minimum size limits are intended to protect the optimal spawning potential of a population in the face of exploitation, the appropriate size may not always be that which protects individuals through only one reproductive season.

Minimum size limits were applied to the 1982 year class of striped bass during the 1980s in order to enhance recovery of the severely depleted Chesapeake Bay stock. After over a decade of reproductive failure, the Chesapeake stock, which normally contributes most of the fishes to the coastwide fishery, produced a moderate-sized year class in 1982 (see Striped Bass in Species Profiles). Once individuals in this year class reached a harvestable size, they were protected by a minimum size limit that was modified several times, each time to protect that year class while allowing larger, less numerous fish to be harvested. In the fall of 1989, the ASMFC, although retaining control over the level of harvest that was allowed, reduced the size limit to allow members of the 1982 year class to be kept when landed. This reduction in size limits was based upon the high reproductive success accomplished largely by that group. Such continued adjustment of a minimum size limit to meet the current needs of a fishery requires persistent attention and analysis by fisheries managers; however, it also represents the most effective means of adjusting allowable harvest to current condition of a population.

Compliance by anglers is essential to the success of any regulation. It is almost inevitable that sublegal fishes will be harvested after minimum size limits are established; this illegal catch may be minor, or it may constitute a substantial portion of the harvest. For example, after a 7-inch length limit on the scup was initiated in Connecticut, less than 1% of the recreational catch was smaller than the legal size (Howell et al. 1984). Alternatively, 5% to 35% of all spotted sea trout harvested by anglers in Texas were smaller than the 305-mm size limit previously established (Meador et al. 1986). Even with sublegal harvest, both the biomass of the former fish stock and the abundance of adults in the latter increased after the size-limit restrictions were established.

Several factors can lead to harvest of sublegal-size fishes, including ignorance of the regulation, inaccuracy in measuring landed fishes, and purposeful violation. Most regulations are easy to understand; the last thing a manager should do is stimulate inadvertent violation of regulations by creating a system that is difficult to follow. As Everhart and his colleagues (1975), somewhat tongue-in-cheek, suggested, going fishing should not require "a lawyer in attendance" to interpret regulations. In inland states, certain freshwater species have fisheries that differ in both fish population and fishing pressure characteristics on practically a watershed-by-watershed basis. The most precise approach toward management of such resources would be to establish specific regulations for each watershed that best fit that fishery. However, managers usually opt for a single, statewide regulation (Hunt 1975), which may not best fit the needs of all stocks but which provides anglers with a simple system that minimizes misunderstanding and inadvertent violation. Most coastal species other than anadromous fishes do not need to be managed by use of a series of regulations that are specific to local areas. Because size-limit regulations are statewide and, in some instances coastwide in their application, there should be minimal angler confusion concerning the length-limit regulations associated with particular species in coastal New England.

However simple length-limit regulations may be, absolute compliance is rarely achieved. Anglers can either inaccurately measure or fail to measure fishes; both of these actions contribute to the harvest of sublegal fishes. Inevitably, creation of regulations requires developing a means of enforcing them in order to minimize the proportion of sublegal harvest that is knowing and purposeful.

When establishing size-limit regulations to protect a fishery, managers must consider death rates of fishes that are landed, handled, then returned to the water. Minimum size limits prevent small fishes from being legally kept but obviously do not prevent them from being landed. Sublegal individuals grab bait or strike lures as readily as larger individuals. Even if they are released after capture, such fishes must endure the physiological strain of the battle and the stress and physical damage caused by handling and hook removal; some fishes inevitably die because of these factors. Numerous studies have investigated the severity of angling mortality and the effect this death rate might have on the ability of size limits to protect the reproductive capacity of a population. Generally, these studies have shown that such mortality is specific to each fishery, depending upon the species of fish, the angling equipment used to land the fish and remove the hook, and the handling technique of the angler. For example, baits may create higher mortality than lures, since baits are swallowed more deeply, causing a greater chance of stress and damage during hook removal. Use of gaffs or similar devices can cause extremely high death rates. Literally keeping a fish under a firmly planted boot while the hook is being removed is obviously more damaging than holding it carefully in one's hand. Equipment or gear restrictions are often applied to freshwater fisheries in which most fishes are expected to be returned to the water after capture (for example, in areas of heavy fishing pressure where populations are managed with very restrictive size and catch limits). Such gear restrictions may forbid the use of baits, treble hooks, and barbed hooks. These regulations are not intended to restrict angler efficiency; they are meant to enhance survival of fishes returned to the water. Such angling restrictions are uncommon in coastal fisheries. Educational programs that instruct the angling public about landing and handling techniques are considered a productive approach to achieving the goal of maximizing survival of small fishes that are landed.

Mortality of landed fishes is perhaps the greatest problem faced by the manager when imposing minimum size limits on commercial fisheries. The otter trawl, the most common means of commercially harvesting a variety of groundfish species, inevitably causes high mortality rates of landed fishes. Thus, minimum legal size limits applied to commercial vessels are often combined with regulations such as restrictions on the use of small mesh openings in trawl netting to reduce the likelihood that sublegal individuals will be landed in the first place.

Catch or "creel" limits. Limiting daily catch, or "possession," is another means of limiting harvest. Daily catch or "creel" limits have been applied to at least one fishery by all of the New England states. A review of 1989 New England state regulations indicates that such regulations were restricted to anadromous species, namely, the striped bass, Atlantic salmon, river herring, American shad, and rainbow smelt. Of these, all but the striped bass are harvested largely within estuaries, river mouths, or the freshwater reaches of rivers during the spawning season. Thus, the striped bass fishery is the only one along the coastline, or seaward from it, to have been regulated in this manner; however, by the end of the 1980s daily catch limits were being considered for other largely recreational coastal fisheries, such as the bluefish. Daily catch limits are not well suited to control commercial harvest, which is typically characterized by participants making multiple-day fishing trips, each day of which produces largely unpredictable catches. Thus, when used, daily catch limits are applied to fisheries in which the greatest harvest is made by anglers.

The major intent of a daily catch limit is to reduce annual harvest below that which otherwise would occur, although nonbiological reasons such as providing anglers with a "satisfaction goal" are sometimes listed as justification for these regulations (Hunt 1975). Daily catch limits do not directly determine the level of yearly harvest, since there are no restrictions on how many anglers might participate in the fishery or on how often each might fish. In order to control total harvest via daily catch limits, the fisheries manager must predict what the fishing pressure on the fishery will be (that is, how many anglers fish and how many total days they will fish). These measures can be complicated by thinking that is characteristic of the angling public. Surveys in freshwater systems have indicated that some anglers become more determined to fish until they have reached a particular catch limit than they might have been before the limit was imposed; evidently, many anglers consider the catch limit a goal rather than a restriction (Hunt 1975). Thus, the psychology of angling can affect the success of trying to predict accurately the intensity of the angler's effort and catch!

Seasonal or annual catch limits, also called quotas, exert much greater direct control on total harvest than do daily catch limits. Daily catch limits attempt to predict fishing effort. Thus, it is hoped that harvest can be controlled by matching that prediction to some desired total harvest. Quotas in theory terminate all fishing once the desired catch has been reached. Thus, the expected total harvest cannot be miscalculated by error in predicting fishing pressure. Quotas are generally applied only to commercial fisheries, since enforcement of seasonal catch limits for anglers would be impossible.

Some gear, such as the otter trawl, will effectively catch a variety of species inhabiting the same areas. Thus, when annual catch limits are reached for particular fisheries, the manager must either (1) close all fishing with gear to which those species are susceptible, even though other coexisting stocks may not have been fully exploited; (2) allow fishing to continue with the requirement that all captured individuals of the target species be returned to the water; or (3) include incidental by-catch predictions in the total harvest that is to be allowed (as done by ICNAF; see Managing Offshore Fisheries Resources in the section on the Evolution of the Management Process). Although effective in protecting stocks, the first option has obvious political and sociological overtones, since the fishing industry feels it is denied the opportunity to pursue its livelihood. The second option, due to the high mortality rate caused by many types of commercial fishing gear, does no more than prevent the fisherman from marketing the catch. Since many fishes returned to the water die anyway, it does not prevent overharvesting of the species being managed. The third option can be viable. Review of the final years of management under ICNAF indicates that quotas can be established that are not disrupted or made ineffective by incidental by-catch (Anthony and Murawski 1986). However, due largely to the perceived intrusion on the rights of individuals to fish for a common resource, quotas have not consistently been applied to New England fisheries resources managed within federal territorial waters since passage of the Magnuson Act.

Fishery closures. Closures are typically established to prevent exploitation at some critical phase of a fish's life cycle. Protection is provided for adult fishes on spawning grounds; adults passing through restricted areas during migration (often spawning migration); and young fishes on nursery grounds or during periods when they are particularly vulnerable to fishing gear (Everhart et al. 1975). Closures usually restrict fishing at specific sites during particular time periods. Such area closures have been an integral part of management strategies directed at specific fish stocks in both state and federal territorial waters off New England, being applied to depleted stocks of anadromous species such as the Atlantic salmon and rainbow smelt and to offshore species such as the haddock and yellowtail flounder.

Closures may restrict activity in fisheries other than the one being directly managed. Two general area closures off the New England coast were established under the Northeast Multispecies Fishery Management Plan developed by the New England Fishery Management Council in 1986 in order to protect the spawning activity of the haddock and yellowtail flounder. Designated spawning areas for haddock on Georges Bank were seasonally closed to all mobile or fixed commercial fishing gear other than scallop dredge gear and hooks of a particular size range. An area of the New England/Middle Atlantic boundary was closed to all mobile gear other than specially permitted midwater trawls, and sea scallop or surf clam/ocean quahog dredges during the spawning season of the yellowtail flounder. Since many fishing gears such as the otter trawl indiscriminately harvest a variety of species, the New England Council found it necessary to close multiple fisheries in order to protect two highly valued, and severely depleted, stocks.

Regulating Effort

Fishing effort is most directly regulated by limiting the number of participants in the harvest. This has historically been a common tactic in wildlife management, where lottery drawings or other approaches are used to select the limited number of hunters allowed to purchase licenses. Regulating effort has also been applied to a variety of fisheries. When effort regulations are established in fisheries management, they frequently control the number of fishing units used by participants, rather than limiting the number of participants; limits on the number of fishing rods, tip-ups for ice fishing, or handlines per angler can be established to restrict fishing effort without denying participation. However, such regulations often simply stimulate the angler to fish longer to accomplish the harvest he or she will be satisfied with; therefore, harvest levels are not necessarily lowered (Everhart et al. 1975).

In New England, participation in recreational inshore shellfisheries managed by local town governments has frequently been controlled, in part to assure that local residents are the major beneficiaries of the harvest. Permit systems are normally used with daily harvest limits to regulate total harvest of these local fisheries.

Limited entry. Limiting effort by restricting participation, commonly referred to as "limiting entry" into the fishery, has often been proposed as the most effective means to insure control of commercial harvest while protecting the health of the fishing industry, especially when combined with other regulatory approaches such as gear and size-limit restrictions. No proposal for control or regulation of commercial harvest raises the ire of the fishing industry nearly as much as does the concept of limited entry. Support for, or more frequently opposition to, limited entry may be related to the "economic self-interest" of those who would be affected by the regulation (Acheson 1980). Those fishermen who favor such regulations may do so more because of their perception that "competitors" would be denied access to the fishery than out of support for sound conservation practices. Most oppose such regulation because "free enterprise," or absolute access to participate in the fishery, is inhibited (Acheson 1980). Although limited entry has been applied to fisheries in other regions under U.S. jurisdiction, such regulations have not yet been established by the New England Fishery Management Council due to the historic overwhelming opposition of the fishing industry. However, with the continued decline of most New England fisheries resources, the council and elements in the fishing industry had become more interested in such regulation by the end of the 1980s.

Regulating Gear

Restricting types of gear. A number of recreational and commercial fisheries have been regulated by restricting the types of gear used. Some of these regulations, such as restrictions on the use of live bait, and treble or barbed hooks, are intended to increase survival of fishes returned to the water. Others, used most often for freshwater recreational fisheries, such as "fly-fishing-only" regulations, have been justified as a way to reduce harvest by decreasing the number of anglers in a fishery. However, such regulations are more often based on social pressures than on conservation needs. Those with the greatest political power gain favorable status by pressuring managers to restrict others from participating in the fishery. Harvest may not be reduced by such restrictions; it may merely be partially shifted within watersheds from protected to unprotected sites. Bluntly put, "purist bait fishermen have as many rights as purist fly-fishermen" (Everhart et al. 1975).

Similar regulations have been imposed upon coastal fisheries. For example, harvests of the striped bass, American shad, and Atlantic salmon were restricted to hook and line only in at least one New England state in 1989. In the instance of striped bass, all commercial harvest was prohibited by the relatively frequent use of a "no-sale" regulation in the coastal fishery. No-sale restrictions have also been considered for Atlantic billfishes. Such restrictions, though they do not favor one type of angling method over another, obviously favor recreational over commercial interests. The creation of no-sale and hook-and-line-only regulations represents the political strength and lobbying power that coastal recreational angling interests have gained in recent decades.

Regulating gear efficiency. Efficiency regulations are commonly applied to commercial harvesting when other means of regulating the fishery are deemed inappropriate. The frequent use of minimum-mesh-size restrictions for otter trawls is intended to prevent the capture of small, young fishes. Thus, it should serve as an indirect means of accomplishing the same result as a minimum legal size limit for the fishes themselves. Mesh regulations are sometimes supported as the only biologically and sociologically suitable means of controlling harvests in fisheries using trawling or gill-netting gear. In such fisheries, minimum legal size limits for fishes may have limited usefulness, since many fishes returned to the water after capture will die anyway due to physiological stress and injury.

Controlling harvest by limiting entry or establishing catch quotas is often unacceptable to the fishing industry, since both methods impede access. Although mesh-size regulations may represent the most acceptable alternative, they do not always work. For example, as trawls are fished, the "spaces" (mesh of the netting) through which smaller individuals are supposed to escape become clogged with fishes. Thus, as a trawl is fished the proportion of the catch made up of undersized individuals will increase. The early (prequota) years of management under ICNAF provide an excellent example of the potential ineffectiveness of mesh-size regulations to control harvest in otherwise unregulated fisheries. Everhart et al. (1975) suggest that gear restrictions that reduce the efficiency of the harvest may really do no more than increase the cost of fishing, since in theory it will take more time to land a particular biomass of fishes if at least some young fishes are escaping. Such restrictions may simply require fishermen to expend greater time and effort, rather than controlling levels of total harvest.

The Use of Multiple Regulations

Generally, no single type of regulation has proven suitable for all management situations, although some have proven to have greater utility than others. The more heavily exploited the fish stock, and/or the more seriously depleted its condition, the greater the likelihood that the management scheme applied to it will be composed of a set of different types of regulations, rather than any single one. Offshore fisheries resources in the New

England region, after a period of management with a focus upon individual regulations (the use of mesh-size regulations, followed by quotas, under ICNAF), are currently under a management regime that simultaneously utilizes minimum size limits, gear regulations, and seasonal and area closures. The state of Maine manages harvest of the Atlantic salmon by the concurrent use of closures, catch limits, size limits, and gear restrictions, on a watershed-by-watershed basis. Collectively, the New England states manage striped bass fisheries using similar regulatory systems. Even as the complexity of management schemes increased in the 1980s, the condition of many New England fish stocks continued to decline, many reaching all-time lows by the end of the decade.

Where Do We Go from Here?

In spite of more than a century of developing the process of managing coastal fisheries resources, New England fisheries are generally in worse condition than at any time since the status of the resources was first monitored. However, the problem is not necessarily due to the lack of a framework through which fisheries conservation can be accomplished. The institutional structure is certainly suitable, and the scientific basis for determining the condition of fish stocks, and the effects that exploitation will have upon these stocks, is sound. The future lies not in developing a system that can work but in those involved (particularly the users of the resource, commercial fishermen and recreational anglers) gaining the resolve to use the available framework to conserve fish stocks properly.

Concern has been expressed that the structure of the regional council system works against conservation, since the majority of voting members have historically come from the fishing industry; it is always difficult for a user group driven by economic considerations to support actions that potentially, in their judgment, cause economic hardship. Emphasis on council appointments that do not exclude commercial users but do include a greater balance of other interests might indeed result in more successful efforts at conservation. However, the Magnuson Act was specifically passed to include sociological and economic factors in the decision-making process. Even if the legislation were not so worded, it would be inappropriate to exclude such considerations from the management process. A publicly owned resource should be available to all; particular users should not be favored, nor should they be excluded from participating in use of the resource if their actions do not threaten the rights of others. Preemptive exclusion of commercial fishing for the benefit of preservation or of recreational interests is no more appropriate than is the favoring of commercial interests to the detriment of others. Indeed, as noted earlier, not all fisheries that are seriously depleted have been extensively harvested by the commercial fishing industry.

The basic objective of the New England Council has always been to minimize regulations imposed upon fishermen. Under the first set of fisheries management plans developed by the council, harvest of haddock, cod, and yellowtail flounder was managed by use of catch quotas. However, generally when quotas were reached, objections of the fishing industry caused the council simply to increase the quotas (Campbell 1989). Such a scheme hardly combines protection of the resource with needs of the users. Harvest limits must become part of the management process.

Both limited entry and catch allocations have been used to manage fisheries outside of the New England region. The New England fishing industry has historically fought such actions on the basis that successful boat owners would not be allowed to expand their fleets and marginal owners would fall into bankruptcy. However, some members of the fishing industry, acknowledging that conditions must change drastically if the industry is to survive, are beginning to speak in favor of limited entry (Campbell 1989).

Similarly, many anglers have argued against regulations on recreational harvest and have fought the establishment of saltwater fishing licenses, the fees from which form a cornerstone of funding for fisheries conservation. Although anglers have accepted regulation better than the fishing industry, particular fisheries will have a much brighter future if anglers become more willing to do so.

Habitat protection may become an even greater battlefield in the future, as development continues to degrade the quality of some habitats while causing others to disappear. Many coastal natural areas are remarkably more valuable when developed than when protected. Regulations controlling harvest will have much less positive effect for many inshore fisheries if habitat deterioration continues to accelerate.

Ironically, there will always be fishes in the sea; the oceans' food webs have the potential to support a tremendous biomass of fishes. However, the species available may not be those that people value, either for recreation or as food; evidence suggests that collapsed fish populations can be replaced by less desirable species that do not suffer the effects of overexploitation (see the discussion of the proliferation of spiny dogfish and skate species in the Northwest Atlantic in the section on Interactions in the Northwest Atlantic in chapter 2). "We have the capacity to destroy [our valued fisheries resources]. . . . We're either going to redefine our relationship with those resources, or we're going to [have to] accept whatever the ecosystem doles out" (S. Murawski, NMFS, as quoted in Campbell 1989).

II

SPECIES PROFILES

The Species Profiles will focus upon particular recreational fisheries and the fish species that support them. Each profile presents standard characteristics of the fishery, such as information on the life cycle and ecological relationships of the species, the condition of its population, and management schemes that have been developed to provide for that species' conservation. In addition, information from the angling perspective, such as how to catch, prepare, store, and cook each species, is included.

The order of the species profiles reflects general evolutionary relationships. For example, all sharks and rays of the Class Chondrichthyes (the cartilagenous fishes) are grouped together, as are the various groups of bony fishes (Class Osteichthyes), such as the flatfishes, members of the cod family, and others. Although this order presents general patterns of evolution, species within these groupings may be arranged to allow the reader to make easy comparisons, rather than to reflect exact evolutionary relationships within groups. Other arrangements, such as an alphabetical one, would separate similar species and prevent easy access to all species similar in appearance and habit.

Species selected for this section come mostly from lists published in the *Marine Recreational Fishery Statistics Survey* (National Marine Fisheries Service, NOAA/U.S. Department of Commerce) and sections 1 and 2 of the *Angler's Guide to the United States' Atlantic Coast* (1974, U.S. Government Printing Office, Washington, DC). Species profiles in this book include not only those fishes avidly pursued by anglers but also a few that are often landed incidentally due to the fishes' abundance and their tendency to strike baits and lures intended for other species. Although several incidental species, such as the spiny and smooth dogfishes, are disliked, they are included here because of the frequency with which they interact with the angler. In several instances, two species are included in the same profile. In each instance, the species are closely related and are part of the same fishery; they are caught in the same general habitats by anglers using particular types of angling techniques and equipment.

Much of the information concerning identification and biology has been taken from the following sources:

Bigelow, H. B., and W. C. Schroeder. 1953. *Fishes of the Gulf of Maine.* U.S. Fish Wildl. Serv. Fish Bull. 74, Vol. 53.

Clayton, G., C. Cole, S. Murawski, and J. Parrish. 1978. *Common marine fishes of coastal Massachusetts.* Mass. Coop Ext. Serv., Univ. of Massachusetts, Amherst. C-132.

Grosslein, M. D., and T. R. Azarovitz. 1982. *Fish distribution.* Mesa NY Bight Atlas Monograph 15. NY Sea Grant, Albany.

Scott, W. B., and M. G. Scott. 1988. Atlantic fishes of Canada. *Can. Bull. Fish. Aquat. Sci.* 219.

References listed at the end of the relevant species profile reflect additional sources from which information was taken.

The section on Status and Management in each species profile presents historical changes in the condition of the stock or population, with reflections on the causes of these changes and a review of the types of management schemes developed to protect the viability of these stocks. Management information is not intended to serve as an updated guide for the reader. Human interactions with some species such as the striped bass, haddock, cod, and some flatfishes are so dynamic that regulations controlling harvest are changing frequently, in response to changes in the condition of the population and changes in the level at which it interacts with its "users" (anglers and commercial fishermen). Thus, if specific regulations updated to 1989 were listed for these species, the reader would be presented with outdated information in subsequent years. Management schemes are presented here with a historical perspective, with the understanding that they reflect the manager's attempts to accomplish desirable conservation of these resources through 1989.

A variety of sources were used to compile information concerning the status and management of each species. Major sources include Bigelow and Schroeder (1953) and the *Marine Recreational Fishery Statistics Survey* (cited above); annual documents on the "Status of the Fishery Resources off the Northeastern United States," published by the Northeast Fisheries Center (NEFC) of the National Marine Fisheries Service (NOAA); unpublished reports of state fisheries agencies, the NEFC, and the regional fisheries management councils; and state recreational and commercial fishery regulations for 1989. Other specific sources are listed at the ends of the profiles.

Angling and Handling Tips, presented by Robert C. Biagi, are reflections of his personal experiences in fishing, for most of the species, and numerous conversations and readings he has done, for the others. He also has tried all recipes described in this section, having either developed or added a personal touch to each.

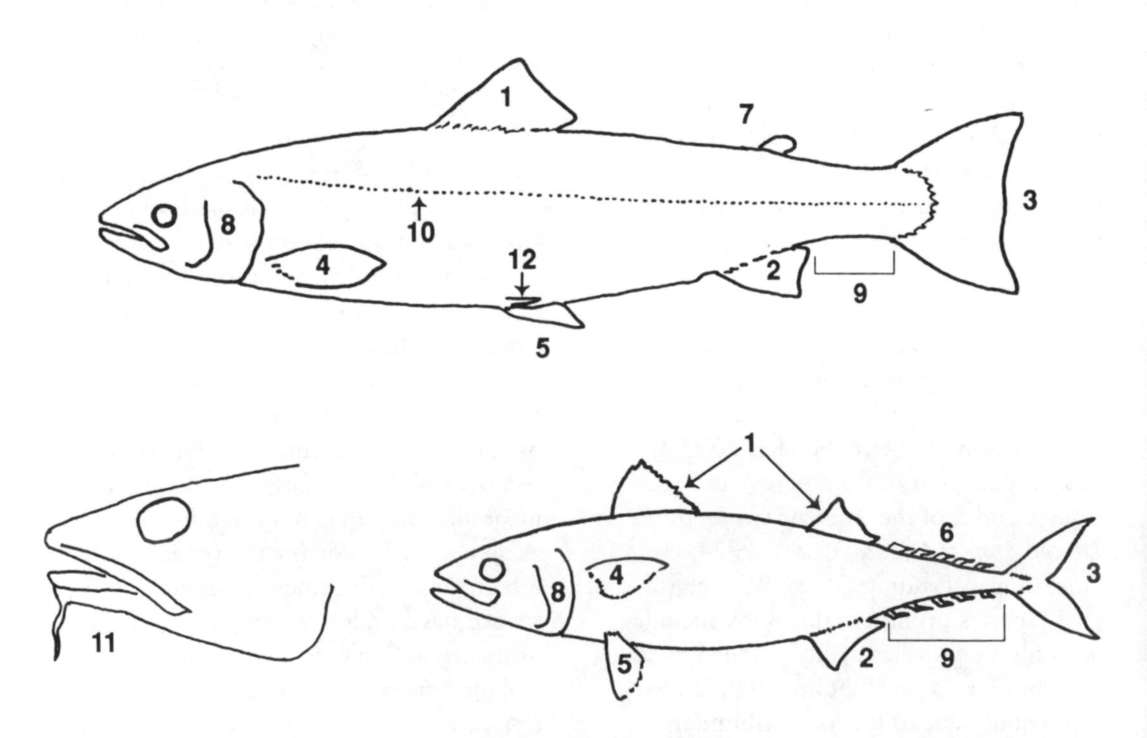

Figure 6. Morphological characteristics used in species profiles to describe fishes. *KEY:* 1, dorsal fin; 2, anal fin; 3, caudal fin (tail); 4, pectoral fin; 5, pelvic fin; 6, finlets; 7, adipose fin; 8, opercle (gill cover); 9, caudal peduncle; 10, lateral line; 11, barbel; 12, pelvic axillary process.

Mako Shark
(*Isurus oxyrinchus*)

The mako shark, a recreational species renowned for the acrobatic battle it provides, is native to the warmer waters of the Atlantic Ocean. It occurs along the Atlantic coast from the Caribbean to as far north as the Gulf of Maine, where it is a summer resident only.

The mako is a member of the mackerel shark family. This group is distinguished from other North Atlantic sharks in having a half-moon-shaped tail with the lower lobe only slightly smaller than the upper and a caudal peduncle that is widely expanded to form a distinctive lateral keel. The mako is most easily distinguished from other members of this family in our region by the shape of its teeth, which do not contain spurs as do those of the porbeagle shark and are more pointed and considerably slimmer than those of the great white shark.

Typically, the mako is grayish blue dorsally, paling to white on the lower sides and belly.

Biology

Mako sharks are oceanic, never occurring within coastal inshore waters less than 9 m (30 ft) deep. They are typically found at or near the surface in offshore areas.

Makos migrate to northerly latitudes during the warm months of summer. As waters cool in the fall, they swim southward along the continental shelf to winter in the Caribbean.

Males become sexually mature when they are 3 to 4 years of age and 1.8 to 2.8 m (71 to 110 in) in length; females mature at about 7 years of age and 2.2 to 3.2 m (87 to 126 in). As in other sharks, fertilization of egg cells occurs internally in makos. Male sharks have two copulatory organs, the claspers, which are specialized pelvic fins. These allow sperm to be passed into the reproductive tract of the female. After eggs are fertilized, they develop within the uterus of the female. Developing young are nourished largely by yolk reserves deposited around the egg cells before fertilization. During later stages of development, they may also feed upon unfertilized egg cells or uterine fluids. After about 1 year of embryo development, a female mako gives birth to from one to several young, each measuring 61 to 79 cm (24 to 31 in) in length. Shark offspring are generally much larger and more highly developed than are newly hatched young in species that reproduce by external fertilization. Thus, the prolonged period of internal incubation furnishes young sharks with a much higher probability of survival than is characteristic of the majority of fish species.

Length at age for male and female makos is presented in Table 4. The oldest female aged in the study from which the growth information was derived was 11.5 years old.

The mako feeds heavily upon a variety of fish species. The bluefish can make up nearly 80% of this species' diet in the southern New England/Middle Atlantic region. The mako also eats small-bodied schooling species such as mackerel and herring and larger fishes such as swordfish, bonito, and tuna species. Squid are often eaten, primarily in offshore areas. Remains of blue sharks, rays, skates, and dolphins or porpoises have been found in a few mako stomachs. However, these are uncommon foods. The dolphin or porpoise remains may have come from animals eaten after they had died. There is no evidence that the mako eats living marine mammals.

Table 4. Calculated length at age in cm (inches in parentheses) of shortfin mako sharks from the New York Bight

Age	*Female*	*Male*
1	116.0 (45.7)	118.0 (46.5)
2	153.0 (60.2)	151.5 (59.6)
3	183.5 (72.2)	180.0 (70.9)
4	210.0 (82.7)	203.0 (79.9)
5	221.5 (87.2)	216.0 (85.0)
6	229.9 (90.5)	231.0 (90.9)
7	246.0 (96.9)	241.0 (94.9)
8	255.0 (100.4)	

Source: Data from Pratt and Casey 1983.

Status and Management

In 1986, an estimated 2.3 million sharks, excluding dogfishes, were landed recreationally along the Atlantic and Gulf coasts of the United States. The Middle Atlantic states landed the highest proportion, approximately 1.1 million; 33,000 sharks were harvested along the northeastern coast. The mako has been one of the most avidly sought shark species. It is pursued in private or charter boats 20 to 60 feet long that are powered by everything from outboards to large diesel engines.

Mako sharks are also harvested by commercial operations, taken either as bycatch by pelagic longline vessels pursuing tuna or by a bottom longline fishery directed specifically toward the mako. The directed fishery for sharks increased in activity during the 1980s. Shark bottom longline gear may be composed of up to 300 baited hooks strung along an 8-mile line, and pelagic tuna longline gear may have up to 2,200 baited hooks spaced along a line up to 60 or 70 miles long.

A preliminary management plan for Atlantic sharks was established in the mid 1980s by the regional management councils in order to reduce the potentially conflicting interests of foreign and domestic fishermen, including anglers. This plan restricted directed landings to domestic vessels and identified seasonal closures for regions off the Atlantic coast.

Angling and Handling Tips

The mako is one of the most exciting recreational species in our region. This species, particularly noted for its spectacular leaping ability, can be pursued by the average angler without a major investment in specialized equipment.

Most mako sharks range from 50 to 300 pounds and can be landed with 30–50-pound test Dacron line. A strong wire leader that will resist the mako's sharp teeth and rough skin is a necessity. The mako prefers baits with a high oil content, such as bluefish and mackerel; however, it will strike at a variety of baits when attracted to them. Bait should be placed whole or in paired fillets on size 8/0 to 14/0 hooks. A float can be used to suspend the bait and hook at desired depths, as long as the float does not impede retrieval of the line.

A variety of tactics are used to attract sharks to baited hooks. Sharks can be attracted from great distances by low-frequency vibrations set up by sound waves in the water. Anglers pound or bang objects off the side of the boat or splash buckets in the water as a means of attracting sharks. A few go to extremes, such as playing music from waterproof speakers placed in the water.

Chumming is considered the most effective way to attract sharks. The simplest method is to ladle a mixture of ground fish and water over the side of the boat, a little at a time, to produce an unbroken chum slick. Occasionally, chunks of fish may be added to the chum slick for better attraction.

When a mako picks up the bait and starts swimming, line should be free-spooled before the hook is set. After a brief run, set the hook hard three or four times and be ready for the battle of your life. Extreme care should be taken when the fish is ready to be landed. Once gaffed, a mako is capable of wrenching the gaff from the angler's grip or bending it out of shape; thus, you need the strongest, longest gaff you can afford. Never attempt to boat the mako prematurely; haste will result in a lost fish, destroyed gear, or possibly severe damage to the boat or angler before the fish is subdued. Many experienced boat captains prefer dispatching the fish with a bullet to the brain. Unless you are planning to eat your shark, releasing it by cutting the line is being encouraged by fisheries managers as an important conservation measure. Regardless, it is best for the novice shark fisherman to accompany someone experienced, so that shark fishing can be learned safely.

Although shark meat was long ago accepted as desirable table fare by many cultures (for example, sharks and skates are used in the English fish and chips and in the raw fish cocktail, seviche, of Latin America), it only recently has been marketed under its own name in the United States. Many Americans are discovering what others have known for some time; when properly handled, shark meat can be as enjoyable as that of more traditional table species. Similar to swordfish, its firm texture, delicate flavor, and moistness yield a fine meal.

Because its skin is so tough, cleaning and cutting steaks from a mako can be a frustrating chore. Try poking a knife through the skin, then turning the sharp edge upward and slicing through the skin inside-out. Keep a stone handy to sharpen the knife often.

Urea, contained in the blood and tissues fluids, must be drained from shark steaks to prevent the flesh from acquiring a strong ammonia flavor. Careful immediate icing is a necessity to keep the flesh from building up ammonia odor. Mako steaks should be soaked in brine made from pickling or kosher salt; do not use iodized table salt. Before cooking, marinate in any citrus juice, vinegar, or milk.

Grilling mako is quick and simple. Marinate 1-inch-thick steaks in orange juice and lemon slices for at least 1 hour. Remove from the marinade, pat dry, and brush with mayonnaise or olive oil, adding some freshly ground pepper. Grill on a hot fire, or broil 4 inches from heat for 5 minutes on a side, and serve with a quartered lime.

References

Pratt, H. J., Jr., and J. G. Casey. 1983. Age and growth of the shortfin mako, *Isurus oxyrinchus,* using four methods. *Can. J. Fish. Aquat. Sci.* 40:1944–1957.

Preliminary fishery management plan for Atlantic billfishes and sharks. 1983. National Marine Fisheries Service, NOAA/U.S. Department of Commerce.

Stillwell, C. E., and N. E. Kohler. 1982. Food, feeding habits, and estimates of daily ration of the shortfin mako (*Isurus oxyrinchus*) in the Northwest Atlantic. *Can. J. Fish. Aquat. Sci.* 39:407–414.

Blue Shark
(*Prionace glauca*)

The blue shark, one of the most frequently encountered shark species of open coastal waters, is truly cosmopolitan in distribution; it is found in all major ocean areas and is widely distributed both inshore and offshore throughout the North Atlantic.

This slender-bodied species has a long snout with a rounded tip. It can be distinguished from other pelagic sharks caught in New England waters by the presence of very long and narrow pectoral fins and an elongate upper lobe of the tail that is about twice as long as the lower lobe. The upper lobe of the tail has a conspicuous notch near its tip.

The blue shark is a dark indigo color dorsally and fades to a bright blue on the sides and white on the belly. The bright coloration fades rapidly to gray shortly after death.

Biology

The blue shark is an open-ocean species. It is a summer visitor to the Northwest Atlantic, occurring from May to October in waters from Cape Hatteras to the Grand Banks off the Newfoundland coast. During this time of year it occurs regularly in the Gulf of Maine and Georges Bank areas and is frequently sighted in waters of 30.5 to 39.6 m (100 to 130 ft) in depth off the southern New England coastline. Large females are typically the first to make the northward and shoreward migration in the spring; smaller females and males follow later in the year. In the autumn this species migrates southward along the outer continental shelf to Cape Hatteras and the margin of the Gulf Stream. Blue sharks occasionally complete trans-Atlantic movements during the annual migration cycle.

Water temperatures apparently influence this species' distribution. Although found in a broad range of water temperatures, 7.8 to 27.2 degrees C (46 to 81 F), blue sharks most commonly frequent temperatures between 12.8 and 17.8 degrees C (55 and 64 F). Temperature preferences influence not only the blue shark's geographic distribution but also its habitat use within seasons. Although frequently sighted at the water's surface in temperate regions, blue sharks most often inhabit deeper, cooler subsurface waters in tropical areas.

Male sharks are at least 175 cm (68.9 in) and females at least 180 cm (70.9 in) long at maturity. Egg cells are fertilized internally, and female blue sharks carry young throughout a prolonged period of development before giving birth. Courtship behavior apparently includes nipping and biting by the male, as most females bear tooth scars on their bodies after mating. Curiously, females as small as 134 cm (52.8 in) and as young as 4 years of age carry distinct mating scars. These females are not yet adult and thus are not ready to bear young in the summer of their first mating. Such females apparently store sperm until after their reproductive system is fully matured and eggs are ready to be fertilized, which normally occurs in the spring of the year following mating. Incubation typically takes from 9 to 12 months; thus, females do not give birth until nearly 2 years after mating. Female blue sharks are viviparous, nourishing young through a placenta attached to the inner wall of the female's uterus. Young are 35 to 44 cm (13.8 to 17.3 in) long at birth. A female may bear up to 82 young, but the average number is much lower. The prolonged period of embryo development and the unusually large size at birth provide young blue sharks with a much higher probability of survival than is characteristic of the majority of fish species (see the section on Parental Care in chapter 2).

The largest blue sharks reach 3.4 to 3.7 m (11 to 12 ft) in length. Table 5 shows average lengths of blue sharks at selected ages. Most blue sharks do not live beyond 13 years of age, although a few large males have been aged up to 16 years.

Species of squid and an open-water octopus are this species' most frequent foods.

Table 5. Length of age in cm (inches in parentheses) of selected ages of blue sharks

Age	*Length*
1	33 (13)
2	47 (18.5)
3	61 (24)
5	210 (83)
8	270 (106)
10	300 (118)

Source: Data approximated from a growth curve presented in Grosslein and Azarovitz 1982.

Squid may be most commonly taken in waters near the surface during the night, when they are migrating vertically to feed. Fishes also constitute an important part of the blue's diet. Bluefish and red and silver hakes are the most important fishes, with mackerel, menhaden, Atlantic herring, and blueback herring also common. Blue sharks feed less frequently on many other fish species, including sand lances, eels, swordfish, goosefish, scup, and yellowtail flounder. They feed at a great range of depths, picking up groundfishes such as members of the cod family and flounders at depths greater than 61 m (200 ft) in Georges Bank, eating midwater squid and fishes such as the bluefish and the spiny dogfish, and aggressively striking at baits such as mackerel and menhaden presented to them at the water's surface. Along with these foods, euphausiid shrimp constitute a major part of the diet of blue sharks on the West Coast.

Status and Management

Interest in shark recreational fisheries has increased markedly in the last 2 decades. In 1970 1.4 million sharks, including the ubiquitous dogfishes, were landed on the East Coast of the United States. Between 1981 and 1986, an average of 2.7 million sharks other than dogfishes were harvested from this same region, with a peak of over 5.6 million in 1983. The New England recreational fishery lands only a small proportion of the recreational harvest, although the percentage of the East Coast landings taken in New England waters increased from an average of 2% in the early 1980s to 14% (33,000 out of 2.3 million) in 1986. The Middle Atlantic states typically land the greatest proportion of sharks.

The blue shark is one of the most commonly caught species of large-bodied sharks. In the New York Bight region, recreational catches of blue sharks are highest in the spring and early summer, with the highest catch rates occurring northward into southern New England as summer progresses.

Like the mako, the blue shark is harvested commercially. The blue shark is taken as by-catch in tuna and swordfish longline fisheries, being utilized mostly by foreign vessels, which market the fins. The blue shark is the most commonly landed shark species in many longline fisheries, comprising 16% to 50% of the total catch during a longline research survey conducted in the New York Bight. Commercial swordfish longlines in the 1970s occasionally caught several hundred blue sharks per set of 1,000 baited hooks.

In order to reduce potential conflicts between foreign fisheries and growing domestic commercial and recreational fisheries, a preliminary management plan for Atlantic sharks has been prepared by the regional fishery management councils (established by the Fishery Conservation and Management Act). This plan restricts harvest to domestic vessels and identifies seasonal closures for regions off the Atlantic coast.

Since this species is largely found in bluewater, offshore areas, no New England state had established regulations to restrict its harvest in state territorial waters as of 1989.

Angling and Handling Tips

Although not as renowned a battler as the mako, the blue shark is a good fighter on lighter tackle and is more readily available than its larger relative. In New England waters, smaller blue sharks are seen as early as May and June, with larger blues up to 200 pounds arriving later with warmer water. Since blue sharks are ocean wanderers, it is difficult to predict where likely sites for fishing might occur. However, this species will strike at any time of day, and a charter boat with experienced crew can maximize the novice's success pursuing this game fish, as well as insure that landing one of these giants will be a safe experience.

Many anglers favor bluefish trolling rods in the 50–80-pound class though lighter rods are preferred by some. Rods should be mounted with a conventional star-drag-style reel, in the 4/0 to 6/0 range. Both Dacron and monofilament line can be used with this gear. Although line as light as 30-pound test is sometimes used, 50–80-pound test is most commonly used.

The best baits include 12-inch fillets of mackerel, bluefish, and bunker, or mullet with the skin left intact, but blue sharks may strike at nearly any bait offered to them. Rigs consist of a single, large hook attached to a heavy wire leader. The hook should be passed through the bait so that the barb is free and the bait will flutter. Some anglers add a plastic skirt or a piece of squid to the front of the bait to add color. If one is a beginner at shark angling, it may be best to purchase premade rigs sold at many tackle shops.

Baited hooks are drifted through the

water while chumming to attract sharks to the bait. Some charter boats prefer to have several baits simultaneously drifted at varying distances and depths behind the boat, from depths of about 30 feet deep just behind the boat to 80 or 90 feet deep several hundred feet behind the boat. Coastal bait and tackle stores often sell 5-gallon containers of frozen ground, premixed chum, which can be hung behind the boat to melt. Anglers can prepare their own chum by grinding or chunking carcasses or bait fish and mixing with seawater. Chum should be ladled into the water frequently to make a continuous chum slick. Some anglers also make as much noise as possible by splashing around the boat with a paddle or bucket. Since sharks seek food mainly by their sense of smell and sound, these actions can improve the catch markedly.

Once the bait is taken, wait 10 to 15 seconds, then set the hook as forcefully as possible. A stand-up fighting harness will be useful if you hook a large shark. Anglers need to keep a careful eye on the drag and prevent the line from tangling or passing under the boat. Although providing a powerful battle, blue sharks often fight more fiercely during landing. Unprepared anglers have been yanked right off their feet by explosive diving of the blue at this time.

Sharks should be landed with utmost care. Never bring a live shark into the boat. Any shark that is to be kept should be shot in the head with a gun or bang stick. Once dead, a shark may sink, so rope and gaff your fish before killing it. Use two lengths of ¾-inch rope and two gaffs to secure the head and tail. Lifting the tail out of the water by pulling on the tail rope will help calm a thrashing shark. Unless they are to be eaten, blue sharks should be released alive by cutting the line.

Sharks that are to be eaten must be bled as soon as possible after being landed. This cleanses the meat of urea that is stored in the shark's tissues. If not bled, urea can give the meat a strong odor and flavor of ammonia. A deep cut just in front of the tail will accomplish proper bleeding.

After cutting off the head, tail, and fins and gutting, the carcass should be flushed in seawater and cut into 12-inch chunks with skin intact to store on ice. Later it can be steaked for eating or freezing.

Blue sharks have snow white, mild-flavored meat when properly handled, and the lack of bone in their skeleton means that more usable meat can be filleted from the cartilage or steaked. Blue sharks can be grilled, fried, baked, and broiled or boiled to flake in salads. In general, cook a total of 10 minutes per inch of thickness, cooking the first side longer than the second and adding 5 minutes if using a sauce. The shark is done when it turns white and flakes.

Baked Shark Italiano is an excellent way of preparing blue shark. Place fillets or steaks in a baking dish and spread with bottled Italian dressing (or your own made from olive oil, vinegar, parsley, oregano, basil, salt, and freshly crushed black pepper). Marinate in a refrigerator for an hour or more, turning several times. Place the steaks in a plastic bag with seasoned bread crumbs, pressing to coat both sides, and put in a greased baking dish. Bake in a 450 degree F oven for 6 minutes, then turn the fish over and sprinkle with grated Parmesan cheese. Return it to the oven for 4 to 5 minutes until the meat flakes.

To grill shark, place steaks in a glass baking dish and add ½ cup of lime or lemon juice. Add freshly ground black pepper, two finely chopped garlic cloves, and ½ cup of olive oil, and grate the rind of a lime or lemon and 1 or more inches of ginger over it. Marinate in a refrigerator for 1 to 2 hours. Grill on a hot barbecue or gas grill 6 minutes on the first side and 4 to 5 minutes on the second. Baste with the marinade.

References

Grosslein, M. D., and T. P. Azarovitz. 1982. *Fish Distribution*. Mesa NY Bight Atlas Monograph 15. NY Sea Grant, Albany.

Harvey, J. T. 1989. Food habits, seasonal abundance, and sex of the blue shark, *Prionace glauca*, in Monterey Bay, California. *Calif. Fish Game* 75:33–44.

Kohler, N. E. manuscript. Food and feeding ecology of the blue shark (*Prionace glauca*) in the western North Atlantic. Narragannsett Laboratory, NMFS/NOAA.

Pratt, H. J., Jr. 1979. Reproduction in the blue shark, *Prionace glauca*. *Fish. Bull.* 77:445–470.

Preliminary fishery management plan for Atlantic billfishes and sharks. 1983. National Marine Fisheries Service, NOAA/U.S. Department of Commerce.

Smooth Dogfish
(*Mustelus canis*)

The smooth dogfish (also called the smooth dog and, improperly, the sand shark), an extremely abundant inshore species in southern New England, ranges throughout coastal waters of the western Atlantic from southern Brazil to Cape Cod. It has rarely been captured in the Gulf of Maine and northward.

The smooth dogfish has two large, spineless, and equal-sized dorsal fins, and its teeth are low and flattened for grinding. These characteristics alone clearly separate it from all other coastal New England shark species. The body is very slender and flattened ventrally, and the posterior outline of the upper lobe of the tail is deeply notched.

The smooth dog is grayish olive, slate gray, or brown dorsally, and its belly is yellowish or grayish white. Unlike most sharks, this species has the ability to blend in with its surroundings, often appearing nearly pearly white when around white sand bottoms and gray or brown over mud or other dark substrates.

Biology

The smooth dog is commonly found in harbors and bays and occasionally journeys into freshwater tributaries of estuaries. This species is closely associated with the coastline, although it can occasionally be found in more offshore waters at depths up to about 165 m (540 ft). It occurs in southern New England waters as a summer transient, arriving by May and leaving by October or November.

The smooth dogfish gives birth to young after up to 10 to 13 months of embryological development in the female's uterus. This species is viviparous; while developing, young are nourished by a placenta. There typically are 10 to 20 pups in a litter, although as few as four have been reported.

Young are about 29.2 to 36.8 cm (11.5 to 14.5 in) long at birth, and growth is rapid throughout the first several years of life. Calculated sizes at age for female and male smooth dogs are presented in Table 6. Females of this species may reach the maximum size of just under 150 cm (about 59 in) in just 6 to 7 years.

The smooth dogfish feeds most heavily upon lobsters and crabs. Its preference for lobsters hardly endears it to fishermen and others who consider the dogfish a competitor for the highly valued lobster resource. Even Bigelow and Schroeder (1953) calcu-

Table 6. Calculated length at age in cm (inches in parentheses) of male and female smooth dogfish

Age	*Female*	*Male*
1	76.5 (30.1)	76.1 (30.0)
2	105.1 (41.4)	95.3 (37.5)
3	122.5 (48.2)	103.7 (40.8)
4	133.2 (52.4)	107.6 (42.4)
5	139.7 (55.0)	108.8 (42.8)

Source: Calculations based upon tooth replacement and body growth rates, as presented in Moss 1972.

lated the consumption rates of lobsters by dogfish and speculated on the "destructiveness" of this species (interestingly, no mention is made of the amount of food that would be available to the smooth dog if humans did not harvest lobsters so heavily). Smooth dogfish also eat squid, small fishes such as the Atlantic herring, menhaden, and tautog, and invertebrates such as the razor clam, moon snail, and mantis shrimp.

Status and Management

Like the spiny dogfish, the smooth dog has never attracted a very active commercial fishery. Although a very abundant inshore resource, it has little market value for human consumption. It is considered a nuisance by commercial netters and a threat by lobster harvesters. Similarly, it holds little aesthetic value to the typical angler, who dislikes it due to its abundance and its tendency to take bait intended for preferred fish species.

Since it is generally abundant and little harvest is directed toward it, the smooth dogfish was not managed by any New England state as of 1989.

Angling and Handling Tips

Many anglers call this species the sand shark, although that is actually the name of a separate, larger-bodied shark species. The smooth dogfish is one of the most common species landed by anglers in the summer along the New England coast. Individuals 1 to 2 feet long are commonly caught in bays and harbors, on jetties, or in the shoreline surf.

Catching these fish is simple. Bottom-fish with any outfit used to catch smaller species such as the scup. A light-to-medium-action surf rod with 10-pound test line works well. Rig the line with a long-shank flounder hook, a 1-ounce barrel or bank sinker, and a swivel to prevent twisting. Nearly any bait will do, including sea worms, squid, and pieces of fishes or crustaceans. Anglers usually do not try to catch the smooth dogfish; they just can't prevent it! Dogfish are great bait stealers with a keen sense of smell for anything edible. Once a school of small pups finds your fishing spot, the only way to avoid catching them is to leave.

This is an ideal species for the weight-conscious American consumer. The smooth dogfish has a higher percentage of protein (26%) and a lower percentage of fat (0.2%) than any other species of coastal New England fishes that has been tested. This species "deserves" to be in our coolers, rather than left to rot or be eaten by crabs on some beach or jetty.

As with all sharks and skates, the smooth dogfish must be bled to remove urea from its tissues. Bleed the fish by cutting off or through its tail, and under its gills, as soon as it dies. After bleeding, wash the fish in salt water to remove ammonia odors. Ice the catch as soon as possible.

Anglers who enjoy eating dogfish whole or steaked usually start cleaning smaller individuals by cutting off their tails and gutting them. To gut, face the shark's belly upward and insert the tip of your knife into the vent. Slide your knife forward along the inside of the belly, cutting to the gills. Slide out the entrails, then cut the head, gills, and intestines off in one cut behind the gills.

Dogfish skins are so tough they were once used for fine wood sanding by furniture makers. Thus, cutting through their skin is a problem. Keep several sharp knives handy when using the following technique: Push the point of a sharp knife through the skin along the side of the dorsal fin and slide the knife under the skin along the length of the spine cutting through the skin from the inside out. After cutting the length of the fish, fillet as you would any other species. Your knife will slide easily along the cartilage. Since it has no bone in its skeleton, this species is easier to fillet than many fishes.

To steak larger dogs, first remove the fins. Next, cut the skin on the top and bottom of a side and pull the skin off with a pair of pliers. Cut the steaks to the desired size right through the cartilaginous skeleton. These steaks can be cooked in any manner, but grilling and broiling are best for larger individuals.

Well-iced smooth dogfish fillets or steaks stay fresh for nearly a week when stored near 32 degrees F. They also improve with time, getting firmer after several days in the refrigerator. Dogfish fillets also freeze very well when wrapped tightly to prevent freezer burn.

The meat of the smooth dogfish is more tender and has a finer texture than that of larger sharks. Small pieces of the smooth dogfish can be sautéed in an iron skillet. After cutting fillets into bite-size pieces, lightly sprinkle with salt while turning them over. After salting, let stand for 15 minutes. Next, cover with milk, buttermilk, lemon juice, or watered-down vinegar, stir, and place in the refrigerator until you are ready to start cooking. The dogfish can be left in the marinade overnight, if you wish, to enjoy a Scandinavian breakfast the next morning.

Remove the fillets from the marinade and drain, bring them to room temperature, sprinkle with black pepper and dust lightly with flour. After dusting, heat a pan and add ¼ cup of olive oil or a stick of butter. Sauté the shark until it is light brown, remove, and keep warm. Add several chopped cloves of garlic and one thinly sliced or chopped red onion to the pan and brown, adding oil as necessary. Next, add 2 or more tablespoons of catsup and stir over heat for 2 minutes. Pour over fish, thinly sliced home fries, and slices of fresh French bread. This exotic and hearty breakfast will have you ready for a new day of fishing, like an old salt from the North Sea!

Dogfish Teriyaki is another unusual yet easy dish to prepare. Mix 2 tablespoons of soy sauce, 1 tablespoon of tamari sauce (optional), ¼ cup of dry sherry, 1 teaspoon of lemon or lime juice, two large cloves of minced garlic, and ½ inch of grated ginger root in a glass baking dish. Add thinly sliced pieces of dogfish and marinate for 1 to 2 hours in the refrigerator. Pour enough oil to coat the bottom of a heated wok or frying pan and raise the heat to 375 degrees F. Add the fish pieces and quickly brown on both sides. Add all of the marinade, reduce the heat, and cook for 5 to 8 minutes. Serve over rice with a green vegetable such as steamed or lightly sautéed broccoli.

References

Faria, S. 1984. *Northeast seafood book: A manual of seafood products, marketing, and utilization.* Mass. Div. Mar. Fish.

Moss, S. A. 1972. Tooth replacement and body growth rates in the smooth dogfish, *Mustelus canis* (Mitchill). *Copeia* 1972:808–811.

Spiny Dogfish
(*Squalus acanthias*)

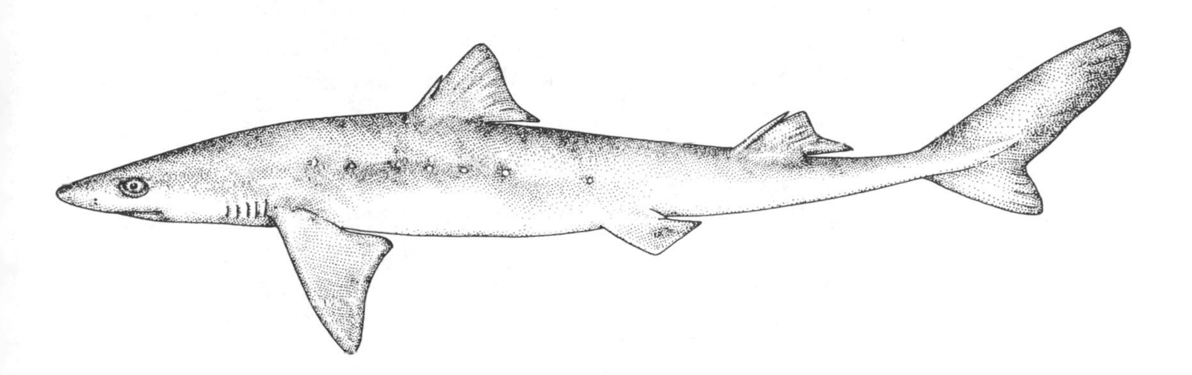

The spiny dogfish, perhaps the most ubiquitous and disliked of species encountered by New England anglers, has a worldwide distribution in the northern hemisphere. It occurs along the East Coast of North America from Newfoundland to Georgia and is abundant from Nova Scotia to Cape Hatteras.

This small shark species can easily be identified by the presence of a well-developed, sharp spine on the anterior edge of its two dorsal fins. It can be distinguished from the similar black dogfish, a deepwater species rarely encountered by anglers, by the presence of a deep notch in the upper lobe of the tail, a characteristic lacking in the latter species. The spiny dogfish has a slender body, a flattened head, and lacks an anal fin.

The spiny dogfish is slate gray dorsally. This color fades laterally to a pale gray or white on the belly. A row of white spots runs laterally from above the pectoral to above the pelvic fins. Several other white spots occur near the base of both dorsal fins. These spots are faded or absent in large individuals.

Biology

The spiny dogfish migrates extensively along the Atlantic coast and is a transient in a large portion of its range. Tagged individuals have migrated over distances as great as from Newfoundland to Virginia, or from the northern Atlantic coast eastward to Iceland. This species occurs chiefly north of Cape Cod in the summer and extends no farther south than Long Island throughout much of the fall. A general southward migration expands the distribution of dogfish to North Carolina for the winter. As they migrate northward in the spring, spiny dogfish first appear in shallow reaches of Georges Bank by March or April and reach the inner waters of the Gulf of Maine in May. They also display an inshore/offshore seasonal movement, tending to overwinter in the deeper waters of the outer continental shelf and to move to shallow waters in the summer. Seasonal migrations move this species to regions with appropriate temperature regimes. Preferred bottom-water temperatures range from 7.2 to 12.8 degrees C (45 to 55

F). The spiny dogfish also exhibits expansive within-season movements, as large schools swim about in search of food. Dogfish segregate into separate schools according to size as juveniles and according to size and sex as adults. Schools of mature females predominate in more inshore areas, whereas juveniles are most abundant offshore.

Spiny dogfish have a prolonged juvenile life stage, with few individuals reaching sexual maturity before the age of 9. The age at which approximately 50% of all dogfish reach sexual maturity is 12 years in the Northwest Atlantic.

As with all sharks, eggs are fertilized internally. Spawning typically occurs during the winter in offshore waters. Spiny dogfish in the Northwest Atlantic give birth to up to 15 (average of 6.6) large, well-developed young after a gestation period of nearly 2 years. The number of offspring produced increases with age and size of the female. Although offspring undergo extensive intrauterine development, there is no placental attachment between embryo and uterine wall.

The average length of dogfish "pups" at birth is 26 to 27 cm (10.2 to 10.6 in). Females grow faster and attain larger sizes than do males after the age of 7 years. Figure 7 illustrates calculated lengths at age for both sexes. Maximum size of females is about 124 cm (49 in) in length and 9.8 kg (26 lb) in weight. Maximum age is about 40 years.

The spiny dogfish is disliked by many anglers because of its voracious appetite; anglers are likely to catch little else on baited lines when their boat settles over a feeding school of dogfish. The dogfish eats many species of fishes, including mackerel, herring, scup, flatfishes, and smaller cod and haddock. Invertebrates including shrimp, crabs, squid, sipunculid worms, and a variety of other taxa are also eaten.

Other sharks are the only natural predators of the spiny dogfish. This species suffers relatively minor rates of natural mortality throughout its life cycle, compared to the majority of fish species that reproduce by releasing eggs and sperm into the water column.

Status and Management

The spiny dogfish had a bad reputation even with early Northwest Atlantic fishing operations. Possessing only modest market value itself, this species caused major disruption of seine and longline operations by feeding upon entrapped fishes. Many commercial operators considered longline fishing to be futile when dogfish schools were in the area. Due to periods of tremendous abundance, dogfish landings occasionally reached high levels; 12,250 mt (27 million lb) were harvested annually off the Massachusetts coastline in 1904 and 1905. Although major fisheries developed in Europe (this species is a component of Great Britain's fish and chips), the spiny dogfish developed only into an industrial fishery in New England. During World War II, dogfish livers were processed to extract vitamin A as a replacement for cod liver oil. However, after the war, the production of synthetic vitamin A replaced this use of dogfish livers. Dogfish were also processed into fish meal products for a time in the first half of this century. As fish meal markets declined, so did domestic harvests. Distant-water foreign fishing fleets accounted for virtually all of the harvest from the early 1960s until the 200-mile fishery conservation zone was established. Since then, U.S. harvest levels have been modest. Landings declined for five consecutive years, until increasing slightly to 2,900 mt (6.4 million lb) in 1988. Due to the lack of foreign harvest, the New England and Middle Atlantic stock of spiny

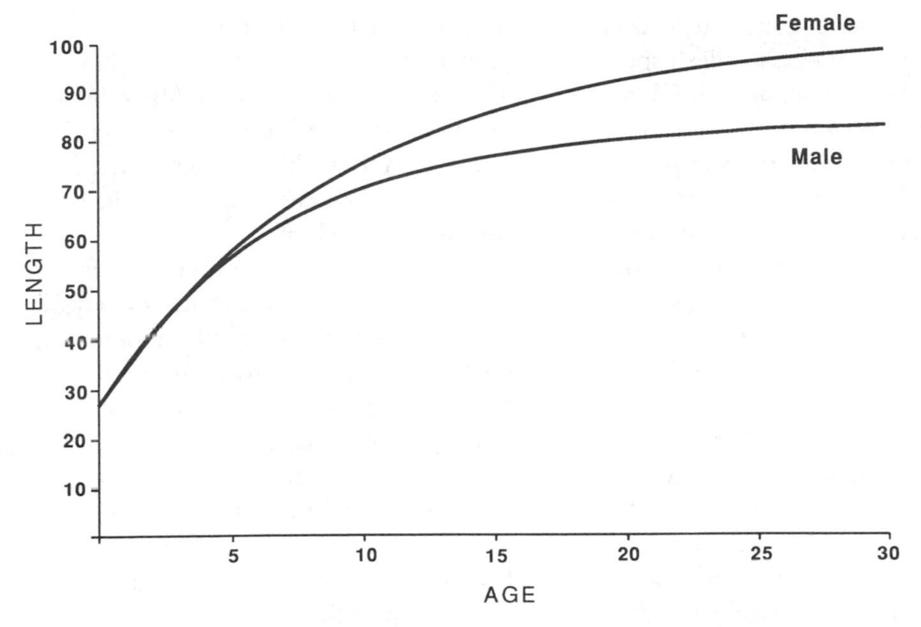

Figure 7. Calculated length at age of male and female spiny dogfish (calculations from von Bertalanffy growth equations presented in Nammack et al. 1985).

dogfish increased dramatically during the 1980s.

Because it reaches periods of almost spectacular abundance, the spiny dogfish is frequently represented as an underutilized species offering great potential for increased marketing. However, were it to gain greater desirability in the seafood markets, it might be very vulnerable to severe population depletion. Its slow growth, low rate of offspring production, and late age at sexual maturity provide the spiny dogfish with only limited capacity to withstand losses caused by increased levels of fishing exploitation. The abundance of dogfish in the Northwest Atlantic was 47% lower in 1979 than the 10-year average between 1968 and 1977. This rapid decline was caused by the intensive foreign fleet activities of the 1970s. Similarly, stocks in the eastern Pacific and the North Sea collapsed due to exploitation in the 1940s and 1950s–1960s, respectively. However, due to its extremely high abundance throughout the 1980s and its modest value to both the fishing industry and anglers, harvest of this species has not been regulated in New England waters.

Angling and Handling Tips

Although this species is considered a nuisance by many, it produces occasional thrills for the novice angler or youngster, who, regardless of what the party boat mate may call it, knows that the 2–3-foot-long fish he or she has just caught is a shark—is a shark—is a shark!

The novelty of catching the spiny dogfish may soon wear off, however, as the angler discovers that, once this species is found, little else may be caught. Spiny dogfish typically travel in large schools, voraciously grabbing all baited hooks offered in their direction, chasing away other species that they are feeding upon, and even attacking fishes already on the angler's hook. Dogfish schools rarely stay in one place very long, but follow the movement patterns of their prey. One hour the angler may be experiencing great success catching any number of groundfish species, the next nothing but dogfish. They are particularly frustrating to the party boat mate, who must undo the mass of tangled lines and rigs created by the spinning, twisting acrobatics of a hooked dogfish. When stumbling across a school of this persistent nuisance, one has two choices: Move to another site, or enjoy catching and eating this small shark species.

Catching a dogfish is no problem; a variety of gears rigged with baited hooks will often attract more than you might wish to handle. Landing a dog can be another matter. The spines on each of this species' dorsal fins can inflict a painful wound if the angler is careless. Use gloves and a billy for protection from wounding and possible infection. For general handling of this species, see the species profile for the smooth dogfish, which is treated in the same manner; the recipes listed in that species profile are also interchangeable with the recipes here. Spiny dogfish are usually bled and cut into steaks or fillets. Some anglers pull the skin off prior to cutting, and others fillet and skin as with any other tough-skinned species.

Although a major component of British fish and chips, and heavily exploited along the European coast, spiny dogfish are usually discarded in this country. One of the problems with this species' image is associated with its name. It may be some time before the American consumer eats anything with "dog" in its name, other than the beloved American hotdog. However, those willing to try this fish find the meat white, firm, and mild-flavored when properly prepared; it even has a slightly sweet flavor when marinated in lightly salted milk.

The spiny dogfish can be broiled, baked, fried, or poached. The traditional way to prepare this species is as fish and chips. This dish consists of dogfish fillets dipped in batter, fried, and served with large french fries. For a hearty beer batter, mix 1½ cups of flour, 1 teaspoon of salt, ¼ teaspoon of black pepper, ¼ teaspoon of garlic powder, 1 tablespoon of cooking oil, and two beaten eggs in a bowl. Slowly add ¾ cup of beer, stirring until smooth. Cover the bowl and allow it to sit in the refrigerator for several hours or overnight. The typical beer batter can be improved by using ale, which adds a nice flavor due to the hops used in brewing, or one of the newer dark beers on the market in New England. Each of these will add a distinctive, enjoyable flavor.

Once the batter is ready, place serving-sized pieces of shark into it. Preheat a deep skillet or wok and add enough oil to cover one layer of fish, raising the heat to 375 degrees F. Keep the fish and batter at room temperature to allow the batter to adhere and to prevent the oil from cooling when the fish strips are added. Fry dogfish pieces for 4 to 5 minutes. As the edges of the pieces turn to a golden brown, turn them once and continue frying for 2 to 3 additional minutes. Place the fish on an oriental frying rack over a wok or on a warming tray until all the pieces are cooked. Sprinkle with white or cider vinegar, serve with large french-fried potatoes, and learn why the British enjoy this dish so much.

For an interesting and nutritious meal, try serving a shark burrito. Cut shark fillets into short strips or 1-inch chunks, soak in salted milk for 1 hour, and drain. Put some flour into a plastic bag, add the shark pieces, and lightly coat. Heat ¼ cup of olive oil in a skillet or wok, shake the excess flour from the shark pieces, and sauté them until lightly browned. After sautéing, sprinkle one pack of prepared burrito, taco, or chili mix over the fish, add 1¼ cups of water, and stir. Next, drain and add one 15-ounce can of pinto or red kidney beans. Bring to a boil and simmer for 15 minutes. Warm tortillas (taco shells or pita breads work as well) in the oven and place several spoonfuls of the mixture on each. Top with finely chopped onions, shredded lettuce, diced tomatoes, and shredded cheddar cheese. Add a dash or two of Tabasco sauce for spice if you like burritos hot. This is a balanced meal that needs only a brewed beverage of your choice to be complete.

Reference

Nammack, M. F., J. A. Musick, and J. A. Colvocoresses. 1985. Life history of spiny dogfish off the northeastern United States. *Trans. Am. Fish. Soc.* 114:367–376.

Little Skate (*Raja erinacea*) and Big Skate (*R. ocellata*)

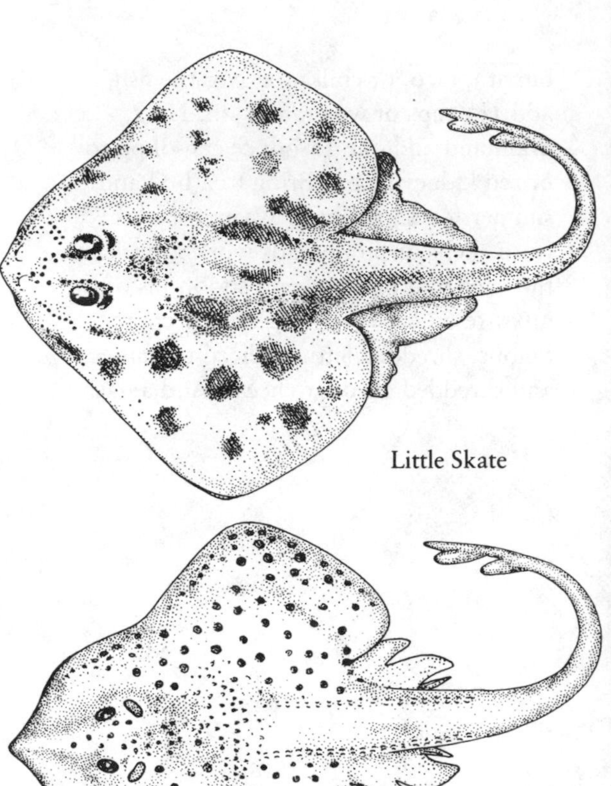

The little skate (also called the summer or common skate) and big skate (also called the winter skate) range from North Carolina to the southern side of the Gulf of St. Lawrence and are abundant off the U.S. coast from Georges Bank to the Chesapeake Bay.

Skates are easily distinguished from other recreational species found in New England. Their bodies are very flattened dorsoventrally. There are five pairs of gill openings, which are located ventrally on the body. The pectoral fins are fleshy and very expanded, and their anterior margin is attached to the head in front of the gill openings. Due to their appearance, these fins are often referred to as "wings."

The little skate and big skate are so similar that few people can consistently identify them to species correctly. The outline of the big skate's head bulges anterior to the eyes. Also, teeth are set in a series of 72 to 110 rows in the big skate, and 38 to 66 in the little skate.

The little skate is gray to dark brown dorsally, with numerous small, dark spots. It is white to light gray ventrally. The big skate is light brown with dark brown spots dorsally; some individuals have a large white "eye spot" patch with a black center on the posterior base of the pectoral fin. When present, this spot readily separates this species from the little skate. The big skate is white ventrally.

Biology

Both species are most abundant from shallow inshore waters to depths up to 110 m (361 ft). The big skate occurs most commonly in water temperatures from 1.2 to 21 degrees C (34 to 70 F); the little skate, 1.2 to 19 degrees C (34 to 66 F). Both species display seasonal movements. The big skate is concentrated on Georges Bank in the summer and fall and aggregates there and along the eastern shore of Long Island in the winter and spring. The little skate is distributed throughout the coast from Chesapeake Bay to Georges Bank in the winter and spring, with major concentrations along Long Island's coast. Some individuals south of Long Island migrate to deeper waters in the late spring, but most apparently move northward. This species is most abundant between Long Island and Georges Bank in the summer and fall.

The big skate matures at a length of about 64 cm (24 in) or larger; the little skate, 40 cm (16 in). Skate eggs are fertilized inside the female's reproductive tract, but unlike their shark relatives, skate species release egg cases into the water where much of the embryological development takes place. Limited evidence suggests that the big skate spawns in the summer and early fall in the Gulf of Maine, and from summer to January in southern New England. The little skate spawns throughout the year, with peak numbers of pregnant females occurring from November through January and from June through July in southern New England.

The egg cases, called mermaids' purses, are formed two at a time, with one egg housed in each case. No counts of fecundity have been made for these species; however, females produce 60 to 150 eggs per year in other, closely related skate species. Sticky filaments on the surface of the egg case allow it to adhere to the substrate, where it stays while the embryo develops. Hatching occurs 6 to 9 months after fertilization. Young are 9 to 10 cm (3.5 to 3.9 in) long at hatching.

Growth rates of the little skate are presented in Table 7. Maximum length for the little skate and big skate is about 54 cm (21 in) and 107 cm (42 in), respectively. The annual mortality rate for juvenile and adult little skate is 33% to 39%.

The diets of these two species are similar. Rock crabs, bottom-dwelling shrimp, and squid are eaten in greatest abundance, with polychaetes, amphipods, razor clams, and a variety of small fishes also frequently being taken. The little skate occasionally depends strongly upon one type of food, as shown in one study in Delaware Bay and another in Block Island Sound, where individuals fed almost exclusively upon sand shrimp and digger amphipods at the former and latter sites, respectively.

Table 7. Average length at age in cm (inches in parentheses) calculated for little skate from the Georges Bank/Delaware Bay region

Age	*Length*
1	16.0 (6.3)
2	25.3 (10.0)
3	32.2 (12.7)
4	37.8 (14.9)
5	41.6 (16.4)
6	45.3 (17.8)
7	47.8 (18.8)
8	50.5 (19.9)

Source: Data from Waring 1984.

Skates are known to be eaten by the monkfish and are probably eaten by other large fish predators as well.

Status and Management

Historically, two fisheries developed utilizing various species of skates, one in which wings were sold for human consumption, and the other in which whole bodies were processed into fish meal. Big skate and little skate landings averaged 217 metric tons (477,400 lb) and 953 mt (2.1 million lb) in the southern New England trawl fishery between 1969 and 1978. These landings represented 6.5% and 17% of the total industrial fish harvest in southern New England during that time period.

There was no fishery directed at skates in New England in the early 1980s. Most of the domestic catch landed while fishing for other species was discarded at sea. However, landings increased from 1,000 mt (2.2 million lb) in 1982 to 5,100 mt (11.2 million lb) in 1987 for all skate species from the Gulf of Maine to the Middle Atlantic states. This increase was due to a broadened market for the skate as a food fish and a bait market that developed in southern New England. The little skate was the principal species harvested in the bait fishery.

Skate species are landed incidentally in New England recreational fisheries; thus, the recreational harvest has always been an insignificant component of the total harvest. Skates were considered underutilized resources throughout the 1980s, and no harvest restrictions for either recreational or commercial fisheries had been established for them by the end of the decade.

Angling and Handling Tips

Skates, long a part of England's famous "fish and chips," are considered "rough" or "trash" fishes in the northeastern United States. Traditionally, skates have held little commercial value and have been treated as discards when landed by recreational anglers and commercial fishermen alike. Although in disfavor with many due to their odd, somewhat ominous appearance, skates can be a culinary delight to those willing to experience them by taste rather than merely by sight.

Skates can be caught from coastal areas seaward to very deep waters. Although large, "door-size" skates are occasionally caught inshore, most encountered by anglers are 1 to 2 feet across. Although they can be landed year-round, spring and fall produce the best catches. Anglers may not actively pursue skates, but few would misidentify these "winged" fishes when they are landed. Skates are caught with nearly any gear capable of handling sinkers that can hold bait onto the bottom. From 12-pound to 30-pound test line can be used, depending upon the habitat fished: jetty, surf, or deeper water. Skates will eat a variety of baits, ranging from shrimp, clams, and squid strips to a variety of bait fishes. When hooked, a skate swims away in one slow pull; the resistance of its large wings during retrieval is similar to trying to pull a kite underwater.

After catching a skate, dispatch the fish by hitting it between the eyes with a billy. Then remove the wings as close to the body as practical. A layer of white meat lies under the skin on either side of a thin, flat piece of cartilage. Fillet the meat from the cartilage and skin it. Wings can also be lightly poached, after which the skin can be easily pulled away, allowing the meat to fall from the cartilage. Europeans often consume the entire skate wing by simmering it in broth or stock long enough for the meat to fall apart and the thinner cartilage to dissolve.

Like all shark relatives, skates store urea in their tissues. Storing this nontoxic waste product aids them in maintaining proper water and salt balances in their body fluids. However, unless the urea is washed away, fillets may have a noticeable odor and flavor of ammonia. If the wings are removed soon after the fish is landed, the bleeding that ensues removes much of the stored urea. After the wings are iced, they can be marinated in water with kosher salt, vinegar, lemon juice, orange juice, or even milk to remove any remaining ammonia flavor. Skates also tend to improve in texture with proper storage. After cleaning and washing, place the wings in a plastic bag and store in the coldest part of the refrigerator for several days. Such storage makes the flesh firmer and more easily handled when cooking. Milk or juice can be added to the bag to marinate while storing. Of course, too long a storage will result in stronger-flavored flesh, as it does with most fishes.

The advantages of preparing skates more than make up for their less than attractive appearance. Their skeleton is cartilaginous rather than bony; thus, they are easier to fillet. They also contain no sharp little bones that discourage diners, particularly children. Since skates contain a significantly higher percentage of protein than beef, and almost no fat, these fishes may become a health food of the near future. Considering the adventuresome direction fish cuisine has taken in recent years, we may soon see specialized dishes such as "skate in black butter sauce" featured in other than just ethnic restaurants.

Skate wings in the past were cut into round, scallop-sized pieces and marketed as sea scallops. Although this "innovative" marketing technique has since been ended, it does provide a hint as to one possible preparation. If you would like to try mock scallops, sauté skate chunks in butter or margarine or bake them in a casserole as you might sea scallops. Poached skate flaked in a cold fish, crab, or lobster salad is also a tasty treat.

To broil, place 2–3-inch strips of filleted and skinned skate in a marinade of commercial or homemade Italian salad dressing. After 10 or more minutes, place slices in a baking dish or pan, pour on remaining marinade, and broil 4 inches from the heat. Cook 10 minutes for each inch of thickness of meat. Since many fillets are thin, as little as 3 minutes of cooking per side may be suitable. Serve topped with some freshly ground black pepper, using the cooked marinade as a sauce.

Do not fillet when poaching, as the meat can be easily separated from cartilage and skin when cooked. First, wash and scrub whole wings with a brush under running water. Cover them with water in a pan and add ½ cup of white vinegar or white wine and some salt. If you wish, add the following: thinly sliced onions, some black peppercorns, and parsley flakes. For variety, add a little tarragon, a teaspoon of Tabasco sauce, or a tablespoon of soy sauce. Bring to a boil, reduce heat, and simmer for about 3 minutes. After poaching, gently place the fish on a plate, skin, and remove the meat from the cartilage. Brown some crushed or minced garlic in ¼ cup of butter or olive oil, arrange the fish on a serving platter, pour on the butter and garlic, and top with fresh parsley.

For a wonderful cold dish, place skate wings in a pan and cover them with white wine. Sprinkle with thyme and flaked parsley and season with a little salt and ground pepper. Bring the wine to a boil and simmer for 10 minutes. After cooking, drain gently, reserving liquid, remove the skin, and place on a glass serving dish. The meat can be left on the cartilage so the diner can remove it while eating. If separated from the cartilage, use a wide spatula to transfer it from pan to dish, so that it can hold its shape. Continue to heat the reserved poaching liquid until it is reduced by half, cool, then pour over the fish. Chill in the refrigerator until it gels. Serve with olive oil and vinegar dressing or use a commercial Italian salad dressing.

Reference

Waring, G. T. 1984. Age, growth, and mortality of the little skate off the northeast coast of the United States. *Trans. Am. Fish. Soc.* 113:314–321.

American Eel
(*Anguilla rostrata*)

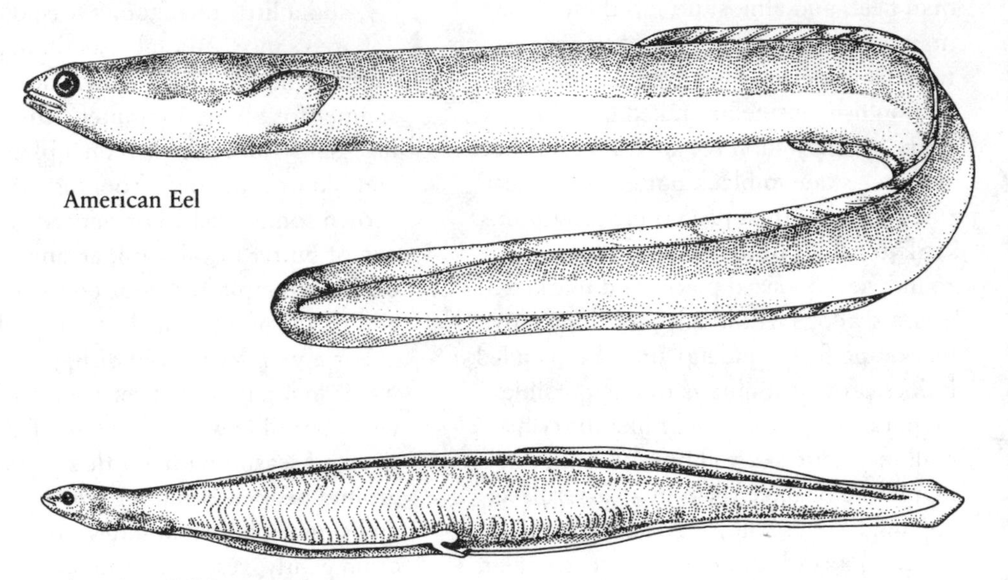

American Eel

American Eel Leptocephalus

American eels commonly occur in freshwater systems and coastal waters from southwestern Greenland and Labrador southward along the Atlantic coast of North America and throughout the Gulf of Mexico to Panama. This species has spread far into inland waters throughout its range, including landlocked lakes and watersheds separated from coastal waters by hydropower or other types of dams.

The American eel is unlike any other fish species that New England anglers are likely to encounter. The term "eellike" probably suffices to conjure up an image that would identify this species to anglers. The body is elongate with long dorsal and anal fins connected to, and appearing continuous with, the rudimentary tail fin. The eel's scales are minute and embedded in the skin. The mouth is large, with bands of small teeth on the jaws.

Biology

The American eel has a unique life cycle that includes several different body forms and color phases.

Each summer and fall adult eels leave the coastal watersheds of North America and migrate to their spawning grounds south of Bermuda and north of the Bahamas in the Sargasso Sea. Age at sexual maturity varies greatly among and within watersheds from which adults migrate to reproduce. Fecundity varies with size, ranging from 500,000 to 4.0 million eggs for most females. The largest females may produce as many as 8.5 million eggs.

After hatching, larvae remain in the leptocephalus stage for about a year. Leptocephali are transported along the eastern coast of the United States by the Antilles Current, the Florida Current, and the Gulf

Stream. During this period, leptocephali change into the glass eel stage, which is similar to the later juvenile (or yellow eel) stage but lacks pigmentation on the body; thus, the term "glass eel." Glass eels actively migrate toward land and fresh water. By the time they have reached coastal areas, the eels are several inches long, fully pigmented, and are called elvers. In New England, elvers migrate into estuaries and up freshwater systems in the spring. They appear in coastal rivers in Rhode Island in April and May and in Maine from late April to June. After entering estuaries, migration up rivers may take up to several years, depending upon the watershed. This species is able to move upstream of barriers such as dams that totally block off migration of other migratory fishes.

Once in coastal watersheds, elvers enter a prolonged stage of the life cycle during which individuals grow markedly and develop either testes or ovaries, although the eels remain immature. This stage, called the yellow eel, is the one that most anglers encounter. Male and female yellow eels are separated geographically. Males tend to be more abundant in estuaries whereas most females migrate farther upriver, and males predominate in the southeastern United States whereas females do so in more northerly regions. Some investigators feel that an elver may possess gonadal tissues capable of differentiating into either testes or ovaries and that environmental conditions may be largely responsible for determining which sex each individual becomes. This unusual means of determining sex is characteristic of a very small number of cold-blooded vertebrate species. Such a system would explain the geographic separation of sexes displayed by American eels.

After a prolonged period of growth as yellow eels, American eels metamorphose into the silver eel life stage and initiate a spawning migration. This metamorphosis includes (1) a color change to a metallic blackish bronze, (2) fattening of the body and thickening of the skin, (3) enlargement of the eyes for migrating at ocean depths, and (4) degeneration of the digestive tract. Spawning migrations occur in the summer and fall, with silver eels from northeastern North America leaving fresh water earlier in the season than those from southeastern U.S. watersheds. The degeneration of the digestive tract and the absence of adult eels after the spawning season indicate that eels die after reproduction.

Table 8. Length ranges at age in cm (inches in parentheses) of American eels from two northeastern U.S. watersheds

Age	*Rhode Island*	*New Jersey*
4	27–46 (10.6–18.1)	29–32 (11.4–12.6)
5	28–51 (11.0–20.1)	
6	28–51 (11.0–20.1)	41–67 (16.1–26.4)
7	29–58 (11.4–22.8)	36–67 (14.2–26.4)
8	33–64 (13.0–25.2)	44–70 (17.3–27.6)
9	38–62 (15.0–24.4)	37–74 (14.6–29.1)
10	37–65 (14.6–25.6)	44–86 (17.3–33.9)
11	46–65 (18.1–25.6)	63–90 (24.8–35.4)
12		67–94 (26.4–37.0)
13		68–98 (26.8–38.6)
14		78–97 (30.7–38.2)
15		78–104 (30.7–40.9)
16		77–100 (30.3–39.4)
17		95–99 (37.4–39.0)

Source: Data from Facey and Van den Avyle 1987.

Table 8 lists lengths of various ages of American eels from two northeastern watersheds. Females typically grow faster and larger than males. Females may weigh up to 6.8 kg (15 lb) and reach nearly 127 cm (50 in) in length, although most individuals do not reach this size.

Yellow eels feed primarily at night, probably depending more upon sense of smell than sight to find food. Feeding activities and movements of yellow eels within estuaries generally follow tidal flows, individuals moving upstream on flood tides and downstream on ebb tides. Eels tend to return to specific daytime resting areas after carrying out nightly foraging excursions away from their resting sites. When feeding, they may swallow food whole or grasp large food items and violently spin in the water to tear off pieces of flesh. In fresh water, eels eat insects, worms, crayfishes, frogs, and fishes. Crustaceans such as blue crabs, bivalves such as soft-shell clams, and polychaete worms are important foods in estuaries.

Eels are eaten by a variety of predatory fishes, such as the largemouth bass and striped bass, and by gulls, ospreys, bald eagles, and other fish-eating birds. Larger eels may eat considerable numbers of elvers in some watersheds.

Status and Management

Europe has been the major outlet for U.S. commercial landings of yellow and silver eels. They are shipped to foreign markets frozen, or densely packed alive in cool, moist crates supplied with oxygen. Historically, commercial landings from the Middle Atlantic (New Jersey to Virginia) have exceeded those from the North Atlantic (Maine to New York). From 1970 to 1973, an average 125.5 mt (over 276,000 lb) were harvested from North Atlantic states. Massachusetts landings plummeted from 100.5 mt (221,000 lb) in 1978 to less than 4 mt (8,400 lb) in 1985. This decline, mirrored throughout New England, was due in large part to the loss of export markets.

A modest elver fishery developed in Maine in the 1970s in response to a European market for these young eels. The market proved to have limited potential at that time. Landings in Maine were 10 mt (22,000 lb) in 1977 and 7.6 mt (16,720 lb) in 1979. During this time period Massachusetts prohibited harvest of elvers except for aquacultural purposes.

Table 9. Total recreational harvest in New England waters from 1981 to 1986, with states exhibiting greatest annual catches

Year	*Recreational harvest (nos. of fishes)*	*State with greatest harvest*
1981	101,000	
1982	178,000	RI (107,000)
1983	76,000	MA (37,000)
1984	119,000	MA (55,000)
1985	482,000	RI (248,000)
1986	42,000	MA, RI, CT (each < 30,000)

Source: Data from the NMFS *Marine Recreational Fishery Statistics Surveys* for the Atlantic and Gulf coasts.

Recreational harvest is modest in New England. Table 9 lists estimates of total recreational harvest in New England states from 1981 to 1986, with states exhibiting the greatest catches each year also noted. Several New England states employ harvest regulations to manage their American eel resources, including minimum size limits and gear and capture method restrictions.

Angling and Handling Tips

Recreational harvests are taken largely by eel pots and traps. These traps consist of cylinders of mesh netting with a funnel entrance that allows individuals in but confuses their exit. Pots are baited with fishes or fish frames, horseshoe crabs, or other crushed shellfishes and fished through the night, since eels are most active nocturnally. Many tend eel traps for bait as well as food, since this species is considered one of the best midsummer baits for striped bass. Eels are also taken in the winter by spearing. Those who spear through the ice often use fish spears or hooked tines attached to a 12–18-foot pole.

Anglers catch eels while fishing for other

species, especially when bottom-fishing in the evening or at night. A medium-weight spinning outfit with 12-pound test monofilament is well suited for this species. Some anglers feel heavier line is a necessity, as a hooked eel's violent twisting wraps the line around pilings, brush, logs, or other submerged structures. Small-shanked hooks baited with a small piece of fish, squid, or sea worm and a fish-finder rig (an egg-shaped sinker through which the line passes) complete the rigging. A small split shot should be placed 10 inches from the hook to prevent the egg sinker from sliding down to the hook. Cast the rig out to deeper waters, along ledges or rock walls, or in narrow channels of an estuary.

Landing an eel is much easier than taking it off the hook. The novice is in for quite a surprise when attempting to pick up and hold an eel. Its smooth, slick skin and its tendency to twist and spin when handled make the eel nearly impossible to hold in the bare hand for more than several seconds. This frustrating problem is easily remedied by using a towel or rough cloth when grasping an eel.

One advantage that the eel has over nearly all other recreational species is that it is easy to keep alive, thus absolutely fresh, until you are ready to clean it. Eels can live for extended periods of time if placed in a mat of wet seaweed within a burlap bag or covered container. Those who grew up in a coastal area may remember when the local "fishman" carried live eels in baskets of wet seaweed for door-to-door sale. Many a wide-eyed child would watch as live eels were dispatched and cleaned on the doorstep as part of the sale.

Eels are relatively easy to clean. Tie a 10-inch looped cord to a nail or hook attached to a post above eye level. Put the loop around the "neck" of the eel, behind the gill covers. Cut the skin around the neck posterior to the loop (do not cut the meat). Grip the skin with pliers, and peel it backward toward the tail. Next, sever the head, then gut, wash, and cut into 2–3-inch lengths. Filleting eels is also possible, but many do not consider the bones a nuisance. Eels can be grilled without removing the skin, but they should be thoroughly washed and rubbed with salt.

Eels, prized delicacies in many parts of the world, are practically unknown to American palates. In Europe, the eel is a popular holiday treat, often served at Christmas (as it once was in many homes of coastal North America). The snakelike appearance of the eel prevents many from eating it, but what it lacks in looks it makes up 10-fold in flavor. The eel's meat is delicate, with a wonderfully sweet flavor. Like species such as the Atlantic salmon, the eel does not freeze well. However, its fat yet firm meat makes the eel well suited for frying, sautéing, grilling, boiling, and smoking. Smoked eel is moist, and smoking results in a lower fat content than the other cooking methods. After smoking, remove the meat from the bones and serve as smoked fingers with wedges of lemon.

Traditionally, 2–3-inch lengths are rolled in flour and fried. Remember, the firm texture of an eel may need a little more cooking than a soft-fleshed species to insure that the inside of the meat is not undercooked.

For a tasty dish, try Eel Risotto. Sauté eel pieces in a large iron frying pan with ½ cup of olive oil for several minutes, turning to color all sides lightly and firm the texture. Remove the eel, add several cloves of finely sliced or minced garlic and one chopped small onion to the pan, and lightly brown. Add ½ pound of sliced mushrooms and

1 cup of white wine, cover, and cook until the wine is reduced by half. Add the eel, 1 cup of tomato sauce, some freshly ground black pepper, ½ teaspoon of oregano, ½ teaspoon of basil, and salt to taste; cover and simmer for 20 to 30 minutes. Add water if necessary.

To serve Eel Risotto, remove the eel and keep it warm. Mix precooked, cooled rice into the sauce until the rice is evenly red. Place several pieces of eel on a bed of risotto and serve with a sprig of parsley and freshly grated Parmesan cheese.

References

Facey, D. E., and M. J. Van den Avyle. 1987. *Species profiles: life histories and environmental requirements of coastal fishes and invertebrates (North Atlantic)—American eel.* U.S. Fish and Wildlife Service Biological Report 82(11.74). U.S. Army Corps of Engineers, TR EL-82-4. 28 pp.

Dutil, J.-D., A. Giroux, A. Kemp, G. Lavoie, and J.-P. Dallaire. 1988. Tidal influence on movements and on daily cycle of activity of American eels. *Trans. Am. Fish. Soc.* 117:488–494.

American Shad

(Alosa sapidissima)

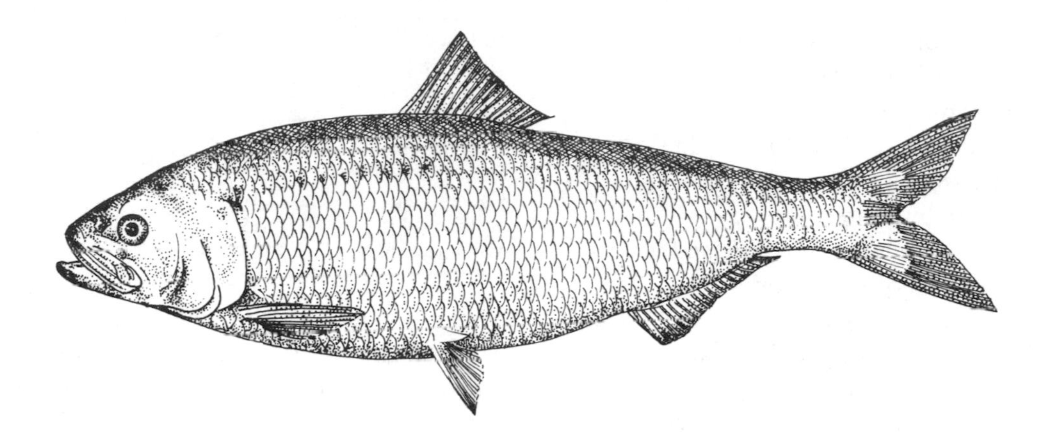

The American shad, noted for the seasonal fishery created by its yearly migrations into coastal rivers, is native to the Atlantic coast from Florida to Newfoundland and is most abundant from North Carolina to the southern New England region. It was successfully introduced into the Sacramento and Columbia rivers of the West Coast in 1881 and is now established from southern California to Alaska and westward to the Kamchatka Peninsula in the Pacific basin.

Members of the shad (or herring) family have laterally compressed bodies with one short-based dorsal fin and a small triangular flap of skin (the pelvic axillary process) at the base of the pelvic fin. The lateral line, a series of sensory pores running laterally along the body of most fishes, is absent. The midline of the belly is narrowed to knife-edge thinness, and the posterior edge of the scales along this edge is slightly elevated, creating a saw-toothed appearance. Although members of this family are easily separated from most coastal species, they are difficult to distinguish among themselves. Anglers can separate the American shad from the other two species they might encounter, the so-called river herring (the alewife and blueback herring), by the presence of one or two series of four to six dusky spots running laterally along the shad's body behind the gill cover and by its larger mouth. The American shad's upper jaw extends beyond a point below the back of the eye, whereas the river herrings' jaws do not extend beyond the middle of their eyes.

The American shad is dark blue or green dorsally, fading to a silvery tinge laterally and white on the belly.

Biology

The American shad is an anadromous species, spending most of its life in oceanic waters but migrating into coastal rivers to reproduce. During the oceanic phase of its life cycle, the American shad is a schooling fish, undertaking extensive migrations along the East Coast. The shad tends to move offshore in winter and return to more coastal waters in the spring. Adults that spawn in New England rivers overwinter in the Middle Atlantic coastal region and migrate northward in the spring, most tending to remain well offshore due to water temperature patterns in the Middle Atlantic Bight. Postspawning adults and immature fish congregate in the Gulf of Maine and the Bay of Fundy in the summer and fall, moving offshore and southward by winter.

Shad spawning in New England waters generally do not live beyond 5 to 7 years of age, although individuals 10 to 11 years old are infrequently caught. Males reach sexual maturity between 3 and 5 years of age, females between 4 and 6. Fecundity of females is generally highest in populations reproducing in the southern portion of this species' range and decreases progressively northward. Within a geographic region, fecundity increases with size of the female. Females spawning in the Connecticut River system release 256,000 to 384,000 eggs per female annually.

Shad return to their natal rivers to spawn. Olfactory (sense of smell) and rheotactic (orientation to currents) cues guide their homing migration to the rivers in which they had hatched. Shad learn these cues during their initial migration to the ocean as young fish. Water temperature appears to control the timing of migration into natal rivers. Shad migrate into the Connecticut River at approximately 16.7 degrees C (62 F) and cease to migrate when the water reaches 20 degrees C (68 F). Migration may extend as far as 500 miles upriver; however, migration distances in most rivers are now severely limited by dams and other man-made obstructions.

After hatching, young fish spend their first summer in the river. During this time they feed heavily on microscopic zooplankton and insects. In the fall, dropping water temperatures initiate downriver migration into brackish water and ultimately into the ocean. Shad remain in the ocean until participating in a spawning migration as adults.

Table 10. Percentage of repeat-spawning shad in watersheds of the Atlantic coast, with rivers listed in order of ascending latitude

River	*Latitude* (N)	*Repeat spawners* (%)
St. Johns (FL)	30	0
Ogeechee (GA)	32	0
Edisto (SC)	33	0
York (VA)	37	24
Potomac (VA)	38	20
Susquehanna (PA)	40	37
Connecticut	41	63
St. John (N.B.)	45	73
Miramichi (N.B.)	49	64

Fish spawning south of Cape Fear (NC) die after reproducing; in rivers north of this region, some individuals survive and return to the ocean, living to spawn in subsequent years. The percentage of adults that live to be repeat spawners increases as one progresses northward along the Atlantic coast (Table 10).

Although shad weighing in excess of 4.5 kg (10 lb) are infrequently captured, adult males typically weigh between 0.7 and 2.7 kg (1.5–6 lb), females between 1.6 and 3.6 kg (3.5–8 lb). Shad may grow to 76 cm (30 in) in length, but the largest individuals caught are usually less than 61 cm (24 in) long.

After migrating to the sea, American shad feed upon mysid shrimp, euphausiid shrimp, amphipods, fish eggs, and small fishes. In fresh water, juveniles are fed upon by American eels, birds, and juvenile striped bass. At sea, shad are eaten by porpoises, seals, sharks, tunas, and kingfish. However, shad do not appear to be the major food item of any predator in either freshwater or marine habitats.

Status and Management

Shad populations spawning in New England watersheds had been severely depleted before the beginning of the 20th century due to the combined effects of damming, pollution, and exploitation, activities that date back to at least colonial times. For example, a dam constructed on the Connecticut River at Turner's Falls, Massachusetts, in 1798 cut off half the watershed originally used by spawning shad. Subsequently, dams constructed downriver at Holyoke, Massachusetts, in 1849 and Enfield, Connecticut, in 1880 further reduced habitat available for spawning and the survival and growth of young shad. The Merrimack River has a similar history. Diversion dams and locks had been constructed in the watershed before 1800. By the mid 1800s, heavy industrialization within the river basin caused major pollution problems. Industrialization also cut off most of the river from shad spawning, particularly after construction of a 30-foot-high dam at Lawrence, Massachusetts, in 1847. Development of this kind occurred in watersheds throughout New England. Also, extensive commercial and recreational fisheries developed for shad along the entire East Coast in the 19th century. The combined effects of exploitation, pollution, and loss of access to spawning and nursery habitat markedly reduced New England shad stocks.

Concern over the loss of migratory fisheries, particularly the Atlantic salmon and American shad, led to efforts to restore these lost resources. The states of New Hampshire and Massachusetts established commissions to address restoration of migratory fish stocks in 1864 and 1866, respectively. Subsequently, these states, aided by legal rulings establishing the responsibility of industrial operators of dams and diversions to provide fish passage over these obstructions, initiated long-term programs in construction of fishways (fish ladders and lifts) and in stocking. These actions produced only modest, inconsistent success until the mid 1900s, when improved knowledge and technology concerning fish-passage facilities produced almost startling results. For example, improvements in fish passage at Holyoke increased the number of shad passed upstream over the dam at that site from 25,000 in 1973 to 380,000 by 1980.

The Atlantic States Marine Fisheries Commission established a coastwide management plan in 1985 to address coordinated management activities directed toward this highly migratory resource. Objectives of this plan include regulating the harvest, improving habitat quality and accessibility, and initiating stocking programs to restore shad to rivers where they historically existed. The New England states establish management plans for shad in their territorial waters. Regulations, including daily creel limits, area and seasonal closures, and gear limitations (several have established rod and reel only), differ from state to state.

Angling and Handling Tips

Shad fisheries date back to precolonial times. Various Indian tribes took great advantage of the readily available food source created by shad spawning runs. Early colonists also utilized this abundant resource, and thousands of barrels of shad were harvested annually to feed troops during the American Revolution. The shad run after the terrible winter of 1776 is believed to have saved George Washington's troops from starvation.

Recreational fishing for shad has

changed very little throughout the 20th century. Migrating shad are not attracted to live bait but will hit artificial lures. The ultimate weapon for taking a shad today, the shad dart, is the same lure used by our forefathers. This is a cone-shaped, lead-head jig with a diagonally flattened face that creates a wiggling motion when retrieved or held in currents. This lure, weighing from 1/16 to 1/4 ounce, has a hook positioned to point upward when the lure is retrieved. This prevents the hook from easily snagging on the bottom. Traditionally, darts are red and/or white, with a tail made of white or yellow hairs. Some anglers prefer other head colors, such as yellow, or tailless darts.

Most anglers use a 6-foot light-action spinning rod to detect the light taps and bumps that telegraph impending action. Up to 8-pound test line is spooled onto the reel, although 6-pound test is preferred; the drag of heavier lines influences the action while drifting the jig, thus affecting the catch.

Shad spawning migrations provide angling opportunities from river mouths and adjacent estuaries upstream to spawning grounds. Some anglers believe that cloudy days provide the best action, since shad generally are most active at night and during other periods of dim lighting. When fishing, cast the lure slightly upstream across the current, let it drift close to the bottom, then slowly retrieve the line. The line should remain tight at all times. The greatest number of strikes will occur during the drift phase of a cast. When fishing in the middle of the current, simply cast downcurrent and retrieve the dart very slowly, bouncing it off the bottom.

Shad usually hook themselves, so hooks do not need to be forcefully set. Patience and a lightly set drag are virtues when fishing for shad. This species has a delicate mouth from which a hook can easily tear if the fish is retrieved through the current too rapidly. The battle is often exciting, as a hooked shad usually swims with the current, adding additional pressure to the rod and action to the angler. Leaps that cause the fish to clear the water's surface are also a common part of the action.

In addition to being a great battler, the shad can provide wonderful table fare. The shad's reputation for boniness is accurate. Prior to modern filleting techniques, it took some skill and patience to feast on the shad. An old Micmac Indian legend tells of the unhappy porcupine asking the Great Spirit to turn it into some finer form of animal. The great spirit grabbed the "porky," turned it inside-out, and tossed it into the water to become the bony but delicious shad! However bony, the sweet flesh of this species makes a delightful meal for anyone.

Fresh shad roe is highly prized. Even some who care little for caviar rave about shad roe, the "New World's caviar." The preferred way to prepare the shad's paired, egg-filled ovaries is to dust them lightly with flour and fry in butter that has browned some onion and garlic. Fresh roe is most flavorful. Roe chilled on ice for 10 minutes and gently poached for several minutes in simmering water holds together better during subsequent cooking.

Due to their extreme boniness, filleting shad for total bone removal is not easily mastered by anglers. Instead of bone removal, many recipes require slow cooking to loosen and remove the rows of "Y bones" that make eating difficult. Baking in an oven at low heat (250 degrees F) for 5 to 7 hours is a traditional method originally practiced overnight on the wood stove. One alternative includes covering the shad with barbecue sauce and onion slices and wrapping tightly in foil. This package can be cooked on a charcoal grill; open the foil after an hour to add a smoked flavor.

Fast pickling is an alternative that eliminates problems with bones while producing a tasty dish. The shad should be cut into steaks or chunks, which may then be placed in a glass or other noncorrosive container. Soak up to 48 hours in chilled brine made from two cups of kosher salt per gallon of water; this removes moisture from the fish and firms the flesh. Drain the brine, cover with white or cider vinegar, and soak up to 24 hours before draining once more. Layer the steaks in a glass container, sprinkling some pickling spices between each layer. Add some of the following: onion rings, bay leaves, dill, mustard seeds, crushed hot pepper, or thin slices of cross-cut celery, fresh red or green pepper, lemons, oranges, or garlic. Mix ½ cup of sugar in 1 cup of vinegar and some white wine or water, pour over layered steaks, and cover. Bake in an oven at the lowest possible setting for 6 or more hours (or overnight). Pickled shad may be eaten hot or cold and will store in your refrigerator for several weeks. Mix fish pieces with sour cream for an interesting spread.

References

Moffitt, C. M., B. Kynard, and S. G. Rideout. 1983. Fish passage facilities and anadromous fish restoration in the Connecticut River basin. *Fisheries* 7:2–11.

Stolte, L. 1981. *The forgotten salmon of the Merrimack*. U.S. Government Printing Office, Washington DC. 214 pp.

Weiss-Glanz, L. S., J. G. Stanley, and J. R. Moring. 1986. *Species profiles: Life histories and environmental requirements of coastal fishes and invertebrates (North Atlantic)—American shad*. U.S. Fish and Wildlife Service Biological Report 82(11.59). U.S. Army Corps of Engineers, TR EL-82-4. 16 pp.

River Herring: Blueback Herring (*Alosa aestivalis*) and Alewife (*A. pseudoharengus*)

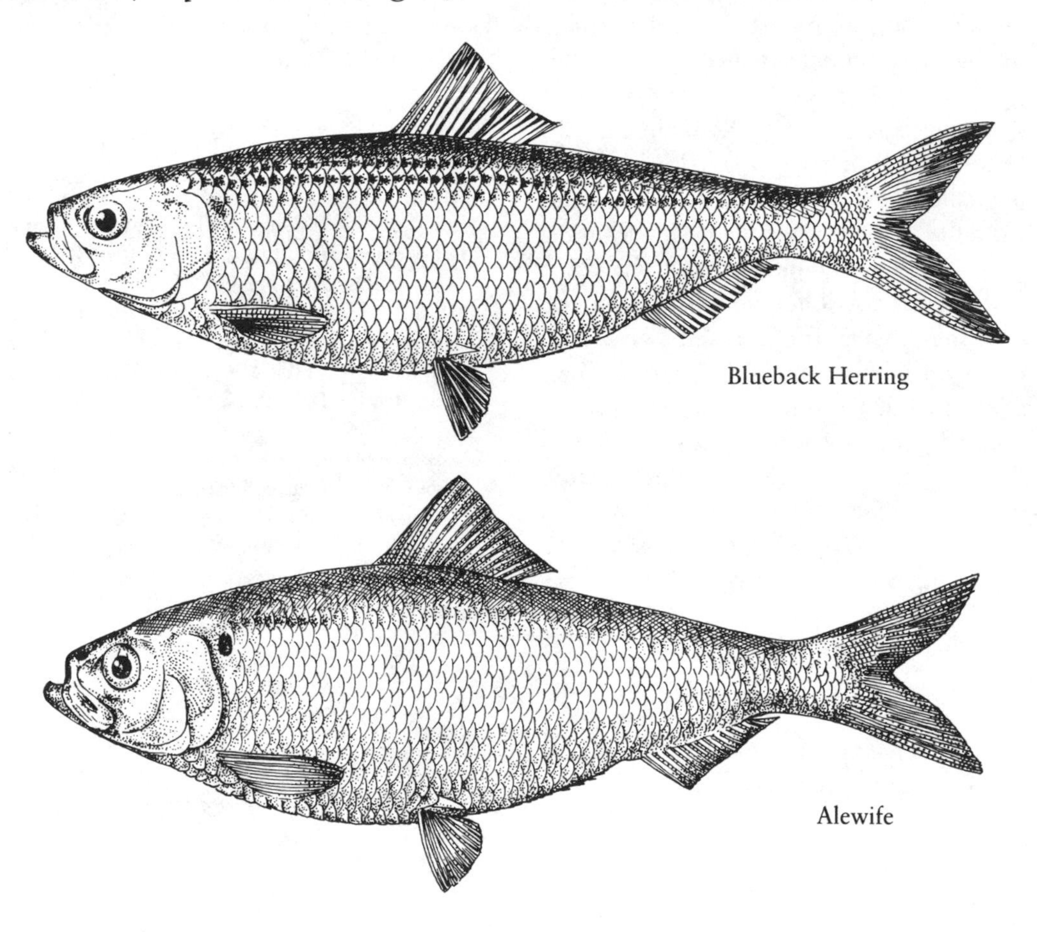

Blueback Herring

Alewife

River herring are actually two closely related anadromous (growing up in marine environments and reproducing in fresh water) species, the alewife and the blueback herring. Both of these very similar species are native to the East Coast of North America, the alewife occurring from Newfoundland to South Carolina, the blueback herring from Nova Scotia to the St. Johns River in Florida.

River herring have laterally compressed bodies with one short-based dorsal fin and a small triangular flap of skin (the pelvic axillary process) at the base of the pelvic fin. The lateral line, a series of sensory pores running laterally along most fishes' bodies, is absent. The midline of the belly is narrowed to knife-edge thinness, and the posterior margin of the scales along this edge are slightly elevated, forming a saw-toothed pattern. River herring differ from the closely related American shad in the size of the upper jaw. The shad's upper jaw extends beyond a point below the back of the eye, whereas the river herrings' jaws do not extend to the middle of the eye. The alewife and blueback are typically managed as a single fishery in part due to the extreme difficulty one has in distinguishing them from each other. The most reliable identifying characteristic is internal. The peritoneum (tissue lining of the body cavity) is uniformly dark brown or blackish in the blueback but is gray or silvery with small dark spots in the alewife. Externally, the species differ only very subtly in scale patterns. Nearly all external characteristics are indistinguishable, or greatly overlapping, in the two species. Some researchers have described differences in dorsal color patterns between the species, but others have felt that color differences are not consistent and that coloration of either species varies greatly with changes in lighting conditions.

Biology

Although many populations of alewives have been successfully introduced into landlocked freshwater systems, both of these species are anadromous, spending all of their lives, after the first year, in the ocean, except during spawning migrations into freshwater systems.

While in the ocean both species display seasonal migrations, possibly in conjunction with changing patterns of water temperature. In summer and fall, both species are confined to areas north of 40° north latitude, particularly Nantucket Shoals, Georges Bank, and the Gulf of Maine. They are distributed over most of the continental shelf in late winter and spring as they migrate toward their spawning rivers.

Fecundity of river herring increases with size and age. The alewife produces 60,000 to 100,000 eggs per season, depending upon body size. A 24-cm (9.4-in) blueback herring produces about 45,800 eggs; a 31-cm (12.2-in) individual produces about 349,700 eggs.

Alewives and bluebacks reproduce from April to mid July, spawning earliest in the southern portion of their ranges. Each year, males arrive in spawning streams before females of the same species. Alewives usually spawn 3 to 4 weeks earlier than bluebacks in the same watershed. Alewives initiate spawning when water temperatures reach 10.6 degrees C (51 F), bluebacks, 13.9 degrees C (57 F). Both species cease spawning when waters warm to 27.2 degrees C (81 F).

Most fishes return to spawn in the watersheds in which they hatched, although researchers have shown that some individuals stray from their natal rivers into adjacent watersheds to reproduce. Olfaction (sense of smell) appears to be the major sensory mechanism used by individuals to find and migrate into their natal watersheds. Eggs are buoyant in flowing water but settle to the bottom in still water. In New England, young-of-the-year remain in their home streams until fall, when they migrate downriver to the ocean. Heavy rainfalls, high waters, and/or sharp declines in water temperatures stimulate the start of this migration.

These species suffer relatively high rates of mortality throughout their life cycles. Fewer than 1% of all eggs survive early life stages to migrate to the sea. Total annual mortality of adults is about 70%. Of all adults, 32% to 90% die annually, during, or due to, spawning migrations and reproduction. Adult spawning mortality is highly variable from area to area and from year to year within specific areas.

Table 11. Length at age in cm (inches in parentheses) of alewives and blueback herring

	Alewife		*Blueback*	
Age	*Male*	*Female*	*Male*	*Female*
3			25.9 (10.2)	26.2 (10.3)
4	26.4 (10.4)	28.4 (11.2)	26.7 (10.5)	27.7 (10.9)
5	27.7 (10.9)	28.4 (11.2)	27.9 (11.0)	29.2 (11.5)
6	29.0 (11.4)	30.0 (11.8)	28.7 (11.3)	30.2 (11.9)
7	30.2 (11.9)	30.7 (12.1)	29.7 (11.7)	31.0 (12.2)
8		32.5 (12.8)		

Source: Data from Fay et al. 1983.

Growth rates, age at sexual maturity, and longevity vary greatly for both species geographically. Few individuals of either species exceed 30 cm (12 in) in length or about 1 kg (2.2 lb) in weight. Lengths of 3–8-year-old Connecticut River specimens of both species are listed in Table 11. Alewives are somewhat longer than bluebacks of the same age throughout their ranges. Within each species females tend to grow somewhat faster than males. Age at sexual maturity for both species is higher in the northern portions of their ranges. For example, most alewives spawning in North Carolina watersheds mature at age 3; in the central portion of this species' range, many do not mature until age 4 or 5; and none matures until at least age 4 in Nova Scotia.

River herring feed primarily upon zooplankton, eating microcrustaceans, fish larvae, fish eggs, insects, insect eggs, and crustacean eggs. River herring are eaten by a variety of predators, particularly schooling species such as the bluefish, weakfish, and striped bass. They are also fed upon by avian predators such as gulls and terns.

Status and Management

River herring populations along the East Coast started to decline in colonial times. The combined effects of overfishing, pollution, and damming of spawning rivers and streams have had drastic long-term effects upon these two species. For example, only 9 of an original 27 streams in the Gulf of Maine coast of Massachusetts that once held major river herring spawning runs still did so by 1920. However, long-term efforts to restore depleted stocks have led to significant recovery in many New England watersheds.

River herring have traditionally supported a modest commercial bait industry in New England. Offshore landings are typically picked up as by-catch while harvesting other species, and inshore and river harvests are directed specifically toward herring spawning runs. Coastwide commercial harvests have steadily declined during the 20th century. In New England, coastal Maine and Merrimack River catches declined by over 76% and 99%, respectively, between the turn of the century and 1950. By the late 1980s, North Carolina remained the only state with a substantial commercial fishery, accounting for 80% of the total coastwide catch.

Recreational fishing accounts for modest harvests, with the greatest effort occurring in the Middle Atlantic states. Much of this harvest is used as bait for other predator sport fishes.

Since most of the harvest in New England traditionally occurred adjacent to or in estuaries and river mouths, management of these species has been focused upon small geographic areas. For example, Maine has a management plan that sets regulations on a county-by-county basis. In Massachusetts, each town may petition the director of marine fisheries to establish local control of management. The state manages runs that are not overseen by town authorities.

In part due to the prolonged depletion of river herring and shad stocks in the Middle and southeastern Atlantic states, the Atlantic States Marine Fisheries Commission established a coastal management plan for the river herring and American shad in 1985. The objectives of this plan include regulating the harvest, improving habitat quality and accessibility, and initiating stocking programs to restore populations in rivers where they historically but do not presently occur.

Angling and Handling Tips

Most herring are caught by dip-and-cast nets. Few anglers purposely fish for river herring. Most are landed incidentally while anglers are fishing in river systems for other species such as the American shad. A herring will occasionally strike at very small shad-type darts or at weighted flies with a tiny leadhead and a thin tail feather. These "miniature shad" provide a scrappy battle to anglers who use an ultralight spinning outfit or fly rod.

Recreational anglers often use river herring for bait rather than table fare. Many striped bass anglers consider live-lining river herring as one of the surest means to hook a large fish. When live-lining, anglers hook live herring either in front of or behind the dorsal fin, then let them drift without any lead weight, which allows them to swim about naturally.

At one time in New England, river herring were a common part of people's diets. Though not considered as flavorful as sardines (small sea herring), river herring were commonly eaten fresh, salted, smoked, or pickled. River herring roe was considered as desirable a delicacy as shad roe.

Roe is excellent when sautéed. After removing the egg-filled ovaries, parboil them to prevent spattering during subsequent cooking. Sauté finely chopped garlic in butter or olive oil at medium heat and add the roe, sautéing until golden brown. Season with salt and pepper and serve New England–style with wedges of lemon, very small boiled potatoes, and white wine.

Cooking and marinating lighten the flesh and flavor of herring, making these fish enjoyable for most who try them. Traditionally, herring were smoked. To smoke river herring, clean and gill them, leaving heads intact, and place in a glass or plastic container. Mix a solution of 1 gallon of water, 1 cup of pickling or kosher salt, and 1 cup of brown sugar. If you like, add pickling spices, hot Tabasco sauce, or soy sauce to liven the flavor. Heat the mixture to dissolve the salt and sugar thoroughly, then let it cool. After cooling, pour over the herring and allow to stand for 20 minutes. Finally, rinse the fish in cool fresh water and hang them for an hour or more until dry. This process removes moisture from the flesh and firms its texture.

Smoke at low heat (120 degrees F) for several hours; then raise the temperature to 180 to 190 degrees F for several more hours. The longer you smoke, the greater the smoked flavor. Cool, wrap, and refrigerate. Smoked river herring may be stored up to several weeks before eating.

For an additional treat, use some of your smoked herring in a dip. Flake 1 cup of fish from the bone and add it to 1 or more cups of sour cream and a packet of dried vegetable soup mix (not onion mix).

Mix, refrigerate for several hours, and serve with crackers.

Reference

Fay, C. W., R. J. Neves, and G. B. Pardue. 1983. *Species profiles: Life histories and environmental requirements of coastal fishes and invertebrates (Mid-Atlantic)—alewife/blueback herring*. U.S. Fish and Wildlife Service, Division of Biological Services, FWS/OBS-82/11.9. U.S. Army Corps of Engineers, TR EL-82-4. 25 pp.

Atlantic Salmon
(*Salmo salar*)

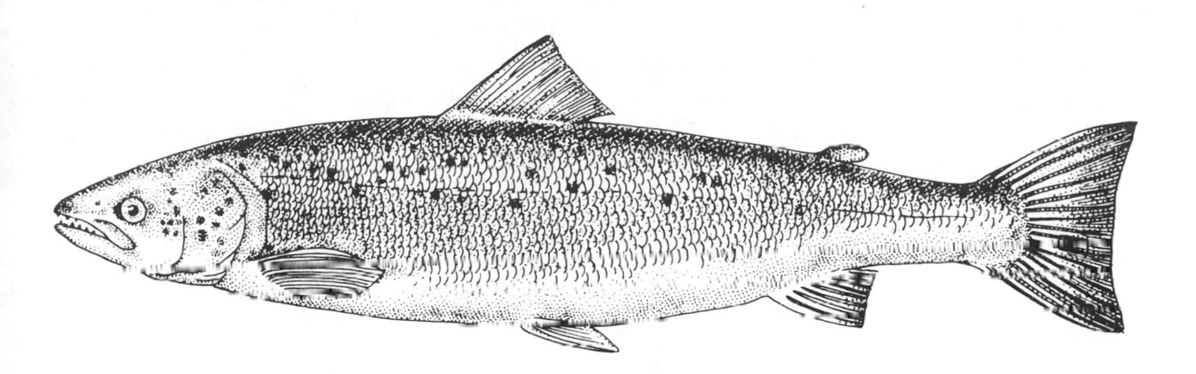

One of the most revered of coastal sport fishes (Scott and Scott 1988 call it "a renowned aristocrat among the game fishes"), the Atlantic salmon occurs on both sides of the North Atlantic Ocean. In the western Atlantic, it is found along most of the Canadian coastline and extends as far southward as Massachusetts. This is a homing species; thus, each coastal watershed where it occurs has a separate, reproductively isolated stock. Most of the New England stocks are presently seriously depleted or extirpated (locally extinct).

The Atlantic salmon can easily be distinguished from recreational species in other fish families by the presence of an adipose fin and a pelvic axillary process. However, separating the Atlantic salmon from other members of the family Salmonidae that occur in New England waters is more difficult. The recently introduced West Coast coho salmon, which was first stocked into the Great Bay and Piscataqua River of New Hampshire in 1969, has more anal fin rays (13 to 19) than the Atlantic (7 to 12). The brook trout has a blackish background color on its back and sides, with whitish ripple markings dorsally from head to tail and numerous white and scattered red spots on its sides. The salmon has a somewhat more deeply forked tail than the rainbow and brown trouts, and its mouth is smaller. An Atlantic salmon less than 15 cm (6 in) long has an upper jaw that extends to below the middle of its eye, while in larger individuals the jaw reaches to below the rear edge of the eye. The upper jaws of both trouts reach the back edge of the eye in fishes less than 15 cm, and well beyond the eye in larger individuals.

Biology

The Atlantic salmon is anadromous, normally living in fresh water at least 2 to 3 years after hatching before migrating to the sea. Large, cool streams with gravel and boulder substrates are essential for young Atlantic salmon. After the young, called parr, grow to about 15 cm in length, they undergo a series of changes, smoltification, that prepare them for life in the ocean. "Smolts" that are undergoing this process migrate to the sea, where they spend 1 to several years before returning to fresh water as adults to spawn.

The Atlantic salmon may undergo extensive migrations while at sea, moving to and from distant feeding grounds, frequently north of the Arctic Circle and as distant as the waters off Greenland. Although many individuals undertake these migrations, others apparently remain near coastal waters until they mature. Upon reaching sexual maturity, nearly all adults return to the stream in which they hatched in order to reproduce (this return to natal streams is called homing). Although most salmon spend up to several years in the ocean, some males return to fresh water as adults after only 1 year at sea. These small males, called grilse, are believed to be a separate genetic strain from males that attain significantly larger sizes while remaining at sea for a much longer time before maturing.

Adult Atlantic salmon enter rivers from spring to fall, but spawning occurs only in autumn. Thus, some adults spend up to half of a year in fresh water prior to spawning, whereas others spawn shortly after completing their spawning migrations.

Once spawning commences, the female salmon builds a gravel nest by vigorously flapping her tail while lying on her side against the gravel substrate. Once nest building is completed, she is joined in the nest depression by a male, who fertilizes the eggs as she releases them. The female may build a series of nests sequentially, spawning in each one, before she has released all of her eggs for the season. During this process, the male will continue to court her and will drive off other males that attempt to do likewise. Small mature male parr and grilse often attempt to rush into a nest in order to fertilize some of the eggs being released by the female spawning with the courting male. Typically, these smaller males will reproduce only if they are successful at this "sneaking" behavior, as they are largely unable to court a female successfully if larger males are present.

Unlike Pacific salmon species, some Atlantic salmon survive the first spawning season and migrate to the sea, returning to reproduce in 1 or occasionally 2 or more subsequent years. The percentage of adults that live to spawn in subsequent years varies from river to river, depending in part upon the amount of energy expended to migrate up specific rivers; repeat spawners occur more frequently in shorter rivers that require less rigorous migrations.

The large eggs, buried up to 25.4 cm (10 in) in the nest's gravel, develop through the winter. The length of time to hatching, which varies with water temperature, is about 110 days at 3.9 degrees C (39 F). Young often remain in the gravel after hatching until their yolk reserves are completely exhausted. At this time, they emerge from the gravel as large, very well-developed fish. Due to their size and advanced stage of development at emergence, young salmon exhibit markedly higher rates of survival than do newly hatched young of most oceanic species. Although it may take as many as several years for parr to reach 15 cm and smoltify, smolts grow rapidly once at sea. A salmon may weigh from 1.4 to 6.8 kg (3 to 15 lb) by its first spawning run. Repeat, or second-time, spawners may weigh up to 9.1 kg (20 lb), although larger individuals are occasionally landed.

Young Atlantic salmon in fresh water eat larvae of aquatic insects and adult terrestrial insects that are caught in a stream's current. While at sea, the salmon eats amphipods, and euphausiids and other shrimp when small, and a variety of fishes such as the herring, alewife, smelt, capelin, mackerel, sand lance, and cod when larger. Adults generally do not eat while in fresh water to spawn.

Young salmon are eaten by birds, such as the American merganser and kingfisher, and by fishes such as the American eel. Predators such as harbor seals, gray seals, sharks, pollock, and tuna species eat sea-run salmon. Gulls and cormorants may eat large numbers of smolts as they migrate downriver to the sea. Spawning adults may be eaten by bald eagles and ospreys.

Status and Management

At least 28 New England rivers contained significant salmon stocks in precolonial times. As many as 300,000 adult fish may have entered these rivers annually, with the Merrimack, Androscoggin, Kennebec, and Penobscot producing the largest spawning runs. The Connecticut River was once thought to possess a significant run; however, anthropological evidence suggests that the American shad was the dominant anadromous species in this watershed, with the Atlantic salmon possibly being represented by only a modest-sized spawning run annually.

Atlantic salmon spawning runs were extirpated (brought to local extinction) from New England watersheds other than some in Maine by the mid 1800s. Populations in Maine rivers were severely depleted by the early part of this century. State efforts at management in the latter half of the 19th century, including construction of fish-passage facilities over dams and stocking, were moderately successful. However, the lack of control of harvest and the continual decline of water quality prevented recovery of the many spawning runs of this species.

By the mid 20th century, restoration of the Atlantic salmon to the New England coast received great attention. A series of legislative actions, including federal statutes, international treaties, and an interstate compact, established the means whereby multiple agencies cooperated in an intensive management program.

The Federal Power Act required hydroelectric licensees to construct and operate, at their own expense, whatever fish-passage facilities were deemed necessary by the secretaries of commerce and interior. The Anadromous Fish Conservation Act of 1965 provided the means for the federal government to share with the states the cost of conservation of anadromous fishes. Fear that stocks would not recover even with careful riverine management because of commercial fishing in international waters led the United States to join the Convention for the Conservation of Salmon in the North Atlantic Ocean, which created the North Atlantic Salmon Conservation Organization (NASCO). Concurrently, the federal government and various New England states entered into a series of compacts to address management issues cooperatively. Among these are:

1. The Connecticut River Atlantic Salmon Commission, and the Policy and Technical Committees for Fisheries Management of the Connecticut, with membership including representatives from Connecticut, Massachusetts, Vermont, and New Hampshire, as well as from the National Marine Fisheries Service (NMFS) and the U.S. Fish and Wildlife Service (USFWS).
2. The Policy and Technical Committees for Fisheries Management of the Mer-

rimack River, with state agency representatives from New Hampshire and Massachusetts, and federal agency representatives from the NMFS, the USFWS, and the U.S. Forest Service.

3. The Maine Technical Advisory Committee, with membership from Maine's three state fisheries agencies, the University of Maine, and the USFWS. This committee recommends conservation action to the Atlantic Sea-Run Salmon Commission, composed of members from Maine's fisheries agencies and from the general public as appointed by the governor.
4. The New England Atlantic Salmon Committee, composed of all New England states and of federal fisheries agency heads. This committee interacts with the U.S. commissioners to NASCO.

This management system, surprising both for its complexity and for the level of interagency and international cooperation, appears somewhat ponderous due to the redundancy of committees and commissions attempting to accomplish management goals; however, since each river system contains a separate stock of Atlantic salmon, and each river system is overseen by different states within whose borders the stock occurs, such a management structure is necessary.

Management programs generated by the participating state and federal agencies had achieved modest success by the end of the 1980s. The focus of the management effort had been upon hatchery production of offspring, construction and operation of fish-passage facilities, severe restriction on harvest in territorial waters, and creation of agreements on harvest in international waters. The fish culture program, consisting of one hatchery in 1970, had grown to 11 hatcheries in all six New England states by 1985. Smolt production from these hatcheries increased from 100,000 fishes in 1970 to 1.3 million in 1985. In 1989, the New England states controlled harvest in territorial waters by using an array of regulations, including size and creel limits, gear restrictions, closed areas, and closed seasons.

Even with this expansive effort, salmon stocks remained low by 1989. Salmon entered 16 river systems in New England during the 1980s. The total yearly number of adults in these watersheds varied from 4,000 to 6,000 fish, with perhaps 1,000 of those resulting from natural reproduction and the rest from hatchery production. Only seven of these rivers possessed any wild salmon, the rest relying wholly upon hatchery fish. Eight of the 16 rivers, all in Maine, provided significant sport fishing, with anglers harvesting from 10% to 25% of these runs annually.

Multiple problems still need to be addressed before restoration and conservation goals can be met. Long-term changes in water quality in specific watersheds such as the Connecticut River may affect the capacity to restore successful salmon reproduction. Enhancing the passage over dams of adults upriver, and young and adults downriver, is still in the developmental stage technologically, with downstream migration still posing the greatest problem; the easiest way for a fish to move downstream past a dam is to pass through its turbines, which may cause high mortality rates.

Biological problems also exist. The restoration of populations that had entirely disappeared may require genetic adaptation of the introduced strains. Although hatchery operations using standard animal husbandry techniques can accelerate the rate of genetic change in a population over nature's slow pace, multiple generations of intensively controlled breeding may be required before strains of salmon well adapted to specific watersheds are developed. Also, although coastal and riverine harvest is well managed, commercial exploitation in international waters is harder to control. Such fishing probably harvested more than one salmon for every salmon that returned to its natal river in the 1980s.

In one sense, managing the Atlantic salmon is less complicated than many other coastal species that are targets of severe commercial and/or recreational exploitation. Harvest in territorial waters is strictly controlled, largely due to the agreement of anglers, the only exploiters along our coast. However, the problems listed above, in combination with the fact that there are numerous separate stocks, each in some stage of serious decline, collapse, or total extirpation, pose monumental challenges to fisheries managers and probably dictate the need to expend persistent energy over prolonged periods in the future before substantial results might occur.

Angling and Handling Tips

Fishing for the Atlantic salmon has long been the sport of the elite in Europe and Canada. Waters and the fishes within them historically were owned by the rich in Europe, and salmon fishing was restricted to those few, based upon ownership laws. Although always a costly venture, fishing has been more a right than a privilege of ownership in the United States, since fishes have been considered a publicly owned resource. Thus, angling for salmon has been openly available to all who wish to spend the money to do so.

Atlantic salmon are available in certain Maine rivers, with the Penobscot, Denny's, East Machias, and Narraguagus rivers being the most productive systems. Even in these rivers, the Atlantic salmon requires such patience that many who have attempted to fish for it leave the fishing site empty-handed. The salmon feeds very little throughout its stay in fresh water; thus, if it strikes a fly it is because the angler has shown greater patience than the fish, repeatedly dropping the fly across the fish's nose over and over until it strikes, perhaps more out of annoyance than hunger. Many anglers joke that angling for Atlantics is more watching than fishing! However, once it is hooked, the thrill and excitement of fighting and landing a large salmon in a fast-flowing river make for an experience that is difficult to match.

Many anglers prefer a moderately stiff fly rod with some backbone, measuring 8.5 to 9 feet long and weighing about 5 ounces. Bigger rods are necessary in faster water, though smaller rods are used by experienced anglers who prefer to provide the fish with additional advantage. Avoid "broomsticks" or you'll miss most of the fun. The rod should have a large reel that holds 150 or more yards of 20-pound test monofilament backing in addition to the fly line and tippet or leader. The fly line should be size 8 or 9, either floating or with a sinking tip. The tippet should be 10-pound test or larger.

Flies of all sizes, shapes, and colors are used. Standard bucktails and nymphs like the silver and the rusty rat and black dose (size 6 to 8) are particularly popular. Large flies are used in some of the bigger rivers. Since colors and patterns depend upon water conditions, visiting tackle shops near the river or visiting with successful anglers can aid in the selection of the best flies.

Salmon are usually found in more protected areas than trout, preferring moderate turbulence and flow. Water temperature can affect their selection of resting sites. Thus, anglers can occasionally find beautiful salmon resting in the smallest of cold-water freshets.

It's a lucky angler who has experienced whole or steaked salmon cooked on the campfire. Whole fish are wrapped in foil, placed in a metal basket or directly on a grill. Once cooked, all the fish needs is a little salt, pepper, and freshly squeezed lemon juice. Though you can't bring the atmosphere of the camp home with you, salmon can be cooked similarly on a barbecue or gas grill. To grill steaks, melt a stick of butter and coat the steaks on both sides. Place on a hot grill and baste with the butter. Cook the fish for 3 to 4 minutes on one side, until the grill wires leave dark lines in the flesh. Turn over, baste once more, sprinkle with dried dill, black pepper, and salt, and cook for another 5 or 6 minutes. Total cooking time should be 10 minutes or less, since the less salmon cooks, the more tender and juicy it will be. Place on a warm platter and serve with large wedges of lemon.

Adventurous salmon lovers, especially those who appreciate lox and smoked salmon, should try "gravelox," an ancient Scandinavian dish. This was originally made by burying cleaned, spiced, and salted salmon in riverbank gravel for several days. Cold weather, salt, and the weight of the gravel cured the fish and removed moisture. (Large rocks were used to keep animals away.) For a modern approach to curing, use the following method: Prepare a mixture of ⅓ cup of pickling or kosher salt, ⅓ cup of sugar, 2 tablespoons of freshly ground black pepper, and 1 tablespoon of dried dill (if possible, a cup of fresh dill can be substituted to enhance the flavor). Place a scaled fillet, skin side down, in a large glass baking dish and spread the mixture over it. Cover the fillet with fresh dill and place a second fillet, skin side up, over it. Cover the dish with foil or plastic wrap and place another baking dish on top. Place a 5-pound weight (unopened cans of vegetables, etc.) in the top dish and refrigerate. After 24 hours, remove the covering, drain the liquid, and turn the fish over. Re-cover and refrigerate. Repeat the draining and turning process for three to four 24-hour periods; the fish should now be cured. Next, scrape the mixture from the fish, and place skin down on a cutting board. Then, beginning at the larger end of the fillet, cut thin slices from top to belly without cutting through the skin. A knife held at a slant will give wider, more attractive slices. Scandinavians eat this cured fish as is for breakfast. You might try the following appetizer: Spread crisp rye crackers or brown bread with butter or light mayonnaise. Place slices of the salmon on the crackers and squeeze on a little lime juice. If wrapped in plastic, this cured fish will stay fresh for up to 2 weeks.

Gravelox can also be cooked like fresh salmon, but it does not need as much heat or the same length of cooking time, since it is already cured.

Rainbow Smelt
(*Osmerus mordax*)

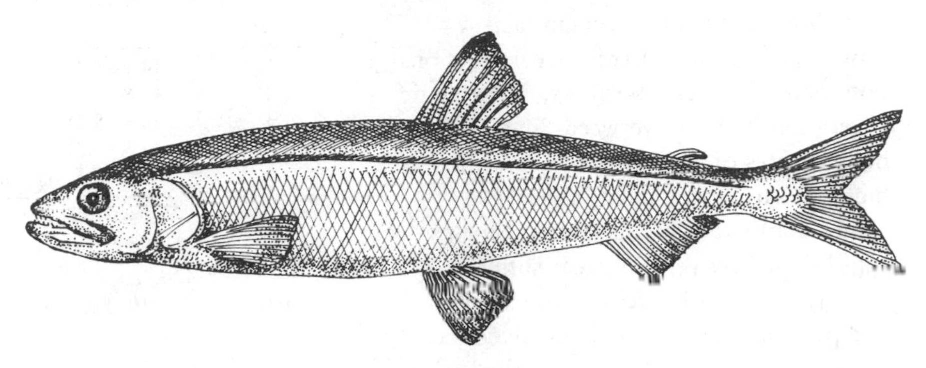

The rainbow smelt, a small fish eagerly pursued by anglers because of its fine flavor, is found along the coastal inshore areas of northeastern North America from Newfoundland to New Jersey but is most abundant from the southern Canadian Maritime Provinces south to Massachusetts. Once exclusively an anadromous species (residing in salt water but entering fresh water to reproduce), the smelt has been successfully introduced into freshwater systems throughout the northeastern and central United States.

The rainbow smelt is slender with a pointed head and a large mouth. This species can be distinguished from other small-bodied fishes along the New England coast by the presence of large teeth on its jaws and tongue, a deeply forked tail, and a small adipose fin that is narrow at its base.

The smelt is green dorsally with slightly paler sides and a silver belly. It has a conspicuous silvery streak running lengthwise along its sides.

Biology

Smelt are pelagic, usually traveling in schools less than 1 mile from shore and in water less than 5.8 m (19 ft) deep. Their movement patterns are associated with seasonal changes in water temperatures. In summer, schools move to deeper, cooler waters, and in the fall they enter bays and estuaries where they actively feed until the onset of winter. During winter months, adults and juveniles are found between estuary mouths and the brackish water areas of coastal streams. The spawning season in New England begins in late winter or early spring. Timing is influenced by water temperatures (about 4.4 to 5.6 degrees C; 40 to 42 F), increasing day length, and the breakup of ice covering the water's surface at the spawning grounds.

Both sexes become sexually mature at about 2 years of age, although some 1-year-olds may participate in spawning. Fecundity ranges from 33,400 eggs for a 16.5 cm (6.5 in) fish to 75,600 for a 22.9 cm (9 in) fish.

Adults spawn only at night, swimming from estuaries into fast-flowing fresh or slightly brackish stream stretches where eggs and milt are released. Upstream movement to the spawning ground typically occurs during flood tides. Spawning fish move back into downstream resting areas in estuaries before sunrise. Fertilized eggs sink and adhere to each other and to any stationary material on the stream bottom. Most spawning occurs in fast-flowing, turbulent water in stream sections dominated by rocks, boulders, and aquatic vegetation such as riverweed. Eggs display highest rates of survival when deposited in high-velocity water and on riverweed rather than bare rock substrates. Smelt eggs laid in low-velocity areas suffer extremely high death rates due to siltation and possibly to inadequate dissolved oxygen supplies. Survival of eggs in the Parker River (MA) was found to be 10% when laid on riverweed in optimal currents and 1% in less suitable conditions.

Larvae, about 0.6 cm (0.25 in) in length when they hatch, are carried downstream into estuarine waters by stream currents. By midsummer, the juveniles reside in the deeper waters of estuaries, particularly during daylight hours. They overwinter with adults in stream mouths of the upper reaches of estuaries.

Large smelt may reach 33 to 36 cm (13 to 14 in) in length, but most adults do not exceed 18 to 23 cm (about 7 to 9 in) in length and 1 to 6 ounces in weight. Females are larger than males of identical age after their first year of life. Table 12 lists average lengths of adults. Few smelt live beyond 5 years of age.

Larvae and juveniles feed upon zooplankton, particularly microscopic crustaceans. Adult smelt feed on small crustaceans such as shrimp and gammarids, on squid, crabs, sea worms, insects, and small fishes, including young sea herring, alewives, cunners, sand lances, and silversides. Smelt in turn serve as major prey for fish-eating vertebrates, including the striped bass and bluefish, and a variety of bird species. Mortality due to predation is quite high for this species; up to 72% of adult fishes die annually.

Table 12. Length of age in cm (inches in parentheses) of rainbow smelt from the Parker River estuary, Massachusetts

Age	*Male*	*Female*
1	14.1 (5.6)	14.0 (5.6)
2	18.8 (7.4)	19.7 (7.8)
3	20.8 (8.2)	21.9 (8.6)
4	23.6 (9.3)	24.5 (9.6)
5	24.2 (9.5)	24.9 (9.8)

Source: Data from Murawski and Cole 1978.

Status and Management

Coastal smelt stocks throughout New England suffered marked reductions in abundance by the 20th century, due to human activity in coastal watersheds. Many spawning runs were depleted because of widespread construction of dams and reduction in water quality. In some areas, after dams cut off traditional spawning grounds, eggs were deposited so densely in downstream sites that massive egg mortality occurred as a result of fungal infections. High egg mortality also occurred due to siltation in watersheds experiencing extensive development. Thus, the loss and/or reduction in quality of spawning habitat caused the loss of many local smelt fisheries.

Attempts to manage smelt stocks occurred prior to the 20th century, when, for example, the Massachusetts legislature initiated legislation to protect smelt in the Commonwealth's watersheds in 1874. Starting about 1910, a widespread stocking program was established in Massachusetts to restore smelt runs. In one method, specially constructed trays were placed in stream sites where extensive spawning was occurring. After the surface was covered with developing eggs, the trays were transferred to streams identified for smelt restoration. Transplanting was also attempted by stocking adults that were ready to spawn. Years of such stocking produced no measurable results, so these programs were ultimately terminated. Although similar management attempts were conducted by other New England states, many coastal populations in New England remained down through the 1980s, although specific watersheds produced occasional pulses of great abundance.

Management of smelt stocks has been a responsibility held entirely by the states throughout New England. Management has included addressing habitat restoration, restricting harvest to particular gear types, and prohibiting harvest during the spawning season.

Angling and Handling Tips

Smelt offer an opportunity for action at the time of year when most anglers hang up their rods, store their gear, and warm the easy chair next to the fireplace, spending leisure time indoors protected from the elements. This species provides a hook-and-line autumn fishery in coastal areas and a winter ice fishery in estuaries, river mouths, and riverine areas downstream from spawning sites.

The smelt's attractiveness to anglers is certainly not because of size, as the length of most landed fish is 7 to 8 inches or less. What the smelt lacks in size, it makes up in numbers. Once a school is located, anglers find the action fast-paced; catches of one to several dozen smelt are possible on a good day.

After ice forms in the winter, smelt begin to aggregate and move into coastal streams toward spawning grounds. Anglers wait for this smelt "run" with great anticipation. Some anglers build or rent smelt huts equipped with wood stoves for a comfortable approach to ice fishing while other hardy souls brave the weather. Conditions may be hazardous as well as uncomfortable, as ice is often unstable in areas of flowing water or areas of mixing between fresh and salt water; thus, some productive sites are not always available to the wise angler. Smelt travel in schools and remain within localized areas for extended periods of time. Once a productive spot is found, it may remain productive for some time.

Many types of fishing gear are used, ranging from 9-foot light glass to small ultralight spinning outfits. Very light line, 4-pound test or less, is a must. Handlines are also popular. Baits such as bloodworms, sand worms, grass shrimp, and small local bait fishes are fished on size 6 to 10 hooks, with a small sinker suitable to hold the bait in current.

The depth that bait is fished is critical to success, since schools of smelt move up and down in the water column as they swim about. Baited hooks should be slowly moved up and down to assure they will pass through schools of smelt. One secret to successful fishing is to keep an eye always on the line, no matter how interesting the conversation or beautiful the scenery around you. Smelt produce such light taps or vibrations when grabbing baited hooks that close attention to the rod and line is a must.

Fresh smelt have a characteristic cucumberlike aroma. This species' meat is white, delicate and sweet-flavored. Bones are soft and edible, although many prefer to remove the bones from cooked smelt before eating them. Preparation for cooking usually includes simply removing the head and entrails.

Frying is an easy way to prepare smelt. Mix 1 cup of flour and 1 teaspoon of seasoning normally used in turkey recipes (commercial brands include some mixture of rosemary, oregano, sage, ginger, marjoram, thyme, and pepper) in a plastic bag. Shake cleaned smelt in this mixture so that they are lightly dusted. Fry the fish pieces in ⅛ of an inch of vegetable or olive oil in a heavy skillet over medium heat. Fry for about 3 minutes on the first side and 2 minutes on the other, until lightly browned. Drain on paper towels and eat while still warm. Fried smelt are finger food, to be enjoyed plain or dipped in tartar or seafood sauces.

If you have eliminated fried foods from your diet, broiling or baking is an enjoyable alternative. For a chilled treat, try spiced smelt. Place enough cleaned smelt in a glass or ceramic baking dish to cover the bottom. In another pan, mix the following: two thinly sliced onions, two cloves of sliced garlic, two grated carrots, two bay leaves, ½ teaspoon crushed black pepper (crushed red pepper or hot sauce is optional), six lemon slices, ½ cup of olive oil, ½ cup of white wine, ½ cup of white vinegar, 1 teaspoon of salt, and 1 tablespoon of pickling spices. Bring this mixture to a boil and simmer for 15 minutes. After simmering, pour it over the smelt and bake in a 400 degree F oven for 12 to 15 minutes. Chill well and serve.

References

Buckley, J. L. 1989. *Species profiles: life histories and environmental requirements of coastal fishes and invertebrates (North Atlantic)—rainbow smelt.* U.S. Fish Wildl. Serv. Biol. Rep. 82(11.106). U.S. Army Corps of Engineers, TR EL-82-4. 11 pp.

Murawski, S. A., and C. F. Cole. 1978. Population dynamics of anadromous rainbow smelt *Osmerus mordax,* in a Massachusetts river system. *Trans. Am. Fish. Soc.* 107:535–542.

Murawski, S. A., G. R. Clayton, R. J. Reed, and C. F. Cole. 1980. Movements of spawning rainbow smelt, *Osmerus mordax,* in a Massachusetts estuary. *Estuaries* 3:308–314.

Sutter, F. C. 1980. *Reproductive biology of anadromous rainbow smelt,* Osmerus mordax, *in the Ipswich Bay area, Massachusetts.* Unpublished master's thesis, Univ. of Massachusetts, Amherst.

Monkfish
(*Lophius americanus*)

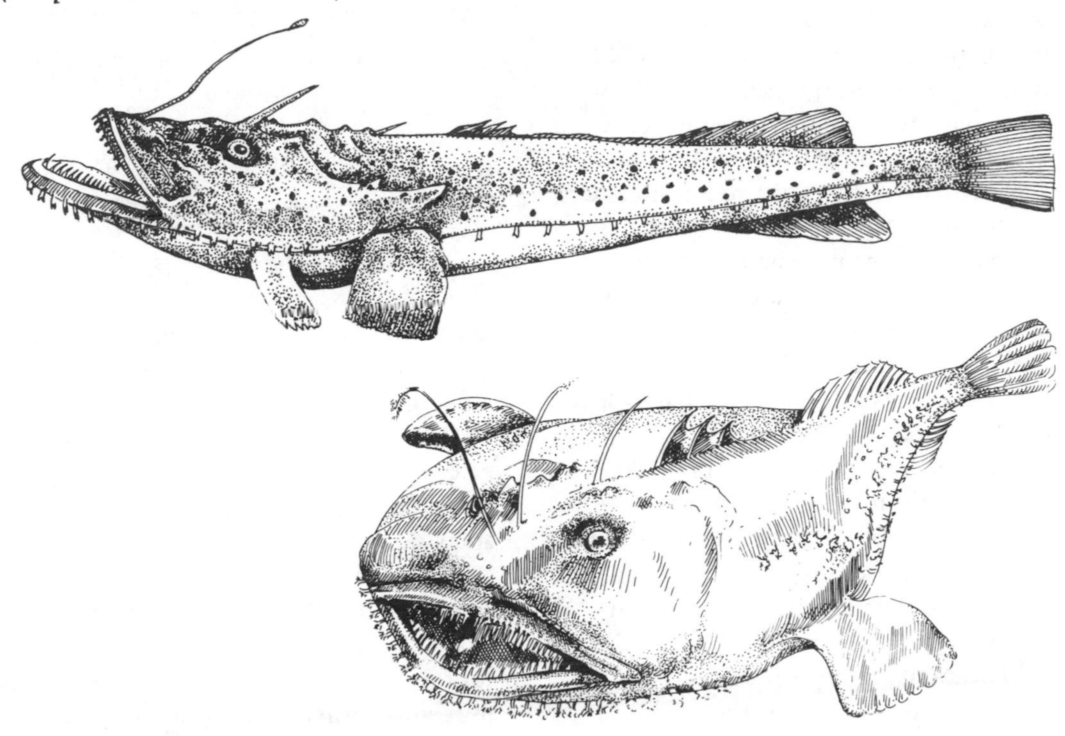

The monkfish, also known as the goosefish or angler, is native to the eastern coast of North America from Newfoundland to North Carolina. It also inhabits the Gulf of Mexico and waters off Brazil.

The monkfish is so unique in appearance it cannot be confused with any other fish of coastal New England. It has an enormous head and mouth and a flattened body of strikingly modest proportion in relation to the head. The eyes are on top of the head, pointing upward. The lower jaw projects so far beyond the upper that its two to three rows of slender, curved teeth are exposed even when the mouth is closed. The upper jaw also has large, curved teeth, and there are several rows of thornlike teeth on the roof of the mouth. The gill openings, positioned behind the pectoral fins, lack the gill covers seen in most bony fishes. The dorsal fin is preceded by a row of three elongate rays, the most anterior of which has a fleshy pad that is used to lure small fishes close to the huge mouth.

A monkfish is usually brown dorsally and pales to white on its belly. The tips of the dorsal fins, pectoral fins, and tail are black.

Biology

In the southern New England region, monkfish are found from the tide line to depths greater than 668 m (2,000 ft) on the continental slope. They live on various types of substrate, including sand, gravel, rocks, beds of broken shells, and mud. Monkfish are tolerant of a wide range of temperatures; they have been collected in waters from as cold as 0 degrees C (32 F) to as warm as 21 degrees C (70 F). However, they prefer winter temperatures of about 3 to 6 degrees C (37 to 42 F) and summer temperatures of 5 to 9 degrees C (41 to 48 F). Monkfish display minimal seasonal movements from the Rhode Island/Nantucket Shoals area northward. South of this region they concentrate in deeper continental shelf waters in the summer and fall (except off Long Island where they occur inshore and offshore year-round), and they spread over the entire shelf from the tide line to 366 m (1,200 ft) in the winter and spring. This movement is probably in response to warming of inshore shallow waters beyond preferred temperature ranges in the summer and fall.

Male monkfish become sexually mature at age 4, females at age 5. Fecundity ranges from 300,000 to over 2.8 million eggs.

Monkfish reproduce in shallow to deep water from spring through early fall. The timing depends upon latitude; individuals in New England spawn from late June to mid September. Large masses of eggs are shed in ribbonlike veils of mucus that may be 7.6 to 11 m (25 to 36 ft) long and 0.6 to 0.9 m (2 to 3 ft) wide. Within each mass, eggs are arranged singly or in small groups in separated, hexagonal compartments. The egg masses float within the water column. Soon after they hatch and their yolk sacs are resorbed, fry begin feeding upon copepods, crustacean larvae, and arrow worms. At this point the fry are about 0.3 cm (0.1 in) long. Fry metamorphose into bottom dwellers by the time they have grown to about 5 cm (2 in) in length.

Monkfish can reach 122 cm (4 ft) in length and up to 23 kg (50 lb) in weight. Length at selected ages is presented in Table 13.

Table 13. Length in cm (inches in parentheses) for selected ages of monkfish

Age	*Length*
1	11 (4)
4	46 (18)
8	74 (29)
9	79 (31)
10	94 (37)
12	102 (40)

Source: Data from Grosslein and Azarovitz 1982.

The voracious monkfish becomes a fish eater by the time it starts bottom-dwelling. Adults feed upon a variety of fishes such as spiny dogfish, skate species, eels, sand lances, herring, mackerel, silver hakes, cod, haddock, flounders, tautogs, and sea bass. "Angler" is a well-earned nickname for this species, which often lies motionless in eelgrass, waving the "lure" at the end of its first dorsal ray. As small fishes approach, the monkfish opens its mouth and literally sucks them in.

The monkfish also eats numerous sea birds such as cormorants, herring gulls, scoters, loons, and diving water fowl. As a result, it has gained the nickname "goosefish," although a full-grown goose has never been found in the stomach of a landed monkfish.

It is not unusual for a monkfish to contain up to half its own weight in its stomach. One specimen taken in a net contained 21 flounders and a dogfish; another had swallowed seven ducks. Its huge mouth allows the monkfish to capture extremely large food items. One specimen had eaten a 5.5-kg (12-lb), 79-cm (31-in) haddock; another had a winter flounder nearly its own size in its stomach. In spite of occasional catches of spectacularly large prey, the bulk of the monkfish's diet apparently consists of large numbers of small fishes.

Status and Management

Historically, monkfish populations along the East Coast of the United States have not been commercially harvested, as this species was not considered to be marketable. This is in contrast to the closely related European monkfish, which was heavily exploited as early as the 1940s due to the high price it brought in English and Scottish ports. More recently, the collapse of populations of high-value fishes in the Northwest Atlantic has led to an increase in the market value of formerly less desirable species. U.S. landings of monkfish increased from 150 mt (330,000 lb) in 1972 to 400 mt (880,000 lb) in 1974. Landings in the New York Bight area increased 10-fold between 1973 and 1978. Although the monkfish now commands a high price because of its greater acceptance as a food fish, it is so sparsely distributed that no fishery has developed specifically targeting it for harvest. The monkfish is typically landed as incidental catch by commercial vessels directing their efforts toward other fish. Likewise, the recreational harvest of monkfish consists mostly of individuals caught by anglers fishing for other species. Therefore, the numbers landed each year are extremely modest. Due to this low rate of harvest, no regulations restricting catch in New England waters had been implemented by 1989.

Angling and Handling Tips

Monkfish are typically captured incidentally by anglers using live bait in pursuit of other fishes. This species is encountered so infrequently that few methods have been developed specifically for pursuing and landing it. Since it typically eats a wide variety of organisms, the monkfish can be hooked with many different types of live bait that are being fished near the bottom for other large predatory species.

Although it is nicknamed "angler," this species is considered by most people to be far uglier than any angler one might encounter! The monkfish's bizarre appearance has led more than one inexperienced person to dump it over the side; however, the joke is on the fisherman, for over the side of the boat went one of the best chunks of thick fillet that the sea can produce. Its firm white meat has been held in esteem in Europe for years. Its texture is similar to that of the lobster, and it is in fact referred to as "poor man's lobster." It is often found on restaurant menus as "lotte" or in supermarkets under its own name, and it is frequently included in dishes like the Spanish paella and French bouillabaisse

After the monkfish is filleted and the membrane (soft tissues just under the skin) removed, its flesh becomes a versatile piece of meat that can be sautéed in a pan or wok; broiled; added to chowder or stew; poached; or cut into strips, dipped in batter, and deep-fried. It is excellent when chunked and added to a kabob, since it stays together like red meat when grilled.

For an interesting appetizer, try monkfish cocktail. Bake a thick fillet in a 350-degree oven for 10 minutes per inch of thickness. After the fillet cools, slice it into short strips or fingers about the size of a shrimp. Chill, arrange on a bed of lettuce, and serve with a commercial seafood cocktail sauce as Monk Shrimp.

Reference

Grosslein, M. D., and T. R. Azarovitz. 1982. *Fish distribution*. Mesa NY Bight Atlas Monograph 15. NY Sea Grant, Albany.

Silver Hake
(*Merluccius bilinearis*)

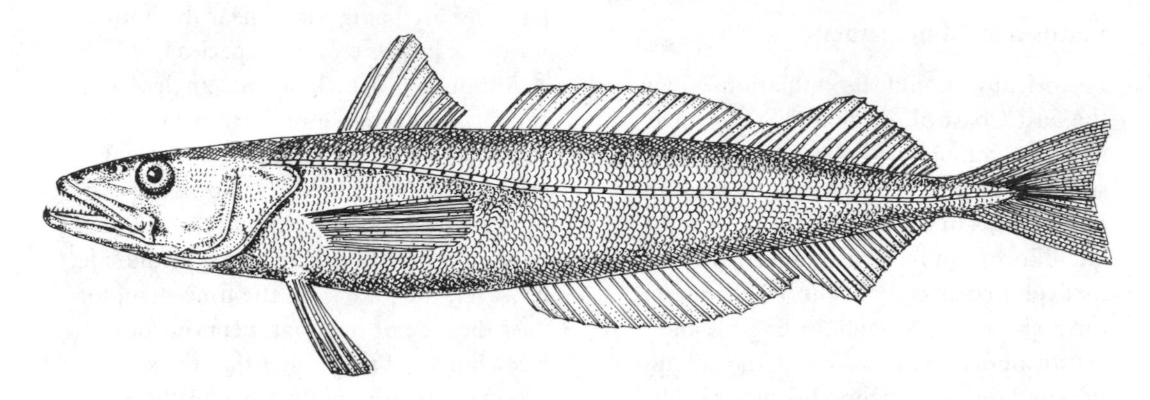

The silver hake, or whiting to many anglers, occurs from South Carolina north to the Grand Banks. It is most abundant between Cape Sable, Nova Scotia, and New York.

The silver hake has two separate dorsal fins, neither possessing spines, with the posterior fin being markedly longer than the anterior one. The pelvic fins are positioned forward on the body, just below the pectoral fins. The anal fin is identical in size and shape to the posterior dorsal. This species lacks the chin barbel and the elongate rays on the pelvic fins that are characteristic of other Northwest Atlantic hake species. It has large eyes and a wide mouth armed with two or more rows of sharp, decurved teeth.

The silver hake is well named, as the dark gray color on the back lightens to a silvery iridescence laterally. However, the iridescence tends to fade rapidly after capture, so most anglers are more familiar with a silvery gray body coloration.

Biology

Silver hakes are divided into two major populations, or stocks, northern and southern groups separated by the shallow waters of Georges Bank. Because they inhabit different geographic regions with varying environmental conditions, northern and southern stocks of whiting display different seasonal migration patterns. Northern whiting overwinter in deep basins of the Gulf of Maine, then move either inshore or to the northern edge of Georges Bank where they spend spring and summer. Southern whiting overwinter along the continental slope south of Georges Bank to Cape Hatteras and in some nearshore areas from southern New England to Chesapeake Bay. Not only does the southern stock move to shallower waters for spring and summer, but individuals overwintering in the southern portion of their range move northward to southern New England in the spring and summer.

Both sexes reach sexual maturity at 2 to 3 years of age. Fecundity of females increases with size and age; a 25 to 30 cm (10 to 12 in) female produces about 343,000 eggs, and a 30 to 35 cm (12 to 14 in) female about 391,700 eggs. Major silver hake spawning areas include the coast of the Gulf of Maine from Cape Cod to Grand Manan Island, the southeastern and southern slopes of Georges Bank, Nantucket Shoals, and south of Martha's Vine-

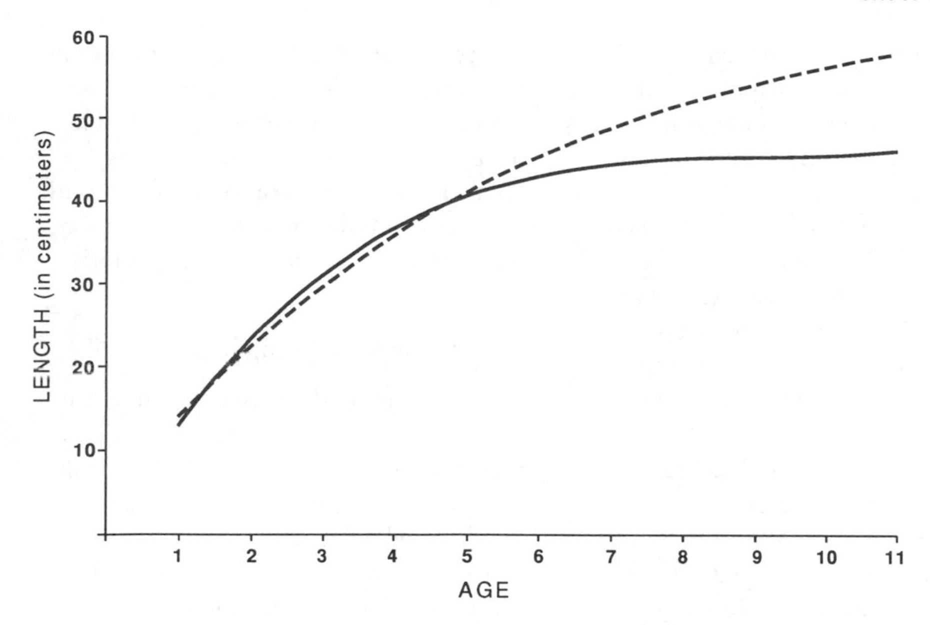

Figure 8. Length at age in cm of the silver hake, showing growth differences that have occurred between the Gulf of Maine (dotted line) and the southern New England Middle Atlantic (solid line) stocks (from data presented in Penttila et al. 1989).

yard. Spawning extends from May to November, although it peaks from June to July on the continental shelf and from July to August in the Gulf of Maine. Individual females generally release eggs during three separate spawning periods during the spawning season; most eggs are released during the first of these periods.

The pelagic eggs hatch in about 2 days. Although occurring over a relatively broad range of temperatures, incubation is most successful when water temperatures are 13 to 16 degrees C (55 to 61 F).

Growth is rapid for the first 2 to 3 years of life but slows markedly after individuals reach sexual maturity. Females generally grow faster and live longer than males. Most males do not live beyond age 6; females may live to be 12 to 15, although individuals this old are rare. Maximum length of females is about 65 cm (25.6 in), and maximum weight is about 2.3 kg (5.1 lb). Rates of growth are highly variable (see Figure 8) and are inversely proportional to population density; highest individual growth rates are generally achieved when population abundance is low.

Juvenile silver hakes feed upon invertebrates, particularly euphausiids and other shrimp. Fishes such as herring and mackerel become an important component of the diet when hakes reach 30 cm (about 12 in) in length and are a predominant food for individuals longer than 40 cm (about 16 in). The whiting is a major predator in the Northwest Atlantic due to the high abundance this species periodically attains.

Whiting are eaten by numerous predators, including the pollock, cod, mackerel, swordfish, spiny dogfish, flounder species and larger whiting.

Status and Management

Because its flesh softens rapidly after capture, the whiting held only modest market interest prior to 1900. Improved freezing methods developed after the turn of the century led to a rapidly expanding commercial fishery. Combined landings in Maine and Massachusetts increased from 909.1 mt (2.0 million lb) in 1902 to 6,363.6 mt (14.0 million lb) by 1919. Harvests generally declined for a decade before once more rapidly increasing. Maine and Massachusetts landings rose from 6,818.2 mt (15.0 million lb) in 1935 to as high as 337,727.3 mt (743.0 million lb) by 1945.

The next major increase in landings occurred with the introduction of the foreign distant-water fleets in the 1960s. Between 1955 and 1961, harvests averaged 49,100 mt (108.1 million lb) and 15,500 mt (34.1 million lb) for the northern and southern populations, respectively. Landings of the northern stock increased to 94,500 mt (207.9 million lb) by 1964, and of the southern stock to 308,500 mt (678.7 million lb) by 1965. Landings declined by 1970, increased somewhat through the early 1970s, then declined to low levels throughout the 1980s. Commercial landings in 1986 totaled about 8,500 mt (18.7 million lb) and 9,900 mt (21.8 million lb) for the northern and southern stocks, respectively. From 1963 to 1977, distant-water fleets were responsible for an average of 49% of the total commercial harvest of the northern stock and 87% of the total harvest of the southern stock.

Between 1962 and 1986, declines in adult biomass corresponded to concurrent drops in harvest. Thus, the reduction in harvest prior to the Magnuson Act was due to population declines rather than reductions in fishing pressure. Further, total exclusion of foreign vessels from the northern fishery and a marked reduction of these vessels from the southern fishery (foreign vessel harvests dropped from an average of 87% of the total before 1978 to an average of 17% after) did not result in recovery of whiting population.

Recreational harvest, being negligible for the northern stock and averaging about 1% of the total harvest of the southern, has had little effect upon population fluctuations noted since the early 1960s.

Angling and Handling Tips

Depending upon the season, whiting can be caught from the coastline to offshore, deep waters, where the larger individuals weighing up to a pound are more commonly caught. The action can be furious when the angler locates a large school of this swift, voracious species.

A 7-foot medium rod with conventional reel spooled with 150 or more yards of 20–30-pound test monofilament is well suited for offshore boat fishing. The mackerel tree, consisting of a 3–4-ounce jig or sinker tied to the end of the line and three to four long-shank hooks attached to the line at 15-inch intervals, is a popular rig for silver hakes. The shanks of the hooks on premade whiting rigs purchased at tackle shops are often covered with bright red surgical tubing. The hooks can carry a variety of baits, including strips of herring and mackerel or squid, clams, and small fishes. This rigging should be jigged with an upward jerk, followed by a lowering of the rod; or, one can still-fish. A slow drift will increase the area covered by the bait, thus possibly increasing the catch. Whiting can be caught at any time, although evening to midnight might be most productive.

When whiting are inshore, use lighter gear (a 5–6-foot light-to-medium-action spinning rod with 6–8-pound test monofilament) and smaller hooks (No. 4 to No. 6), with enough weight to hold the bait in the currents. Some anglers use floats to keep the bait off the bottom and moving. Grass shrimp may be added to the baits listed above when fishing inshore. Some anglers use streamers tied in tandem or small silver lures and bucktail jigs. Add a strip of bait when using these lures to increase the action. After casting and allowing to settle to the bottom, retrieve such lures with gentle jerks of the rod.

In winter, silver hakes may move toward the shoreline, chasing bait fishes in the surf at dusk. Occasionally, during a winter night's feeding foray, hakes stray too close to shore while feeding and are tossed up onto the beach; such events led anglers to refer to this species as the "frostfish." Those that get stranded are left for the hardy angler braving winter fishing or for lucky gulls that find them the next morning.

The whiting is known as the sweetest-flavored member of the cod family and is preferred over the cod and haddock by some gourmets. Before the advent of freezing, the whiting had little value because its delicate flesh spoiled easily. As freezing became common in the early part of the 20th century, whiting were commonly shipped inland to fish-and-chips restaurants. Although it freezes well, a whiting should be defrosted in a colander so it won't get waterlogged.

The whiting is a good frying fish, whole or filleted, when lightly browned and doused with white wine, and is tasty when brushed with spiced tomato sauce and grilled or broiled. To broil, simply brush with melted butter or olive oil and place 3 inches from the heat. Turn the fish after three minutes, brush once more, sprinkle with a little paprika, and finish broiling. Broiled whiting is best served with lemon wedges.

Grilled whiting is considered a gourmet treat. First, make a marinade of ¼ cup olive oil, ¼ cup soy sauce, ¼ cup lemon juice, 1 inch of grated ginger root, two large minced cloves of garlic, and 2 teaspoons of sugar or sweet sherry. Place whole headed and dressed whiting in the marinade and store in the refrigerator for at least 1 hour, turning several times. Use a wire grilling basket, a flat porcelain grilling rack, or a grill brushed with oil to prevent sticking. Place the whiting on the hot grill, skin side down, for 5 minutes, baste with the marinade, and turn. Baste again and cook for several more minutes until the fish begins to flake. Serve with lemon wedges and a sprig of parsley or a length of chive.

References

Almeida, F. P. 1987. Stock definition of silver hake in the New England–Middle Atlantic area. *N. Am. J. Fish. Manage.* 7:169–186.

Penttila, J. A., G. A. Nelson, and J. M. Burnett, III. 1989. *Guidelines for estimating lengths at age for 18 Northwest Atlantic finfish and shellfish species.* NOAA Tech. Mem. NMFS-F/NEC-66. 39 pp.

Ross, M. R., and F. P. Almeida. 1986. Density-dependent growth of silver hakes. *Trans. Am. Fish. Soc.* 115:548–554.

Tomcod
(*Microgadus tomcod*)

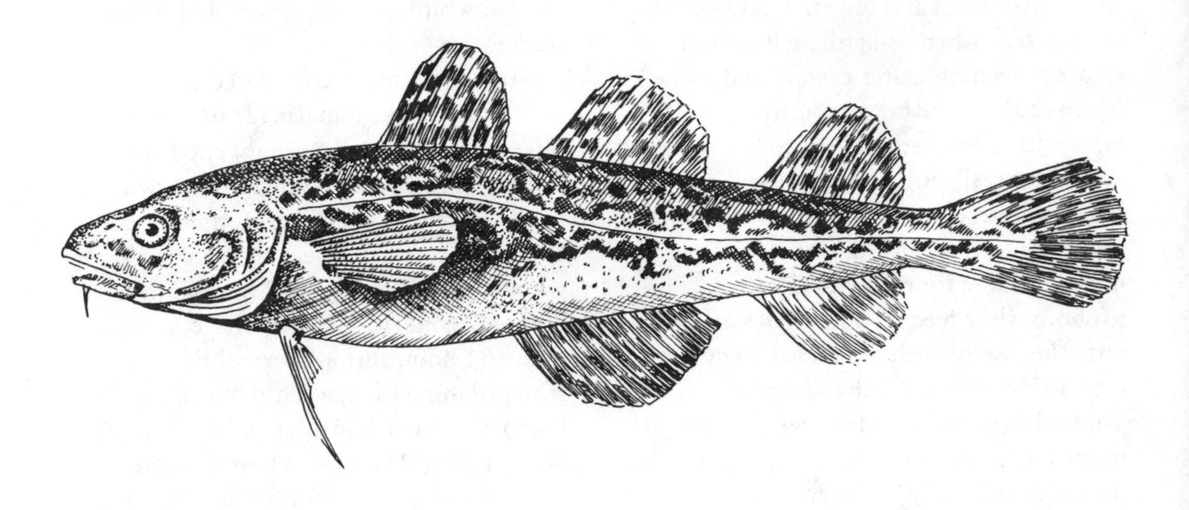

The tomcod, a popular winter recreational species along the New England coastline, is native to most of the northwestern Atlantic, from the Canadian Maritime Provinces south to Virginia. Its center of abundance extends from coastal Newfoundland to the New York Bight region. A typical inhabitant of brackish water, this fish is also found landlocked in several freshwater lakes of coastal Canada.

Like some other members of the cod family such as the haddock, cod, and pollock, the tomcod has three dorsal and two anal fins. The tomcod can be most easily distinguished from these other species by the presence of elongate rays in its pelvic fins and a rounded tail.

The tomcod is olive or brownish green on the back and upper sides, paling to white or yellowish gray on the belly. The back and sides are tinged with yellow and are darkly mottled. The dorsal fins and tail are the same color as the back, and the anal fins are olive or pale gray.

Biology

The tomcod is a nearshore inhabitant along the Canadian and Gulf of Maine coasts; south of Cape Cod it can be found somewhat farther from the coastline, although it never occurs more than 1 mile from shore or in water deeper than about 6 m (20 ft). Tomcod are sometimes caught by anglers on open, exposed shores, but most are found in shallow, protected bays and the mouths of rivers with expansive eelgrass beds.

Like the silver hake, tomcod are called "frostfish" because they appear in very shallow water during the fall and winter. They exhibit no migratory pattern in the Gulf of Maine, but south of Cape Cod they tend to move into deeper waters in the spring and summer and return to shallow bays and estuaries in the late fall and winter. Tomcod are extremely tolerant of the sudden changes in temperature and salinity that are characteristic of shallow estuarine waters.

Both males and females reach sexual maturity by the end of their first year of life. Two- and three-year-old females from the Weweantic River in Massachusetts carried 6,000 to 30,000 eggs. In analyses of Canadian populations, the largest females (36 cm, or 14 in, long) carried as many as 65,750 eggs.

Tomcod reproduce in brackish or freshwater areas of estuaries and streams from November to early February, with peak spawning activity occurring from December to early January. The eggs sink to the bottom and adhere to clumps of seaweed, rocks, or other projections of the substrate. Eggs hatch 22 to 35 days after fertilization, depending upon water temperature; eggs in warmer waters hatch more rapidly than those in colder waters. Within several days after hatching, larvae begin to feed upon microscopic invertebrates.

Larvae normally drift downstream into the estuary, where they stay throughout their first summer of life. Juvenile tomcod are found in greatest numbers in beds of eelgrass, which provide them with shelter and food resources. While in estuaries and quiet bays from October to May, adult tomcod exhibit activity patterns keyed to tidal fluctuations. During flood tides they actively feed over subtidal flats, and during ebb tides they move to deeper channels nearby.

The tomcod can weigh up to 0.6 kg (1.3 lb) and measure 38 cm (15 in) in length. However, most individuals weigh less than 0.5 kg (about 1 lb) and measure less than 30 cm (12 in) in length. Although some tomcod may reach 4 to 5 years of age, few exceed the age of 3.

Tomcod eat many types of fishes and invertebrates. Shrimp and amphipods make up the greatest proportion of their diet. They also eat small mollusks, squid, polychaete worms, and small fishes such as smelt, sticklebacks, alewives, shad, herring, cunners, mummichogs, sand lances, and even young striped bass. Mainly bottom feeders, tomcod use chin barbels and the long rays of the pelvic fins to detect food on the substrate over which they are swimming.

Status and Management

The tomcod supported a very modest inshore commercial fishery through the early 20th century. In 1929, about 12.7 mt (28,000 lb) of tomcod were harvested and marketed in Massachusetts, Maine, and the Canadian shores of the Gulf of Maine. In Massachusetts, the commercial harvest was largely a hook-and-line fishery north of Plymouth and a trap-and-weir fishery to the south. By 1942, the commercial fishery had generally disappeared from Massachusetts but persisted in Maine and Nova Scotia, where 12.5 mt (27,500 lb) were reported landed. No commercial harvest of this species has occurred since the early 1950s.

The recreational fishery for tomcod is most active in the winter and early spring in estuaries and in the mouths of coastal streams. This species is frequently caught incidentally during the active recreational smelt fishery that is conducted along parts of the New England coast. Due to modest harvest levels, no New England state regulated harvest of this species in 1989.

Angling and Handling Tips

Since the most active fishery for the tomcod occurs during this species' spawning migrations into estuaries and river mouths in the winter, most anglers have long since traded their fishing poles for a warm spot near the fireplace. Thus, two of the most important prerequisites for success are selecting a mild sunny day and wearing warm clothing.

Anglers fish for the tomcod from docks, bridges, and banks of shallow harbors, estuaries, and tidal rivers, or from small boats. One cardinal rule is generally followed when fishing for tomcod: Use light gear. Use a light-action spinning rod and a reel spooled with 6-pound test or lighter line. Although tomcod will strike small artificial lures that resemble shrimp or bait fishes, most anglers use bait on small hooks. Clams, shrimp, chunks of cut fishes, or sea worms placed on small hooks are popular baits. A sinker just heavy enough to hold the bait near the bottom should be tied to the line.

The tomcod is a panfish, normally less than a foot long, with a high meat-to-bone ratio. After a cold day of fishing, the tomcod is an "out of the water and into the pan" treat that will delight those who brave the elements to catch it. Tomcod are usually fried whole and served with tartar sauce or lemon; tommies can also be broiled for a nutritious and tasty meal. When broiling, use a whole, gutted fish; leave the head on for moister flesh after cooking. Dust with flour, dot with butter or a little olive oil, and broil in an oiled baking pan 3 inches from the heat for 3 to 5 minutes per side, depending upon size, until golden brown. After broiling, serve with lemon wedges and the vegetable of the season, winter squash.

Reference

Howe, A. B. 1971. *Biological investigation of Atlantic tomcod,* Microgadus tomcod *(Walbaum) in the Weweantic River estuary, Massachusetts.* Unpublished master's thesis, Univ. of Massachusetts, Amherst.

Atlantic Cod
(*Gadus morhua*)

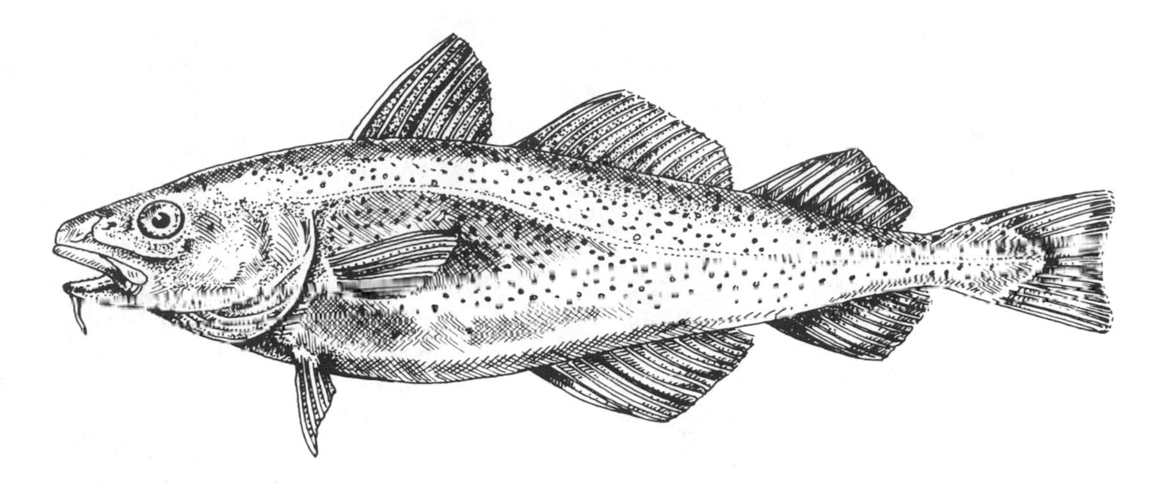

Among New England's rich array of natural resources, the Atlantic cod is perhaps the premier symbol of the region's natural heritage. So entwined in the early history of the settlement of coastal Massachusetts that a model (referred to as the "Sacred Cod") hangs in that Commonwealth's statehouse, this species is native to most of the North Atlantic Ocean. In the Northwest Atlantic it inhabits waters from West Greenland south to Cape Hatteras, North Carolina, and is most abundant from the coast of northern Labrador to the Nantucket Shoals region off of Massachusetts.

The cod is easily distinguished from all recreational species other than several members of its own family (the Gadidae) by its three rounded dorsal fins and two anal fins that are mirror images of the second and third dorsals. The cod also has a prominent barbel (whisker) on the chin.

The cod's upper jaw overhangs the shorter lower jaw, distinguishing it from the pollock, whose lower jaw extends beyond the shorter upper jaw. Also, the pollock's chin barbel is either very small or absent. The cod lacks the large dark blotch on the side that is characteristic of the haddock. The square or indented outline of the cod's tail differs from the rounded tail of the tomcod. The cod also lacks the slender, elongate extensions of the pelvic fins characteristic of the tomcod.

Individuals vary widely in color. Most are grayish green to reddish brown on their backs and sides and white on their bellies. They are speckled on the upper portion of their bodies, the sides of their heads, and their fins and tails. The lateral line is conspicuously lighter than the dark sides of the body.

Biology

The Atlantic cod prefers water temperatures less than 10 degrees C (50 F) and depths up to 366 m (1,200 ft). It is most frequently found at depths of 61 to 110 m (200 to 360 ft) and in temperatures ranging from 1.1 to 7.8 degrees C (34 to 46 F) in the summer, and at depths of 90 to 134 m (295 to 440 ft) and in temperatures of 2.2 to 3.9 degrees C (36 to 39 F) in the winter.

Cod undertake seasonal migrations in the more northerly and southerly reaches of their range in the Northwest Atlantic. Those inhabiting polar waters in the summer and autumn migrate to more southerly and deeper waters of the Northwest Atlantic coast in winter and spring; those summering in the Nantucket Shoals region overwinter nearby, along the New Jersey coast. Cod inhabiting the region between coastal Nova Scotia and Cape Cod do not exhibit predictable seasonal migrations. In New England there are three separate stocks of cod, one inhabiting the Gulf of Maine, one on Georges Bank, and one south and west of Nantucket Shoals. Although some cod move considerable distances in search of food or in response to overcrowding at certain spawning grounds, generally in the New England region they remain within limited areas of uniform physical conditions. Cod do not swim about in large schools, but they do travel in small groups when searching for food.

Both sexes usually reproduce for the first time when 5 or 6 years old. The fecundity of females increases with size and age. A 102-cm (40-in) female may lay about 3 million eggs, and a 127-cm (50-in) female up to 9 million eggs in one spawning season.

The cod reproduces in the winter, spawning in water temperatures of 5 to 7 degrees C (41 to 45 F). Adults inhabiting inshore areas generally move offshore to reproduce. There are several major spawning grounds off the New England coast, but little spawning occurs to the south of this region.

Table 14. Length at age in cm (inches in parentheses) of cod from Georges Bank

Age	*Length*
2	51 (20.1)
3	64 (25.2)
4	71 (28.0)
5	80 (31.5)
6	84 (33.1)
7	96 (37.8)
8	99 (39.0)

Source: Data from Scott and Scott 1988.

Larvae measuring about 0.5 cm (0.2 in) hatch from 10 to 40 days after spawning, depending upon water temperatures. The larvae inhabit the open water column, feeding upon microscopic copepods for 2 to 3 months after hatching, after which they become demersal (bottom-dwelling) for the rest of their life cycle.

Cod may reach lengths in excess of 152 cm (5 ft) and weights up to 96 kg (211 lb). However, they rarely weigh more than 45 kg (about 100 lb), with the 22–27-kg (50–60-lb) range representing the largest sizes typically caught. Offshore cod tend to be larger than inshore ones, the former frequently reaching sizes of 11.4 kg (25 lb) and the latter usually weighing 2.7 to 5.5 kg (6 to 12 lb). Length at age for cod collected from Georges Bank are shown in Table 14. Maximum age is 26 to 29 years.

Smaller juvenile cod feed mainly upon small crustaceans such as shrimp, amphipods, clams, crabs, and mussels. Adults will eat almost anything small enough to fit into their mouths, including clams, cockles, mussels, and other mollusks, as well as crabs, lobsters, and sea urchins. Adults also pursue schooling fishes, eating substantial numbers of herring, capelins, shad, mackerel, silver hakes, young haddock, and other species. Voraciously pursuing a variety of potential food, cod will occasionally dine upon some very exotic items; ducks, shoes, jewelry, and rope have been found in the stomachs of captured cod.

Young cod are eaten by many predators, including pollock, larger cod, and squid. Predator-caused mortality decreases once juveniles grow to about 20 cm (8 in), as they become better able to escape attacks from many of their potential predators. Adults occasionally fall prey to spiny dogfish and other sharks, as well as harbor, gray, and harp seals, and occasionally pilot whales.

Status and Management

The cod has been an extremely valuable resource for several centuries in New England. Its extensive use as a food dates back to the earliest period of European settlement in coastal New England. In colonial times it was deemed so important that in 1639 the General Court of the Massachusetts Bay Colony ordered that cod could no longer be used as fertilizer by farmers. This action was one of the first recorded attempts at natural resource conservation and management on this continent.

Although one of the earliest fisheries resources to be broadly utilized after European settlement in New England, cod populations proved to be very resilient; after more than 3 centuries of exploitation, the cod still provided about 60% of the total groundfish catch from Georges Bank during the 1980s. Annual cod landings exceeded 50,000 mt (110 million lb) several times in the 20th century. Levels of exploitation since the mid 20th century have markedly influenced the abundance of Atlantic cod stocks. A combination of advances in fishing technology and the introduction into the Northwest Atlantic of distant-water foreign fishing fleets during the late 1950s led to a period of reduced abundance and major annual fluctuations in population size. Reduction in foreign harvest as a result of passage of the Magnuson Act (the 200-mile-limit law) in the mid 1970s provided little relief for cod stocks. During the mid 1980s domestic vessels captured mostly 3-to-5-year-old cod, an indication that few larger, older individuals remain along the North American coast. Harvest levels began to drop in the mid 1980s, with projections for continued decline into the 1990s. Even as harvests dropped, fishing pressure went up. Over 20,000 fishing-vessel days were logged on Georges Bank in 1988, resulting in nearly 70% mortality for that stock. Population abundance declined throughout the 1980s, reaching all-time lows for the Gulf of Maine and Georges Bank. By 1988, the biomass of adults had dropped to one third of that seen in 1980.

When more abundant, cod populations are composed of at least modest proportions of a variety of age groups. However, the harvest on Georges Bank in 1988, up nearly 25% from that of 1987, was highly dependent upon the 1985 year class, the only abundant age group in the population. By 1988 it was felt that, in spite of a good 1987 year class, high levels of fishing pressure that removed either immature or newly mature fish would prevent a recovery of the cod stock from the record low levels.

Recreational harvest constitutes a modest portion of the total cod landed. In 1970, recreational landings were estimated to be 40% of the total harvest in New England. However, from 1979 to 1986, recreational harvest averaged about 11% to 12% of the total cod harvest in the Gulf of

Maine and from Georges Bank southward. The recreational harvest of 2,900 mt (about 6.4 million lb) taken from Georges Bank in 1987 constituted 25% of the total harvest for that region.

Cod harvest in the fisheries conservation zone falls under the Northeast Multispecies Fishery Management Plan of the New England Fishery Management Council. Regulations under this plan include minimum legal size limits for commercial and recreational harvest, area closures, and mesh-size regulations for commercial trawl nets. All New England states managed cod stocks within their territorial waters in 1989, utilizing the same basic types of regulations as in the council's multispecies plan.

Angling and Handling Tips

Many anglers fish for cod on offshore grounds in private or party boats. A 7½–9-foot medium to stiff rod with a conventional 4/0 reel is required when pursuing this species offshore. The reel should be spooled with 50-pound test Dacron line. Many of the most successful anglers use jigs with teasers; 10–20-ounce Vike or Norwegian-style jigs are popular. Some anglers prefer to replace the hook on such lures with a 10/0 or larger treble hook that has a red surge tube over the shank. Jigs should be tied to a monofilament leader fastened to the rod's line by a black swivel. The lure is completed by attaching a red, green, black, or white tube teaser worm on a large 8/0 hook to the swivel. Such a rigging resembles a large fish chasing a smaller bait fish, an effect that causes many cod to strike at the teaser being "chased" by the jig. To fish, lower the jig and teaser to the bottom, free-spooling the line while holding your thumb on the spool to prevent the line from forming into a "bird's nest." When you feel the bump of the bottom, push your clutch lever and raise your rod tip. Take two full turns, wait a minute, and lower the jig to the bottom. Next, slowly raise your rod tip high and let the jig settle back to the bottom. When a cod hits, hold your rod high for a moment. The teaser may flutter in a way that will attract a "doubleheader."

Although jigs hook large cod, bait also brings good luck to the angler. When rigging for bait, attach a 10–20-ounce sinker to a three-way swivel using line that is lighter than that on the reel; if the sinker snags the bottom, all of the rig except the sinker can be recovered. Eighteen inches of 80-pound test monofilament with one or two 4/0 or 5/0 snelled sproat hooks tied along its length should be attached to the second swivel leg. A commercial "Scotsman" or double-hook cod rig can be substituted for the homemade rig. Sand eels, mackerel, strips of herring and other fishes, or crabs can be used as bait. The favored bait of many anglers is a large piece of clam with its entrails trailing from the hook. Clams should be replaced when they turn pale. While fishing, let the sinker stay in place on the bottom; cod strike more frequently at still baits. When you bring your cod to the surface, don't lift it out of the water; the hook will tear from its mouth. Instead, keep your line tight and gaff the fish.

In the early spring, as water temperatures are beginning to rise, cod can be fished along the shoreline during early morning or from late evening until night. Typical gear includes a rod and reel with 15–20-pound test line rigged with a 2-ounce or larger pyramid sinker on a fishfinder. Sea worms or clams on 3/0 hooks are used as bait.

Cod should be iced after capture to retain their delicate flavor; if they are iced in a large cooler, the melt water should be drained occasionally so the fish do not soak in warming water.

The white, firmly textured, and flaky meat of the cod has traditionally been Massachusetts' equivalent to "a chicken in every pot." This flavorful fish can be baked, broiled, boiled/poached, fried, made into cakes or chowder, or salted for long-term storage without loss of flavor or nutrition. The roe (eggs), tongues, and especially cheeks are considered delicacies by many. Internationally, the traditional codfish is usually salted cod (*baccalà*/*bacalao*).

Poaching is a universal method for cooking this species. Boil lightly salted water with lemon slices and add cod fillets or steaks. When the water resumes boiling, remove the pot from the stove and let it stand for 5 to 10 minutes until the meat flakes. Drain and cover with a sauce, or add melted butter for a delicious meal.

For an excellent baked dish, stuff a cod with a "hot" breakfast sausage roll mixed with Italian-flavored bread crumbs or mashed potatoes. Bake at 350 degrees until the cod flakes.

References

Scott, W. B., and M. G. Scott. 1988. Atlantic Fishes of Canada. *Can. Bull. Fish. Aquat. Sci.* 219.

Serchuk, F. 1988. *Updated Georges Bank and Gulf of Maine cod assessments*. Unpublished Reports, NEFC/NMFS, Woods Hole, MA.

Haddock

(*Melanogrammus aeglefinus*)

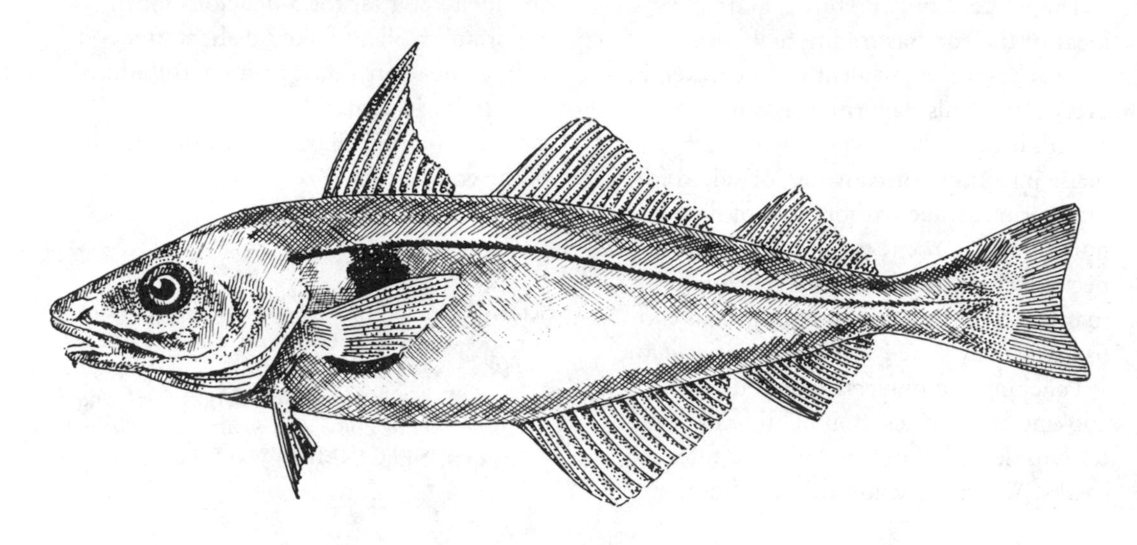

The haddock, a member of the cod family renowned as splendid table fare, inhabits both the American and European coasts of the Atlantic Ocean. In the Northwest Atlantic, it ranges from the southern end of the Grand Banks to Cape Cod in the summer and extends its range southward to Cape Hatteras, North Carolina, in the winter. The greatest concentrations off the New England coast occurs in the southwestern Gulf of Maine and on Georges Bank.

The haddock, like the closely related cod, pollock, and tomcod, is easily distinguished from other coastal New England fishes by its three dorsal and two anal fins. The front dorsal fin is triangular in shape and taller than the following two. The posterior two are squarish, the middle dorsal being slightly larger than the last. Of the two anal fins, the second or posterior one is a mirror image of the third dorsal fin. The haddock can be distinguished from other closely related members of the cod family by a black lateral line and a large spot on each side of the body over the pectoral fins.

The top of the head, the back, and the upper sides are a dark purplish gray. The lower sides are shiny gray tinged with pink, and the belly and lower head are white. The haddock has dark dorsal fins, pectoral fins, and tail; the anal fins are pale and spotted with black at the base.

Biology

Haddock inhabit deep, cool waters, rarely entering estuaries or river mouths. They are primarily found at depths of 45 to 135 m (140 to 450 ft) and generally avoid depths of less than 9 m (30 ft). Haddock prefer substrates of gravel, smooth rock, or sand littered with shells and water temperatures of 2 to 10 degrees C (36 to 50 F). They migrate seasonally to areas that provide optimal habitat conditions. In winter, haddock move to deep water where the temperature is warmer and more constant than that in shallower areas. Most overwinter offshore from southern New Jersey to Cape Hatteras. By early spring they seek more northerly areas in New England, moving into shallower waters of the Gulf of Maine and Georges Bank, where they remain all summer.

Males reach sexual maturity at age 2, and females do so at age 3. Females weighing 1 kg (2.2 lb) produce about 170,000 eggs, and the largest females may release as many as 3 million eggs in one spawning season.

Haddock off the New England coast reproduce on sandy, rocky, or muddy bottoms from January to June, showing the greatest activity in March and April. Spawning occurs offshore at depths of 30.5 to 183 m (100 to 600 ft) and in temperatures of 1.7 to 7.2 degrees C (35 to 45 F). Eastern Georges Bank is the most productive spawning area for haddock in the Northwest Atlantic, with lesser spawning also occurring east of Nantucket Shoals and off the coast of Maine.

The buoyant eggs drift in the water, hatching in approximately 15 days. Young haddock are pelagic for several months after hatching, drifting with the prevailing currents. Subsequently, they move to the ocean floor, where they stay for the rest of their lives.

Like many species, haddock suffer extremely high death rates during early life stages. The number of larvae that survive in a given year is often chiefly determined by their location when they are ready to become bottom dwellers. If the currents in which they have been suspended have carried them far offshore from the continental shelf, few larvae will survive. Thus, the number of fish surviving early life stages is highly variable and unpredictable from year to year. Haddock populations characteristically suffer through extended series of years when few survive early life stages. Recreational and commercial harvests have a great effect upon this species because individuals removed from the population by fishing are not necessarily replaced by reproduction.

Table 15. Average length at age in cm (inches in parentheses) of Georges Bank haddock collected during the mid 1970s

Age	*Length*
1	19.9 (7.8)
2	36.8 (14.5)
3	48.4 (19.1)
4	57.3 (22.6)
5	61.8 (24.3)
6	65.6 (25.8)
7	68.2 (26.9)
8	69.9 (27.5)

Source: Lengths calculated from growth equations presented in Clark et al. 1982.

Maximum age is about 15 years. Although maximum size is approximately 16.8 kg (37 lb) and 112 cm (44 in), few haddock exceed 50 to 60 cm (about 20 to 24 in) in length, 1.4 to 2.3 kg (3 to 5 lb) in weight, and 9 to 10 years of age. The growth rate of haddock is inversely proportional to population abundance, with the average size at age being significantly greater during periods of low abundance than during periods of high abundance. Table 15 shows lengths at age for Georges Bank haddock collected during the mid 1970s.

Before descending to the ocean floor, larval haddock feed upon microscopic copepods. Bottom-dwelling juveniles and adults feed upon almost any slow-moving invertebrates, including small crabs, sea worms, clams, starfish, sea cucumbers, and sea urchins, and occasionally squid. Herring, sand lances, small eels or other young fishes only rarely occur in their diet.

Young haddock are eaten by cod, pollock, and white hakes. Adults are eaten by harbor and gray seals.

Status and Management

Historically, the haddock was abundant throughout the open waters of the Gulf of Maine and on all offshore banks, especially Georges Bank. The Georges Bank/Great South Channel area once was one of the most productive haddock grounds in the world. This species also occurred in many areas of the coastal belt within 15 to 20 miles of land.

Prior to 1900, the haddock possessed only minor importance as a fishery. Between 1880 and 1903, an average of 24,500 mt (53.9 million lb) of haddock were landed annually from the New England region. The commercial fishing industry, boosted by new markets for fresh fish and frozen fillets, expansion of ice-making and dockside freezing facilities, and improved fishing technology, increased harvests markedly during the 1920s. Landings rose from 41,200 mt (90.6 million lb) in 1921 to a peak of 132,200 mt (290.8 million lb) in 1929. Subsequently, harvests declined sharply, then stabilized at a yearly average of 70,300 mt (154.6 million lb) from 1935 to 1960. Concern over this reduction in harvest was a major impetus in establishment of the International Convention for the Northwest Atlantic Fisheries (ICNAF), a multinational attempt at managing fisheries resources. By 1960, foreign countries had major distant-water fishing fleets operating in the Northwest Atlantic, a development that further depleted haddock populations. Landings on Georges Bank peaked at 150,400 mt (330.9 million lb), then declined to 4,300 mt (9.5 million lb) by 1974. ICNAF initially attempted to protect the haddock by establishing minimum mesh sizes that allowed young fish to escape capture. By 1970, regulations setting yearly total allowable catches were established, and spawning grounds were closed to fishing during April and May.

Implementation of the Fishery Conservation and Management Act eliminated the distant-water fishing fleets from the fishery in 1977. From 1977 to 1982 haddock populations within the U.S. 200-mile fishery management zone were managed under a plan developed by the New England Fishery Management Council (NEFMC). This plan included regulations similar to those developed previously by ICNAF. The Northeast Multispecies Fishery Management Plan of the NEFMC was implemented in 1986. Minimum legal size limits for haddock harvested both commercially and recreationally, and an increase in the time covered by the spawning area closure, were added to the original regulations in the updated plan. In addition, all New England states regulate inshore harvest via use of minimum legal size limits.

In spite of these harvest regulations, domestic fishing pressure caused the haddock population in the Gulf of Maine to decline by more than 80% from 1977 to 1986. Combined effects of U.S. and Canadian fishing caused a similar trend on Georges Bank, where the number of haddock dropped from an estimated 95 million in 1980 to 9 million in 1986 (Figure 9). The 1986 level was comparable to the all-time lows observed in the early 1970s, before elimination of the distant-water fleet. The United States accounted for 69% of the Georges Bank harvest from 1977 to 1984, but the percentage declined to 49% by 1986, after the World Court established the U.S. and Canadian territorial bound-

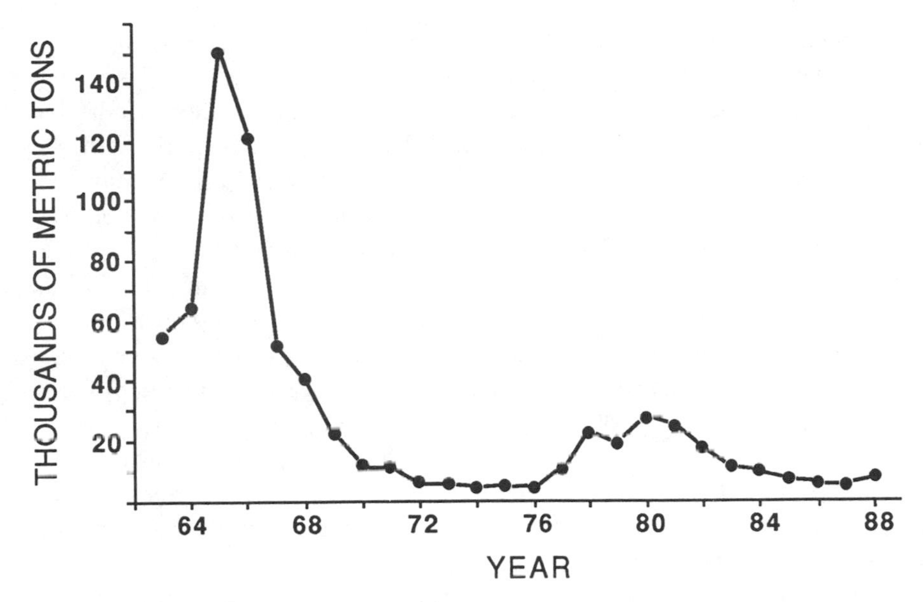

Figure 9. Estimated biomass of age-2 and older haddock from the Georges Bank region (from *Status of the Fishery Resources off the Northeastern U.S. for 1989,* NOAA Mem. NMFS-F/NEC-63).

aries on Georges Bank. Not surprisingly, landings from Georges Bank declined every year from 1980 to 1986.

The extremely low population density of the latter half of the 1980s was a result of continued fishing pressure being imposed upon an already depleted resource that was not being replenished by reproduction. Throughout this time period, the Gulf of Maine fishery was dependent upon haddock from the 1982 and earlier year classes. Most fish in the Georges Bank fishery were from the 1978 and 1983 year classes. Numbers of offspring produced by the Georges Bank stock in 1981, 1982, 1984, and 1986 were among the lowest on record. Haddock populations typically display great variability in offspring production from year to year. During time periods when population abundance of Georges Bank haddock was relatively high, strong year classes were produced on an average of every 3 years. However, this stock and others in Canadian waters have shown above-average reproductive success much less frequently when they have been overfished. Recovery from a depleted condition can only be accomplished through reproduction. Unfortunately, the limited reproductive capacity of a depleted population reduces the chance of timely recovery of the stock. Haddock populations along the New England coast will not be easily rebuilt under management strategies employed in the 1980s.

Harvest from recreational fishing in both the Gulf of Maine and Georges Bank fisheries accounted for a very minor proportion of haddock landings during the 1970s and 1980s. Thus, this activity had little impact upon the condition of the stocks during that time period.

Angling and Handling Tips

Few fishes are more delicately flavored or more finely fleshed than the haddock. Haddock typically brings the highest market price of all the members of the cod family. Traditionally, the skin is left on haddock fillets when marketing, so that the distinctive black lateral line and "devil's thumbprint" (the dusky blotch on the fish's side) will let consumers know that the high price they are paying is indeed purchasing this highly valued fish.

Haddock are caught from spring to fall, with fishing activity generally greatest in August and September. Anglers pursue this deepwater species from private, charter, or party boats.

A medium-action 8-foot boat rod with a fast-tapering tip is preferred by many anglers. A sensitive rod is necessary to feel the light bumps the haddock creates when it grabs a baited hook. Forty-pound test monofilament line on a high-speed conventional reel is usually recommended. Heavy line is necessary even though the haddock is a modest-sized species because anglers fishing in deep waters cannot predict what other larger fishes might grab the bait. Old-timers often favor braided line because it will not stretch as much as monofilament, thus making it easier to hook a haddock in deeper waters.

In a typical haddock rig, a swivel (to prevent twisting) is tied to one end of a 4-foot piece of 50-pound test leader, and a bank sinker is looped to the other end. No. 6/0 or smaller hooks with a short piece of yellow surge wound over their shanks are attached to the leader by two 6–10-inch-long "droppers," or loops. Ten to 20 ounces of sinker are needed to hold the rig on the bottom, depending upon currents and depths. Fresh clams or squid are normally the only successful baits. If you are on a party boat fishing for cod, keep an eye on the old-timers in the back of the boat. If they switch to clam bait on droppers, follow their lead; they're going after haddock.

After the baited rig is lowered to the bottom, all slack should be retrieved. Unlike the cod, which gives a sharp yank, the haddock bites in a series of light bumps. These slight taps can best be felt when the line is held between the thumb and fingers. Because haddock have soft mouths, they are easily lost if not properly played after being hooked. When a haddock taps the bait, the hook should be set with a steady pull rather than a jerk, and the fish should be steadily retrieved without pumping the rod. Just place the rod on the boat's railing and reel steadily.

Haddock flesh is lean and white. It is less firm than cod and flakes beautifully when cooked. Haddock is excellent baked, broiled, poached, microwaved, or used in chowders and stews. Traditionally, New Englanders fry haddock fillets or bake them whole with a bread-crumb and spice stuffing. For a change of pace, try the following simple New England style of poaching. Rub the inside of a dressed haddock with salt and wrap it in cheesecloth, leaving long ends of cloth so that the fish can be lifted from the water after boiling. Cook in boiling water or bouillon for 25 to 30 minutes, or until the fish flakes. Remove to a platter, garnish, surround with alternating boiled potatoes and cooked beets, and serve with mushroom soup as a sauce.

Reference

Clark, S. H., W. J. Overholtz, and R. C. Hennemuth. 1982. Review and assessment of the Georges Bank and Gulf of Maine haddock fishery. *J. Northw. Atl. Fish. Sci.* 3:1–27.

Pollock
(*Pollachius virens*)

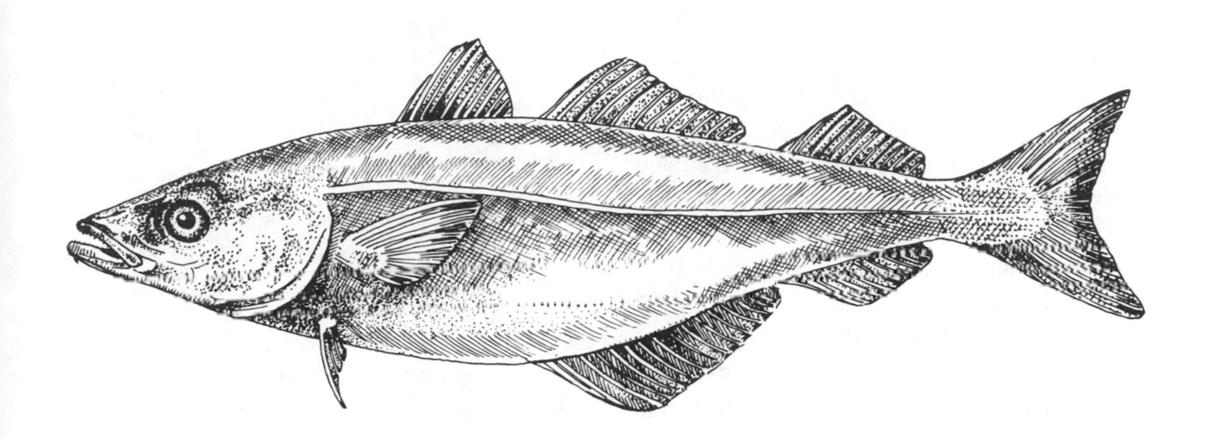

The pollock (also called Boston bluefish), a popular fish available to anglers from inshore bays and estuaries to offshore banks, occurs on both sides of the North Atlantic. In the northwestern Atlantic, it is found as far south as the Chesapeake Bay and as far north as the southern Labrador and western Greenland coasts. The major center of abundance in the western Atlantic occurs in the Gulf of Maine on the offshore banks and along the entire coastline from Nantucket Shoals and Cape Cod to Cape Sable, Nova Scotia.

This species is one of the more deep-bodied members of the cod family. The lower jaw, projecting beyond the upper, has a small chin barbel (whisker) in a young fish that is usually missing in its larger counterpart. The pollock has three separate dorsal fins and two separate anal fins, as do the cod, haddock, and tomcod. The first and the second dorsal fins are triangular in shape, with the second noticeably longer than the first. The tail is indented along its posterior edge, with both lobes of the fin being pointed or angular. This species lacks the dark lateral blotch and black lateral line of the haddock, the rounded tail and equal-sized dorsal fins of the tomcod, and the overhanging snout and shorter lower jaw of the cod.

Pollock range from olive green to brownish green dorsally and pale gray to yellow laterally. The belly is silvery. The lateral line is lighter than the upper sides of the body. Young pollock are darker and yellower on the lower sides than are the older, larger pollock.

Biology

Like other members of the cod family, pollock live on or near the bottom in areas of rocky substrates. They are found from shallow waters to depths as great as 183 m (600 ft), depending upon water temperatures and food availability. Pollock can tolerate temperatures near 0 degrees C (32 F) but are usually absent from surface waters when water temperatures exceed 11 degrees C (52 F). Pollock have been found in waters up to 20 degrees C (68 F) when inhabiting estuaries.

Large schools of pollock migrate inshore during spring and move to offshore waters in the colder months. Large schools of younger fish called harbor pollock move into estuaries and shallow bays in the spring. They remain there until dropping inshore temperatures in the fall force them to move offshore to deeper and, at that time of the year, warmer waters.

Some males and females reach sexual maturity at age 4 and age 5, respectively; most individuals of both sexes have done so by age 6. The average age and size at maturity are inversely proportional to population density; pollock mature at a younger age and smaller size when populations suffer major reductions in abundance. Average fecundity of females is approximately 250,000 eggs. Large females may produce up to 4 million eggs a year.

The western Gulf of Maine region is a major pollock spawning area. Pollock generally spawn during the autumn and early winter in water 27 to 91 m (about 90 to 300 ft) deep. Spawning begins when the water cools to 8.3 to 9.4 degrees C (47 to 49 F) and ends when temperatures reach 4.4 to 6.1 degrees C (40 to 43 F). The buoyant eggs hatch 6 to 9 days after fertilization, and the larvae remain near the surface for at least 3 months before becoming bottom-dwelling juveniles.

Pollock may reach lengths up to 107 cm (42 in) and weights up to 32 kg (70 lb); however, average weight is about 1.8 kg (4 lb). Table 16 lists average lengths for pollock from the Cape Cod coast and the Bay of Fundy. The oldest fish reported is 19 years.

Pollock are largely daytime sight feeders. Young-of-the-year eat crustaceans such as copepods, mysid shrimp, and amphipods.

Table 16. Average length in cm (inches in parentheses) of pollock from two locations

Age	*Cape Cod*	*Bay of Fundy*
1	12.7–17.8 (5–7)	21 (8.3)
2	30.5–33.0 (12–13)	38 (15.0)
3	43.2–45.7 (17–18)	44 (17.3)
4	53.3–55.9 (21–22)	54 (21.3)
5	63.5 (25)	61 (24.0)
6	68.6 (27)	69 (27.2)
7	71.1 (28)	
8	73.7 (29)	
9	76.2 (30)	

Source: Data from Bigelow and Schroeder 1953; Steele 1963.

Adults feed on large pelagic crustaceans such as shrimp, and small fishes such as herring, sand lances, cod, haddock, and hakes. Juvenile pollock occasionally are seen chasing schools of smelt through estuaries in the fall. Unlike the more demersal (bottom-dwelling) cod, pollock will pursue schools of small fishes at any depth, occasionally driving them to the surface of the water where frantic splashing can be seen as the prey attempt to escape. Pollock are noted for their voracious behavior while feeding; one 23-cm (9-in) specimen had 77 5-cm (2-in) herring in its stomach when it was captured during a feeding episode.

Adult pollock are occasionally eaten by harbor seals.

Status and Management

Historically, the pollock had modest market value and was commercially harvested largely as by-catch. As other coastal fisheries have declined in abundance, the pollock has developed a higher market value and has become the target of moderate commercial fishing effort. Commercial landings from the Scotian Shelf to Georges Bank gradually increased from an average of 38,200 mt (84.0 million lb) in 1972–1976 to 66,600 mt (146.5 million lb) in 1986. Total stock size increased throughout the late 1970s and early 1980s, then declined precipitously. The catch per unit effort declined sharply in the mid 1980s. The increase in harvest during the 1980s, coupled with the decline in catch per unit effort and in stock biomass, indicates the stock was being exploited above the level that would allow population maintenance.

Recreational harvest varied from less than 1% to 3% of the total yearly harvest of pollock from 1979 to 1986. During that time period, the total yearly recreational harvest peaked at about 1,300 mt (2.9 million lb) in both 1982 and 1983.

Pollock within U.S. territorial waters are managed under the Northeast Multispecies Fishery Management Plan of the New England Fishery Management Council. Regulations controlling commercial harvest under that plan included minimum legal size limits, mesh-size regulations, and an area closure that was established to protect spawning of the depleted haddock fishery. By 1989, all coastal New England states other than Massachusetts regulated harvests in state territorial waters by means of minimum legal size limits.

Angling and Handling Tips

Unlike the closely related cod, pollock are aggressive, strong fighters that frequently strike at fast-moving lures. Action can be furious when pollock are feeding. Pollock can be pursued from party boats, private boats, or the shoreline. Anglers can catch 25–40-pound jumbos in deeper water, and ½–5-pound schoolies or harbor pollock in inshore areas. Harbor pollock move inshore as the water warms in the spring, retreat to cooler waters in the heat of summer, and return inshore in the fall in large enough numbers to produce the best fishing of the year for this species. Anglers can fish for harbor pollock into the very late fall, depending upon water temperature, the presence of bait species, and the fortitude of the angler. The larger pollock tend to gather in deeper, more offshore waters and are available for party boat anglers in the summer at depths of 100 to 250 feet.

In deeper waters, pollock are taken with the same tackle and rigs as those used for cod, such as a medium/heavy 7–9-foot "cod rod" and a 4/0 conventional reel spooled with 40–50-pound test Dacron line. Lures are especially effective on pollock. A 10–20-ounce Vike or Norwegian-type jig with a dropper/teaser tied about 3 feet above the jig is a particularly popular rigging. Attaching the teaser to a two-way swivel by a split ring or a bead chain helps make the action more effective and does not weaken the line as a dropper can. When jigging with this rig, allow it to settle to the bottom; then retrieve some line, jig the lure, turn the reel handle 5 to 10 times, pause, and jig again. Repeat, and then allow it to flutter downward before repeating the sequence. Pollock most frequently strike during the flutter downward. When bringing the jig to the surface, try a rapid retrieval, since pollock can be hooked anywhere in the water column. When you find a school of feeding pollock, you may be in for additional excitement. The occasional 30-pounder that is accidentally snagged will provide an unforgettable battle all the way to the boat.

While most pollock are caught on jigs and teasers, they are also taken with bait, such as clams with entrails hanging off the hook or 1-by-3-inch strips of fishes.

In inshore areas, pollock are particularly active around breakwaters and other structures during a moving tide. Early morning and evening produce the best results, but pollock can be caught throughout the day. Shoreline anglers can spot a pollock by looking for bait fishes chased to the surface or for actively feeding gulls or other sea birds. Smaller inshore pollock are often pursued with lighter spinning outfits spooled with 12–15-pound test line. One-quarter-ounce to 2-ounce lures such as streamers, leadheads, mackerel jigs, Kastmasters, and small plugs that resemble sand eels all provide successful fishing. A small strip of squid or other bait added to a metal lure can increase the angler's chance of success.

This species may be handled exactly like the cod or haddock. A pollock should be iced at capture to preserve its mild flavor and delicate texture. The pollock's flesh is off-white, with a thin layer of darker flesh under the skin; the color lightens when cooking. The nickname "Boston bluefish" reflects this species' appearance and fighting ability, not its culinary uses. Its light, flaky flesh can be substituted for cod or haddock in most recipes. Pollock can be poached, baked, broiled, grilled, or put into fish chowders. For an easily prepared meal, put a large pollock fillet into a baking dish greased with margarine and cover it with a can of undiluted cream of mushroom soup. Sprinkle with a little paprika and parsley. Bake at 400 degrees for 20 minutes or more, depending upon the thickness of the fillet. The fillet is ready to eat when it begins to flake easily. Serve the fish and sauce on a bed of rice and enjoy a delicious meal.

References

Bigelow, H. B., and W. C. Schroeder. 1953. *Fishes of the Gulf of Maine*. U.S. Fish Wildl. Serv. Fish. Bull. 74, Vol. 53.

Steele, D. H. 1963. Pollock (*Pollachius virens*) in the Bay of Fundy. *J. Fish. Res. Board Can.* 20:1267–1314.

Red Hake
(*Urophycis chuss*)

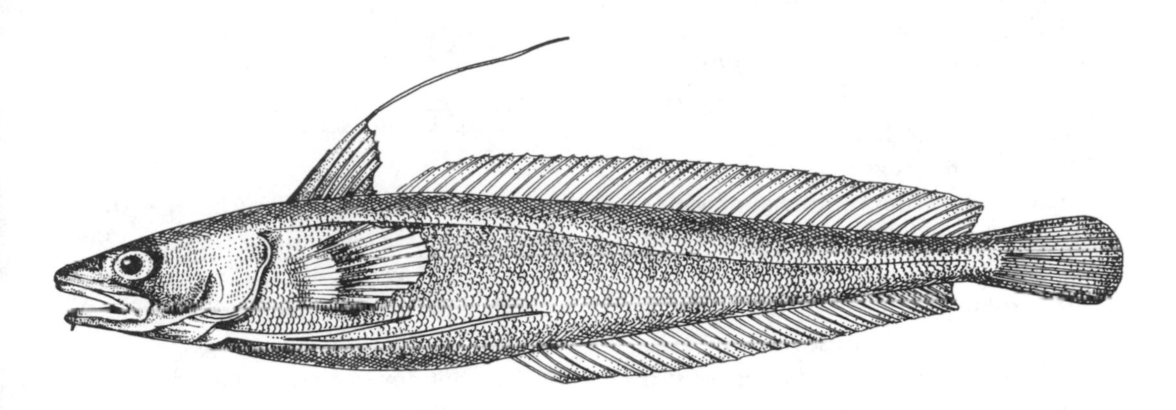

The red hake (also called squirrel hake and ling) occurs from the Gulf of St. Lawrence to North Carolina and is most abundant between Georges Bank and New Jersey. This species so resembles its larger relative, the white hake, that most anglers are unable to separate the two. The red hake has somewhat larger scales; it has 100 to 110 scales along its lateral line, whereas the white hake has about 140. Also, the posterior edge of the upper jaw occurs directly below the back edge of the pupil of the eye in the red hake, whereas it occurs below the rear edge of the eye in the white hake.

Both species are easily distinguished from other recreational species in New England. Red and white hakes have two dorsal fins, the posterior one being more than several times the length of the anterior. They have a long anal fin, a small chin barbel, and pelvic fins that extend into two elongate filaments of unequal length.

The red hake is red, brown, olive-brown, or nearly black dorsally and laterally. Its lateral line is paler than the coloration on its side. The lower body may be tinged with yellow, and the belly and chin are white, yellow, or light gray.

Biology

The red hake is a bottom-dwelling species, occurring in greatest abundance in habitats with sand or mud substrates and temperatures of 5 to 12 degrees C (41 to 54 F). This species displays seasonal movements, in part to seek areas in its preferred temperature range. In the winter, red hakes aggregate in deep offshore areas, residing in greatest abundance at depths of 183 to 457 m (600 to 1,500 ft) off the southern New England coast. By late spring, adults migrate to more inshore, shallow waters. In the summer, the greatest concentration of adults occurs between Martha's Vineyard and Long Island and in the southwestern area of Georges Bank, in waters less than 110 cm (360 ft) deep. Adults are distributed widely both inshore and offshore in the fall but are concentrated in offshore waters once more by winter.

Most red hakes become sexually mature at age 2. Spawning starts in May off southern New England and in June in the Gulf of Maine, peaks in June and July, and extends to September. Red hakes on Georges Bank spawn in 110 to 120 m (360 to 394 ft) of water. The pelagic eggs hatch in about 36 hours at 21 degrees C (70 F) to 90 hours at 15.6 degrees C (60 F). Larvae drift southwest with prevailing currents, becoming bottom dwellers 2 to 3 months after hatching.

Young-of-the-year red hakes live in the mantle cavity of the sea scallop and occasionally in other bivalves such as the surf clam, where they stay until they are forced to leave because of low water temperatures, or they become too large to remain. Red hakes as large as 13 to 14 cm (5.1 to 5.5 in) in length have been found in scallop mantle cavities. This behavioral adaptation is believed to protect them from predation. Although some red hakes have been found in eelgrass, or hiding among rocks or debris, the association with sea scallops is felt to be highly important for survival since the distribution of young red hakes closely coincides with that of scallops. Of all scallops examined from particular areas, 5% to over 80% have contained red hake juveniles. While associated with scallops, juveniles restrict their feeding activity to the substrate immediately surrounding their scallop host. Red hakes remain around scallop beds until the fall of the second year of life, when they move inshore to depths as shallow as 55 m (180 ft), then move offshore to overwinter.

Red hakes normally do not exceed 55 cm (21.7 in) in length and 2 kg (5 lb) in weight, although individuals as large as 76 cm (29.9 in) and 2.7 to 3.2 kg (6 to 7 lb) have been reported. The average fish landed by anglers or commercial boats is about 0.5 to 1.4 kg (1 to 3 lb). Females grow more rapidly and reach larger sizes than males. Maximum age is about 12 years, but few fishes exceed 6 years. Average lengths at age of red hakes are shown in Table 17.

Table 17. Length at age in cm (inches in parentheses) of red hakes

Age	*Length*
2	30.8 (12.1)
3	33.0 (13.0)
4	35.4 (13.9)
5	38.3 (15.1)

Source: Data from Scott and Scott 1988.

The bottom-feeding red hake has taste buds on the surface of its elongate pelvic fins, which it uses to locate food items in the substrate. Juveniles eat amphipods, sand shrimp, polychaetes, and crabs. Adults eat a variety of foods, with amphipods being the most common item. They also feed extensively on squid, shrimp, and fishes, including the alewife, tautog, butterfish, cunner, eel, herring, mackerel, sand lance, and others.

Status and Management

The red hake historically had only modest market value as a food because it has soft meat and poor storage properties even when refrigerated. The development of an industrial use in mink and poultry feeds and the introduction of the foreign distant-water fleets led to markedly increased commercial harvests in the 1960s and 1970s. In the Gulf of Maine, annual commercial landings that had ranged from 1,000 to 5,000 mt (2.2 million to 11.0 million lb) from 1960 to 1971 increased to as high as 15,300 mt (33.6 million lb) between 1972 to 1976. In the much larger Georges Bank/Middle Atlantic population, harvests increased from 4,600 mt (10.1 million lb) in 1960 to 108,000 mt (237.6

million lb) in 1966. Harvests in this region oscillated up and down from 1966 to the mid 1970s. During the 1960s to mid 1970s, foreign distant-water vessels accounted for 83% and 93% of the total commercial landings in the Gulf of Maine and Georges Bank/Middle Atlantic, respectively. Exclusion of foreign vessels from the fisheries after passage of the Magnuson Act led to declining catches through the 1980s. Population levels, which appeared to oscillate with the combined effects of varying fishing pressure and reproductive success throughout the 1960s and 1970s, recovered somewhat during the 1980s. The combination of modest fishing pressure and average or better reproductive success since 1980 led to an increase in stock size in the Gulf of Maine. A similar trend was noted for the Georges Bank/Middle Atlantic stock, although slight declines were recorded in the latter half of the decade. Three or four age groups were consistently well represented in both stocks, indicating that young individuals were not rapidly removed from the population after reaching a size suitable for harvest.

Recreational fishing has been an insignificant part of the total harvest in the Gulf of Maine, and of only minor importance in the more southern population. For example, in the Georges Bank/Middle Atlantic region in 1987, recreational landings comprised only 3% of the commercial harvest.

Red hake fisheries had been managed under a preliminary management plan of the New England Fishery Management Council after 1977. The red hake was not included as a target species in the Northeast Multispecies Fishery Management Plan implemented in 1986. No New England states regulated harvest of red hakes within their territorial waters as of 1989.

Angling and Handling Tips

Red hakes are usually taken by anglers fishing for haddock or other members of the cod family. They are smaller than their nearly identical close relative, the white hake, with most landed fish weighing less than 5 pounds. Their small size and feeding habits make them a common catch when one is fishing with haddock gear.

In deeper waters, a medium saltwater outfit consisting of a 6½–7-foot rod, a conventional 3/0 to 4/0 reel, and 40-pound test line, is typically used. The weight of this outfit is dictated more by the chance that larger-bodied members of the cod family may strike the bait than by its necessity for landing a red hake. Lighter rods and 20-pound test line are favored by many in shallower waters.

The rig consists of a bank-type sinker, which is tied to the end of the line, and a 2/0–3/0 haddock hook attached to the line by a three-way swivel. Most anglers tie the swivel so that the bait is off the bottom by just 1 to 2 inches. A second hook is frequently attached 15 inches or more above the lower hook. This second hook is intended to attract other species that are feeding somewhat higher off the bottom than the red hake. Red hakes will take most types of fresh or frozen baits. Clams are very successful bait, as are cut baits such as mackerel, herring, whiting, and squid.

The meat of the red hake is coarser and less bland than that of the cod. Its flesh is soft and white, and it needs to be handled carefully and prepared while fresh to preserve its pleasant, mild flavor. If red hake meat is packed directly on ice that melts, the meat will become waterlogged; thus, it should be placed in plastic bags when stored on ice. The red hake is best when eaten fresh, as the texture of the meat will toughen if frozen at home. If freezing is necessary, fillets should be stored at temperatures under 0 degrees F to minimize this change in texture.

Red hake fillets can be used in any recipe calling for cod, haddock, or other types of hake. For a spicy meal try Red Cajun Hake. Lightly brown one minced onion and two cloves of chopped garlic in a little olive oil or butter in a deep iron skillet. Add ½ cup of chopped green pepper, ½ cup of thinly sliced celery, ½ cup of diced zucchini, 1 can of crushed tomatoes, 1 teaspoon of prepared dry Cajun spices, and/or 1 tablespoon of red hot pepper sauce (Tabasco). Bring to a boil and simmer for 45 minutes. Mix 2 tablespoons of cornstarch with a little cold water, stirring until the cornstarch is dissolved. Add this to the sauce, sprinkle in parsley flakes, and add salt and freshly ground pepper to taste. Stir until the sauce thickens, then add small whole red hakes, or fillets, and cook at medium heat 10 minutes per inch of thickness. Gently remove with slotted spoon or spatula and serve with pasta or rice.

References

Garman, G. C. 1983. Observations on juvenile red hake associated with sea scallops in Frenchman Bay, Maine. *Trans. Am. Fish. Soc.* 112:212–215.

Scott, W. B., and M. G. Scott. 1988. Atlantic Fishes of Canada. *Can. Bull. Fish. Aquat. Sci.* 219.

White Hake
(*Urophycis tenuis*)

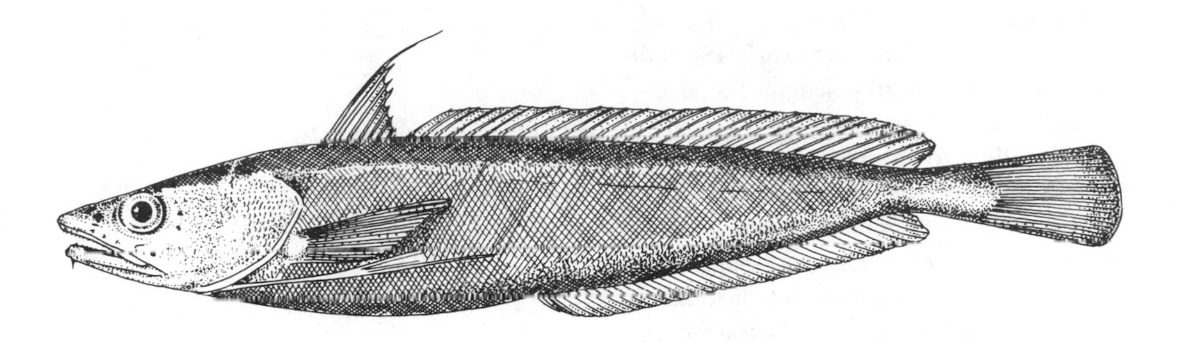

The white hake (also called ling or hake) occurs along the continental shelf and slope of the Northwest Atlantic from Labrador and the Grand Banks of Newfoundland to North Carolina and is most abundant from the Grand Banks to the Gulf of Maine and northwestern Georges Bank.

This species is easily separated from all New England recreational fishes other than its sibling species, the red hake. It is so similar to the red hake that most anglers are unable to separate the two species (see Red Hake for identifying characteristics).

The white hake is muddy to purple brown dorsally. The sides are occasionally bronzed, and the belly is yellowish or dirty white with numerous minute black spots.

Biology

The white hake is more northerly distributed and is more abundant in deeper waters than the red hake. White hakes have been taken in water as deep as 997 m (nearly 3,300 ft). The greatest concentrations are usually in waters deeper than 183 m (600 ft), although they can be caught in much shallower water. In the Gulf of Maine adults move toward shore in the summer and disperse to the deepest areas of the Gulf in the fall and winter. The white hake occurs only along the continental slope south of Georges Bank.

As in many species, seasonal movements are keyed to water temperature changes. All white hakes collected in an extensive Canadian survey inhabited temperatures between 0.6 and 5.5 degrees C (33 to 42 F). White hakes generally avoid temperatures lower than 0 degrees C (32 F) and move to deeper waters in the fall when shoreward waters drop below 5 degrees C (41 F). This species seeks sand or mud bottom habitats throughout its yearly movements.

Fifty percent of male and female white hakes in the Gulf of St. Lawrence are sexually mature when they reach lengths of 40 cm (16 in) and 47 cm (18.5 in), respectively, at about 3 to 4 years of age for both sexes. Female white hakes are among the most fecund of all groundfishes in the Northwest Atlantic. Female fecundity increases markedly with size. Gulf of St. Lawrence females average nearly 640,000 eggs at 50 cm (about 20 in), over 4.1 million at 70 cm (27 in), and nearly 16.5 million at 90 cm (35 in) in length. Females over 100 cm (39 in) may produce as many as 25 million to 30 million eggs.

Spawning occurs in early summer south of Cape Cod and in midsummer to fall in Canadian waters. Larvae are pelagic for several months after hatching, then metamorphose into bottom-dwelling juveniles. Young-of-the-year hake remain in relatively shallow water for a while after becoming bottom dwellers. Juveniles occasionally reside in scallops, but this behavior is not predominant for this species, as it is for the red hake. Larger juveniles are generally distributed across a broad range of water depths.

Growth of the white hake is inversely related to latitude, being slower in the more northerly reaches of its geographic range. Table 18 presents length-at-age data of white hakes from two regions within the Gulf of St. Lawrence. Maximum size is about 135 cm (53 in) in length and 22 kg (48 lb) in weight. Most landed individuals weigh between 0.5 and 9.1 kg (1 to 20 lb), with an average of about 3.6 kg (8 lb). A 30-cm (12-in) individual weighs about 0.2 kg (0.4 lb), a 60-cm (24-in) about 1.7 kg (3.8 lb), and a 90-cm (35-in) about 6.2 kg (13.7 lb).

White hakes eat a variety of invertebrates and fishes, the latter having increasing dietary importance with increasing size of the hakes. Large adults feed almost exclusively upon fishes. Euphausiid shrimp are often a major component of their diet. White hakes off Martha's Vineyard also feed heavily upon small squid, crabs, and butterfish, and those captured from the Middle Atlantic feed upon small mackerel, flounders, crabs, and squid.

Table 18. Length at age in cm (inches in parentheses) of white hakes from two regions in the Gulf of St. Lawrence

Age	*Region A**	*Region B***
3	41 (16)	42 (17)
4	46 (18)	46 (18)
5	53 (21)	51 (20)
6	57 (22)	64 (25)
7	62 (24)	72 (28)
8	66 (26)	76 (30)
9	69 (27)	82 (32)
10	76 (30)	92 (36)

*North and west of Prince Edward Island and Nova Scotia.

**Southwest of Nova Scotia.

Source: Data from Beacham and Nepszy 1980.

Status and Management

White hakes historically supported only modest recreational and commercial fisheries, the latter based upon their moderate value as human food and upon their use in mink and poultry feeds. In the early 1900s, combined landings of white and red hakes from the Gulf of Maine averaged 11,364 to 13,636 mt (25 million to 30 million lb). Harvests increased little from the 1940s to the early 1970s. However, with the collapse of traditional groundfish fisheries in the 1970s, white hakes became increasingly more marketable through the 1970s and 1980s. Since 1968, U.S. vessels have landed about 94% of the total Gulf of Maine/Georges Bank annual harvest. Due to the increasing fishing power of the New England fleet, and an increasing effort directed toward harvest of white hakes, landings rose from an annual average of less than 1,000 mt (2.2 million lb) in the late 1960s to 7,500 mt (16.5 million lb) by 1984. Although population levels remained moderately stable through 1987, landings dropped slightly during the latter half of the 1980s. The reduced commercial catch indicated that the fishery might be unable to sustain continued harvests at the mid-1980s levels without suffering reductions in abundance. The level of recreational harvest, accomplished mostly by offshore party and private boats, constituted an insignificant portion of the total harvest throughout the 1980s.

White hakes in U.S. territorial waters have been managed under the Northeast Multispecies Fishery Management Plan since 1986. Since most of the harvest is taken outside of state territorial waters, no New England state regulated harvest as of 1989.

Angling and Handling Tips

Most hakes run about 5 to 8 pounds, but jumbos of 20 to 30 pounds are occasionally taken by New England anglers. The white hake is usually taken in deep water while fishing for cod; thus, typical cod gear is standard equipment when fishing for this groundfish species. A medium- to heavy-duty rod and a 4/0 to 6/0 conventional reel are preferred by many anglers. A bank sinker is typically tied to the end of the line, and a 2/0 to 3/0 cod hook and leader attached to a three-way swivel are tied far enough up the line so the hook will rest just at the bottom when fishing. A second hook can be placed farther up the line, if desired.

Hooks can be baited with nearly any cut bait, although clam is used most often by party boat anglers. Raising and lowering the bait every few minutes, and changing it frequently to keep it fresh, improve the frequency of the catch. Most baits, particularly clams, turn soft and lose their odor after a time in salt water. Thus, even if fishing at 150 to 200 feet, one should periodically retrieve the bait and change it.

White hakes can be caught bottom-jigging with skirted leadheads baited with clams, as well as still-fishing. When jigging, raise the rod tip so that it lifts the leadhead several feet off the bottom, then let it flutter back down.

The white hake should be handled just like the red hake; it should be eaten fresh or stored at temperatures below 0 degrees F. Freezing above this temperature causes the texture of the hake to become rubbery. Some anglers who freeze white hake fillets believe that freezing in a block of ice provides good storage without loss of quality.

When cooked, the flesh of the white hake is white and mild-tasting. The white hake can be baked, broiled, or poached, and its large fillets lend themselves equally well to chowders and cold salads.

Cold salads are an excellent way to use leftover fish that has been cooked in almost any standard way. To cook fresh, cover fillets with waxed paper and cook 4 to 6 minutes per pound on the high setting of the microwave oven. Mix 6 tablespoons of mayonnaise, ½ cup of prepared green relish, ⅓ cup of very thinly sliced celery, ⅓ cup of minced red onion, and 2 tablespoons of prepared seafood seasoning (or substitute ½ teaspoon of dill, ¼ teaspoon of thyme, ¼ teaspoon of garlic powder, ¼ teaspoon of basil, and several turns of freshly ground pepper) in a bowl. Flake the cooked fish and add to the bowl, mix gently, and chill. Serve on lettuce as a cold summer salad. If you whip the fish, it can also be spooned onto crackers.

For a quick, easily prepared dish, try baked white hake. Place the fillets in a greased baking dish and cover with a layer of Italian salad dressing and seasoned bread crumbs. Top with several slices of tomato or green pepper and bake in a preheated 400-degree oven for 10 minutes per inch of thickness.

Reference

Beacham, T. D., and S. J. Nepszy. 1980. Some aspects of the biology of the white hake, *Urophycis tenuis*, in the southern Gulf of St. Lawrence. *J. Northw. Atl. Fish. Sci.* 1:49–54.

Cusk
(*Brosme brosme*)

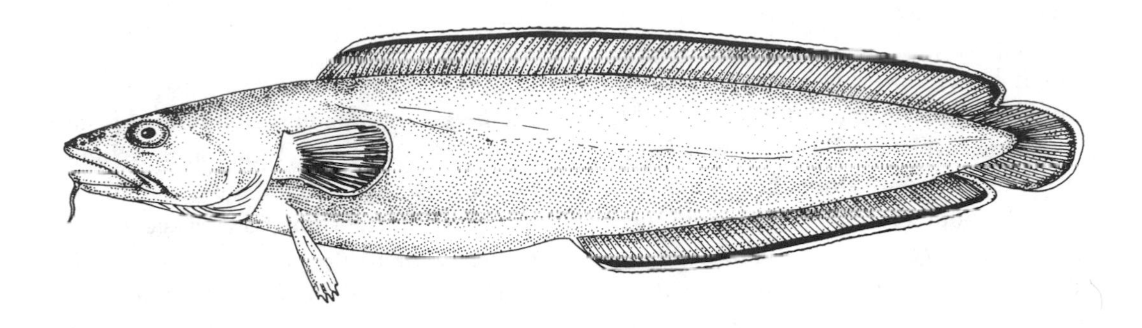

The cusk is a deepwater species found on both coasts of the North Atlantic. In the Northwest Atlantic it occurs from Newfoundland and Nova Scotia southward to Cape Cod and, infrequently, off the southern New England coast to New Jersey.

The single, elongate dorsal and anal fins, attached at the base to the rounded tail, the elongate, slender head, and the chin barbel distinguish this species from all others that New England anglers might land. The pelvic fins are small, and both they and the pectoral fins are rounded, lacking the elongate rays seen on the hakes. The scales are small, and the lateral line is noticeably arched above the pectoral fins.

The cusk's color ranges from a slate gray or a dark reddish or greenish brown to a pale yellow dorsally, usually conforming to its surroundings. The lower sides are gray, and the belly is dirty white. The dorsal, anal, and tail fins match the general body color at their bases and have a black stripe and narrow white edge along their margins.

Biology

The cusk inhabits rough, rocky bottoms with hard substrates at depths of 20 to 900 m (66 to 2,952 ft), although most individuals are found at depths of 73 to 363 m (240 to 1,190 ft). They have been collected in water temperatures from 2 to 12 degrees C (36 to 54 F) but prefer temperatures between 6 and 10 degrees C (43 to 50 F). Cusk are solitary in habit and are weak, sluggish swimmers that exhibit very limited movement within the habitats they are occupying.

The average ages by which 50% of the cusk in a Canadian population reach sexual maturity are 4.7 and 6.5 years for males and females, respectively. All individuals of both sexes are mature by the age of 10 years. Female fecundity increases with size, ranging from about 100,000 eggs for a female 56 cm (22 in) long to over 3.9 million for one 90 cm (35.4 in) long.

Spawning typically occurs from May to

June, although it has been noted as early as April in the Gulf of Maine. Eggs are buoyant, and larvae hatch soon after fertilization at about 0.4 cm (0.2 in) in length. Larvae remain in the water column until they have grown to about 5.0 cm (2 in) in length, when they become bottom-dwelling juveniles.

Only limited information is available concerning growth rates of the cusk. Table 19 lists several lengths at age from a study on a Canadian population. Maximum length is about 110 cm (43.3 in), although few adults exceed 76 cm (29 in). The oldest specimen captured in the Canadian study was 14 years of age and weighed 9.0 kg (20 lb).

Few studies have detailed cusk food habits. This species generally feeds on crustaceans, particularly crabs, and on mollusks and occasionally starfish. The hooded seal feeds at least occasionally upon this fish species.

Table 19. Selected lengths at age in cm (inches in parentheses) for a Canadian population of cusks

Age	*Male*	*Female*
5	45.0 (17.7)	45.3 (17.8)
7	58.8 (23.1)	52.7 (20.7)
9	63.3 (24.9)	61.9 (24.4)
11	72.0 (28.3)	68.9 (27.1)

Source: Scott and Scott 1988.

Status and Management

Due to its preference for rocky habitat and its somewhat solitary behavior, the cusk is not harvested in large numbers by commercial fishermen. However, there has always been a ready market for those that are landed. This species was fished relatively heavily prior to the 1940s, with 909 to 3,182 mt (2 million to 7 million lb) landed by U.S. fishermen from the Gulf of Maine alone. When otter trawling replaced longlining as the major means of commercially harvesting deeper waters off New England, U.S. landings of cusk dropped; landings by U.S. vessels in the Gulf of Maine averaged from 727 to 1,000 mt (1.6 million to 2.2 million lb) from 1945 to 1947. Landings gradually increased as fishing intensity increased over the next several decades. Between 1977 to 1987, annual landings from the Gulf of Maine/Georges Bank region ranged from 1,400 to 4,000 mt (3.1 million to 8.8 million lb), averaging 2,300 mt (5.1 million lb) per year. The United States accounted for more than 75% of these annual landings, with Canadian vessels harvesting the remainder. Due to its preference for deep waters, the cusk is not readily available to many anglers; thus, recreational landings typically constitute an insignificant portion of the total harvest.

Catch frequencies in the Gulf of Maine/Georges Bank region by research bottom-trawl surveys of the Northeast Fisheries Center (of the NMFS) generally decreased through the 1980s, although catches did increase from 1984 to 1986. This trend suggests that the cusk resource is being fully exploited and may be entering a period of decline. However, as of 1989, the New England Fishery Management Council was not regulating harvest of this species.

Angling and Handling Tips

This deepwater species is caught by fishing directly on the bottom. Typical groundfish equipment, such as a 7–9-foot boat rod and conventional reel spooled with 40–50-pound test monofilament line, suits this species well. A cod or haddock rig should be attached to the end of the line by a black swivel. Keeping the baited rig on the bottom is a must when fishing for the cusk, as it is with numerous other members of the cod family. If fishing in a heavy sea, you may need a bank sinker from ½ pound to as much as 1 pound in weight. Heavier sinkers are needed in strong currents to keep your bait on the bottom and to keep your line from tangling with those of your neighbors.

Hooks are usually baited with clams or any other bait used for groundfish. Replace the bait on the hook frequently to keep it fresh. The angler who decides it is too much trouble to reel in 200 feet of line to change bait may spend idle hours waiting for action while a washed-out piece of clam or an empty hook sits in the depths below.

Finally, the angler must remain attentive to his or her pole and line. The cusk takes bait somewhat sluggishly; thus, the angler that is too involved in conversation may miss the gentle tugging of the cusk on the baited hook. Once hooked, this species uses its powerfully muscled body to twist, turn, and pull while the angler is attempting to land it. If you don't use a swivel, this twisting may result in the line tangling into a "bird's nest." Individuals that weigh 5 to 10 or more pounds will provide a nice workout on their way to the boat and the dinner table.

Filleting the cusk can be a task due to its tough skin. To fillet, follow this simple technique: While holding the cusk by its head, poke the point of your fillet knife through the skin along the backbone behind the head and run the knife, blade side up, underneath the skin along the length of the back. Once the skin is cut, remove the fillet as you would from any other fish species.

The cusk has lean white flesh that is not as dry as cod. At one time, many anglers discarded this strange-looking "pollywog." The commercial fisherman, however, was aware of the pleasant flavor of this species and often kept a cusk for his own table rather than sending it to market. The cusk was considered to be one of the finest chowder fishes due to its moist, succulent flesh. When marketed in the past, its white-fleshed fillets joined those of the cod, haddock, and pollock under the market name of "scrod." The cusk has only recently been sold under its own name.

The cusk can be prepared using any recipe for cod or haddock. For a tasty treat, try this species in a traditional chowder. Traditional chowders are made from a variety of items available to the cook, including many species of fishes, all types of shellfishes, vegetables, spices, milk or cream, and even some tomatoes to blend flavors together. Making good chowder takes more of a "feel" than it does a particular cooking skill.

Fish stock can be used in the chowder, but water is quicker. If you want to use stock, put gutted frames, heads without gills, and tails in a piece of cheesecloth, tying it to make a pouch. Place the pouch into a pot containing 2 quarts of water and add one celery stalk, one carrot, one bay leaf, and some peppercorns. Bring to a boil and simmer for 30 minutes. Skim off and discard the foam along with the cheesecloth full of bones.

Meanwhile, sauté ¼ pound of salt pork or bacon in a large kettle or pot, add two medium chopped or sliced onions, and lightly brown. Next add 2 or more pounds of cusk fillets cut into slices or chunks and cook until they flake. Remove the fish and add your choice of the following ingredients (and whatever else seems appropriate):

3 to 5 diced or sliced boiled potatoes
3 to 4 cups milk or half-and-half, or 2 cans cream soup
1 tablespoon of butter
1 teaspoon of dill
1 teaspoon of garlic powder

¾ cup tomato paste
8 peppercorns
Pinch of basil
Pinch of oregano
Several bay leaves
⅔ cup white or red wine
Chopped parsley
1 to 2 quarts water or fish stock
Salt and pepper to taste.

Simmer for 30 minutes. Add the fish and allow to simmer for another 5 to 10 minutes. Remove from the heat and allow to sit (the longer the better), reheating before serving. Serve with oyster crackers for a wonderfully tasty and hearty meal.

Reference

Scott, W. B., and M. G. Scott. 1988. Atlantic Fishes of Canada. *Can. Bull. Fish. Aquat. Sci.* 219.

Striped Bass
(*Morone saxatilis*)

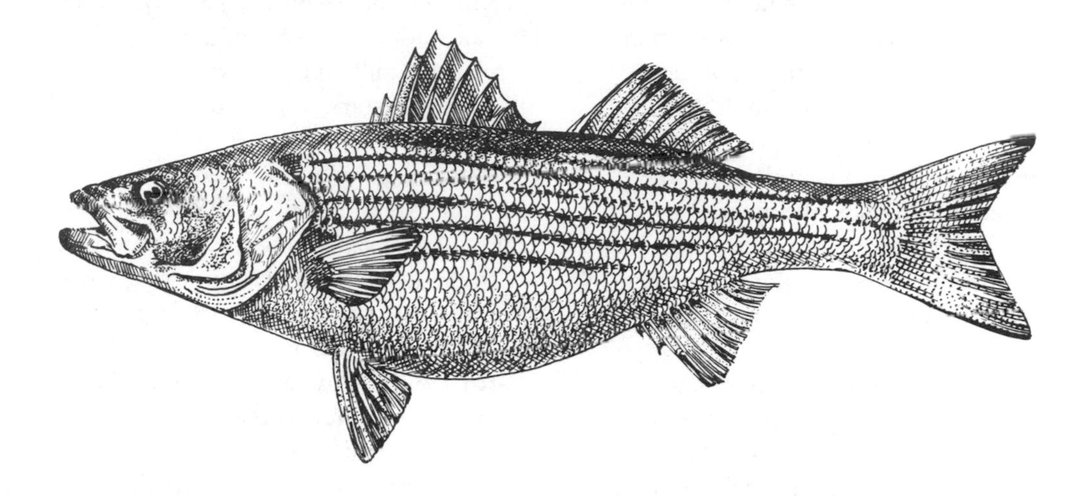

The striped bass, or striper, one of the most avidly pursued of all coastal sport fishes, is native to most of the East Coast, ranging from the lower St. Lawrence River in Canada to northern Florida and along portions of the Gulf of Mexico. The striped bass has been prized in Massachusetts since colonial times. The Massachusetts Bay Colony prohibited its use as fertilizer in 1639; shortly thereafter, this colony established the first public school in America, and financed a fund for widows and families of those who died while serving the colony, with income earned from striped bass fisheries. The unique angling characteristics of this trophy species and its adaptability to freshwater environments have led to a major North American range expansion within the last 100 years. A valuable fishery has been created on the West Coast, and inland fisheries have been developed in 31 states by stocking the striped bass in lakes and reservoirs.

Several characteristics distinguish the striper from other fishes found in coastal New England waters. The striped bass has a large mouth, with jaws extending backward to below the eye. It has two prominent spines on the gill covers. The first (most anterior) of its two well-developed and separated dorsal fins has a series of sharp, stiffened spines. The anal fin, with its three sharp spines, is about as long as the posterior dorsal fin. The striper's upper body is bluish to dark olive, and its sides and belly are silvery. Seven or eight narrow stripes extending lengthwise from the back of the head to the base of the tail are the most easily recognized characteristic of this species.

Biology

Striped bass are rarely found more than several miles from the shoreline. Anglers usually catch stripers in river mouths, in small, shallow bays and estuaries, and along rocky shorelines and sandy beaches. The striped bass is a schooling species, moving about in small groups during the first two years of life, and thereafter feeding and migrating in large schools. Only females exceeding 13.6 kg (30 lb) show any tendency to be solitary.

Schools of striped bass less than 3 years of age (sometimes called schoolies by anglers) occasionally travel far upstream into rivers such as the Hudson, Connecticut, and Merrimack. Although adult striped bass move into rivers to reproduce, fish less than 3 years of age probably make such journeys to take advantage of a river's abundant food resources, including anadromous members of the herring family (the American shad, blueback herring, and alewife).

Striped bass normally do not migrate during the first 2 years of life, remaining within their natal estuary. However, adult stripers generally migrate northward in the spring and summer months and return south in the fall. Stripers produced in the Hudson River generally do not migrate beyond Cape Cod to the north and Cape May to the south. Fish hatched in the Chesapeake Bay exhibit more extensive migrations, some being captured as far north as the Bay of Fundy in coastal Canada. Roanoke River/Albemarle Sound stripers display restricted movement within the Sound and adjacent waters. Fish from this river system contribute very little to fisheries outside of North Carolina.

Stripers are strictly spring-to-fall transients in New England. Only a few fish inhabiting coastal New England waters in the summer have been known to overwinter in the mouths of southern New England streams. Some stripers frequenting coastal New England in the summer will overwinter in the mouth of the Hudson River, and many spend the winter along the New Jersey coast or in the Delaware and Chesapeake bays.

Male striped bass reach sexual maturity at 2 or 3 years, but females do not do so until at least 4 to 6 years of age. The number of eggs produced by a female is proportional to its size; a 1.8-kg (4-lb) female produces about 426,000, a 5.5-kg (12-lb) female about 850,000, and a 25-kg (55-lb) female 4.2 million eggs. Evidence suggests that at least some large, older females do not spawn every year.

Stripers reproduce in rivers and the brackish areas of estuaries. Spawning takes place from spring to early summer, with the greatest activity occurring when the water warms to about 18 degrees C (65 F). The eggs drift in currents until they hatch 1½ to 3 days after being fertilized. Because newly hatched larvae are nearly helpless, striped bass suffer their highest rate of natural mortality during the several weeks after hatching.

Although stripers at one time might have spawned in most East Coast watersheds, the major spawning activity for the East Coast fishery now occurs in the Hudson River, the Chesapeake Bay, and the Roanoke River/Albemarle Sound watershed.

Striped bass can live up to 40 years and can reach weights greater than 45 kg (about 100 lb), although individuals larger than 22.7 kg (50 lb) are rare. After reaching sexual maturity, females grow more rapidly, and weigh more at a given length, than males. Thus, females reach significantly greater sizes than males; most stripers over 13.6 kg (30 lb) are female. The term "bulls," originally coined to describe extremely large individuals, has been more accurately changed to "cows" in recent times. Figure 10 illustrates lengths at age of female striped bass.

Larval striped bass feed upon small, mobile zooplankton. Adults eat a variety of foods, including fishes such as alewives, flounders, sea herring, menhaden, mummichogs, sand lances, silver hakes, tom-

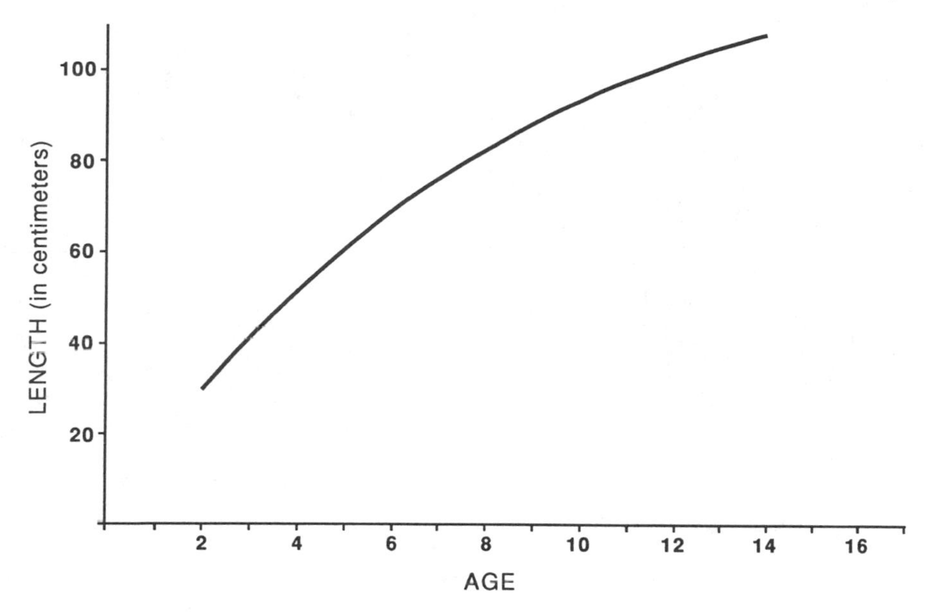

Figure 10. Calculated growth curve for female striped bass from the Chesapeake Bay stock (curve calculated from mean length-at-age data for bass captured in the Potomac River, as presented in ASMFC 1987).

cod, smelt, silversides, and eels, as well as lobsters, crabs, soft clams, small mussels, polychaetes, and squid. They feed most actively at dusk and dawn, although some feeding occurs throughout the day. During midsummer they tend to become more nocturnal. Stripers are particularly active in areas with tidal and current flows and in the wash of breaking waves along the shore, where fishes, crabs, and clams become easy prey as they are tossed about in the turbulent water.

Status and Management

Striped bass populations have a history of periods of abundance interspersed with periods of scarcity. A major coastwide reduction in abundance occurred at the end of the 19th century. No catches of stripers were reported north of Boston for 30 years after 1897. Populations had recovered somewhat by 1921, and an unusually successful year of reproduction in 1934 was followed by 6 years of markedly increased abundance. Great numbers of juveniles were recorded in New England waters in the mid 1940s, and high numbers of increasingly larger individuals followed for several years. This suggests that striped bass populations are dominated for extended periods by fish hatched during occasional years of unusually successful reproduction. Also, a year of successful reproduction is often followed by a series of years when spawning fails or is so limited in success that relatively few new fishes enter the population. Striped bass have typically been most abundant in the New England and Middle Atlantic states following years when reproduction in the Chesapeake Bay has been particularly successful, which suggests that much of the East Coast is dependent upon the success of spawning in that one watershed.

The last peak year of reproductive suc-

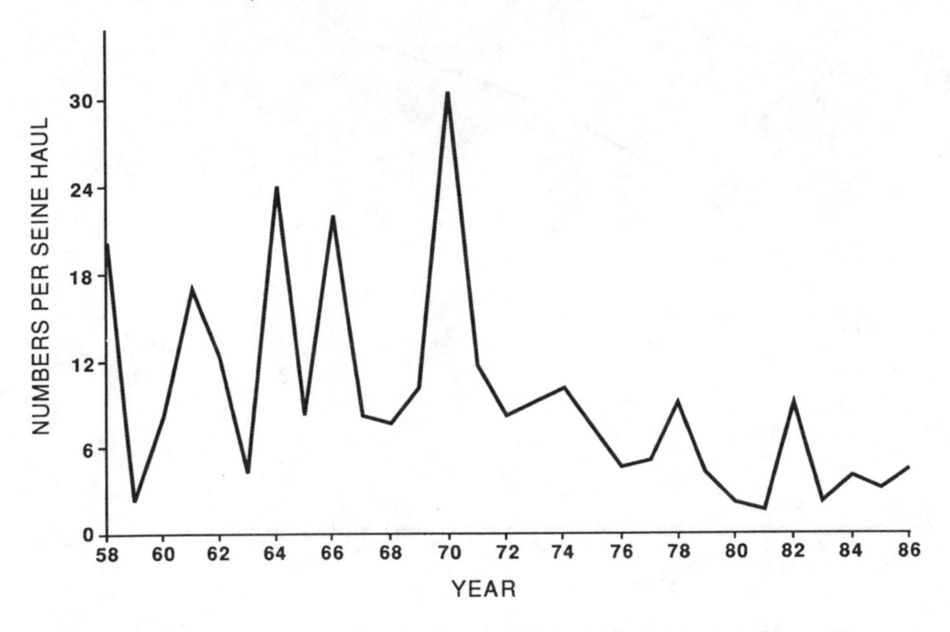

Figure 11. Relative abundance of juvenile striped bass in Maryland waters of the Chesapeake Bay. Catch-per-effort values are calculated from the number of juveniles captured per beach-seine haul during summer and fall survey sampling (ASMFC 1987).

cess in the Chesapeake Bay was 1970 (Figure 11). Levels of reproduction in the Chesapeake Bay have been consistently low since then; 1982 was the only year in the 1980s before 1989 in which even modest numbers of juveniles were produced. When the dominant 1970 year class grew large enough to enter the coastal fishery, nearly 90% of the harvest came from the Chesapeake Bay stock. From 1974 to 1976, 55% to 75% of all spawners in the Potomac River came from the 1970 year class; by 1982, females hatched in 1970 or earlier still outnumbered younger fish in the spawning population, even though the older stripers had experienced more than a decade of natural and fishing mortality. Likewise, the 1965 year class provided 40% to 50% of the Hudson River fishery in following years, whereas the 1966, 1968, and 1969 year classes contributed less than 10% once they grew to harvestable size. However, reproductive success in 1978 and the early 1980s was high in the Hudson. This, coupled with the collapse of the Chesapeake Bay stock, resulted in a much greater contribution of Hudson River fishes to the harvest in New England waters (Table 20).

The recent prolonged period of reproductive failure in the Chesapeake Bay occurred concurrently with a steady decline in striped bass abundance. This decline is felt to have been caused by a combination of factors, including the presence of a variety of pollutants in spawning grounds, fishing pressure, and feeding and nutritional problems of larvae. This decline was reflected throughout the 1970s and 1980s in decreasing success by both the commercial fishery and anglers. Coastwide commercial landings dropped from 79% to 93% between 1973 and 1983, although this was due in part to fishing re-

Table 20. Percentage of the harvest in southern New England states represented by Hudson River bass and more southern stocks, 1984–1985

State and year	*Hudson River fishes* (%)	*Southern stocks* (%)
Massachusetts		
Fall 1984	30	70
Fall 1985	54	46
Rhode Island		
Fall 1984	15	85
Fall 1985	14	86
Connecticut		
Fall 1984	23	77
Fall 1985	52	48

Source: ASMFC 1987.

strictions (closure of the Hudson River and James River fisheries in 1976 due to excessive PCB and kepone contamination, respectively). Commercial landings in 1986 totaled 136 mt (300,000 lb), the lowest harvest on record. The estimated catch by anglers from the Gulf of Maine to the Middle Atlantic region fell from 3,000 mt (6.6 million lb), or about 2 million stripers, in 1979 to 773 mt (1.7 million lb), or 600,000 stripers, in 1985.

A rapidly changing management plan developed in response to the severely depleted status of the striped bass. Prior to the mid 1970s, management of striped bass was carried out more or less independently by each coastal state. In 1979, Congress amended the Anadromous Fish Conservation Act to create the Emergency Striped Bass Study Program. In 1981, the Atlantic States Marine Fisheries Commission (ASMFC) adopted a coastwide management plan, to be acted upon by each coastal state. This plan recommended minimum size limits for fishes caught in nursery rivers and in coastal areas and restricted fishing on spawning grounds during the spawning season. In 1984, Congress passed the Atlantic Striped Bass Conservation Act, from that time on requiring all coastal states to abide by provisions of the ASMFC management plan. In response to constantly dwindling numbers of stripers on the East Coast, the plan was amended in 1984 to reduce the number of fishes harvested by an additional 50% and further amended in 1985 to initiate protection of females from the 1982 year class until they had spawned at least once. This last amendment established a minimum size limit that was increased each year through 1989, in order to protect this 1982 year class from harvest while allowing harvest of the less numerous, older fish. By 1989, regulations included minimum legal size limits, creel limits, closed seasons, gear restrictions (hook and line only), and no-sale provisions. The protection of the 1982 year class led to a measurable increase in abundance of, and high reproductive success by, the Chesapeake Bay stock by 1989. Because of this, in October of that year, the Atlantic States Marine Fisheries Commission chose to modify the Striped Bass Fishery Management Plan to allow a controlled harvest of that age group of fish.

Angling and Handling Tips

Many anglers consider the striped bass to be the premier New England coastal game species. Anglers hike miles along sandy shores to reach the best spots of rough surf, where they then spend hours waist-deep in water fishing for this species. Although some fishes are larger, more beautiful, or harder to catch, the striper is highly prized because few species rank so well in all of these categories.

This species can be found throughout southern New England's coastline from May to November. Angling at dusk or dawn provides the greatest success during most of the season, but night fishing is often best during the midsummer doldrums. Anglers are most successful when fishing the shoreline in areas with drop-offs, holes, sandbars, cuts, or gullies, where tidal rips, strong currents, or wave action create turbulent, "live" water. Surprisingly, many stripers are taken in knee-deep surf.

Use an 8–10-foot surf rod and conventional reel spooled with 30-pound test line to ply the beaches with swimming plugs and live eels. However, a medium to heavy spinning rod with 12–20-pound test line is considered ideal by many anglers for plugging, jigging, or offering bottom-fished baits. Lures are attached directly to the line with a snap swivel. When bait-fishing, the preferred rig consists of a pyramid sinker attached as a fish-finder and a long leader with a brightly colored float attached close to the hook. The float keeps the bait away from bottom-dwelling crabs and skates.

Though artificial lures produce frequent strikes, live and natural baits prove most successful when this species is actively feeding. Live squid, chunks of menhaden, mackerel, or herring, and crabs are all favored baits, but the most successful might be an eel fished on an evening tide. Striped bass of all sizes, including large cows, will even strike a hook baited with sea worms that is drifted over the bottom.

When trolling for bass adjacent to shoreline areas, the rod should be equipped with Carboloy guides to prevent line wear and a high-ratio conventional reel. By choosing among monofilament, lead-core, or wire lines, depths from the surface to the bottom can be trolled. Many lures, including swimming plugs, jigs, tubes, and umbrella rigs, as well as live herring and menhaden, lend themselves to trolling.

Large bass can be steaked and either broiled or grilled. Medium-sized stripers are traditionally stuffed and baked. Split a whole striped bass and stuff it with slices of bacon, onions, tomatoes, green peppers, parsley, apples, and even cranberries. Spice with salt, freshly ground pepper, and a little tarragon and sliced garlic if you choose. Dot the fish with butter or rub with olive oil and lightly flour. Place in a foiled baking dish, add 1 cup of mild red wine, and bake at 400 degrees F until the flesh flakes—25 to no more than 30 minutes, depending upon the size of the fish. Baste and serve with tomato sauce mixed into the cooking liquid.

References

ASMFC. 1987, April. *Interstate fisheries management plan for the striped bass of the Atlantic coast from Maine to North Carolina.* Atlantic States Marine Fisheries Commission Spec. Rep. No. 7: 240 pp.

Boreman, J., and H. M Austin. 1985. Production and harvest of anadromous striped bass stocks along the Atlantic coast. *Trans. Am. Fish. Soc.* 114:3–7.

Boreman, J., and R. R. Lewis. 1987. Atlantic coast migration of striped bass. *Am. Fish. Soc. Symp.* 1:331–339.

Rago, P. J., R. M. Dorazio, R. A. Richards, and D. G. Deuel. 1989. *Emergency striped bass research study: Report for 1987.* Unpublished Rep. of the U.S. Departments of Interior and Commerce to the U.S. Congress.

Sport Fisheries Institute. 1988, May. *Striped bass status report.* Bull. No. 394:7.

White Perch

(*Morone americana*)

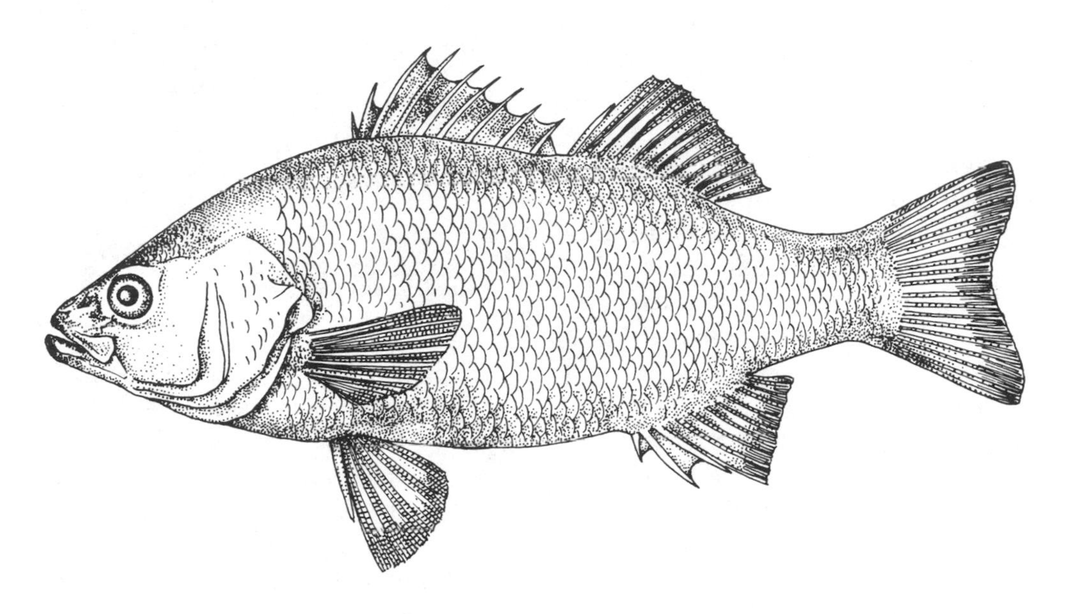

The white perch, a little fish that can reach spectacular levels of local abundance, occurs from New Brunswick and Nova Scotia southward to South Carolina. This species is most abundant in the Middle Atlantic states and occurs in relatively few estuaries north of Cape Cod. Coastal populations are too sparse to have commercial or recreational importance in the Gulf of Maine. However, white perch have been introduced extensively into lakes and ponds in northern New England.

Several traits easily distinguish the white perch from all recreational species of New England other than the striped bass. The white perch has two dorsal fins, the anterior fin having a series of stout, sharp spines. It has a stout spine on the anterior margin of the pelvic fin and a spine near the posterior margin of the gill cover. The white perch differs from the striped bass by having (1) dorsal fins that are continuous rather than separated by a space; (2) second and third anal spines that are nearly equal in length rather than of markedly different lengths; and (3) no prominent lateral stripes.

The white perch is variously colored, with individuals displaying silvery greenish, metallic gray, or nearly black color on the upper part of the body. Large individuals may have a bluish luster on the head. The sides are usually a pale silver, and the belly is silvery white. The underside of the mandible, or chin area, usually has a bluish or pinkish iridescent tinge.

Biology

Coastal populations of white perch exhibit modest seasonal migratory patterns, spending most of the year in estuarine waters and migrating into brackish or freshwater habitats to reproduce. White perch prefer compacted, fine-particled substrates, such as clay, mud, and sand. They often exhibit daily localized migrations, moving shoreward toward shallow water at night and offshore to somewhat deeper water in the day. Coastal populations may overwinter at depths up to 40 m (about 130 ft), occurring most frequently in 12 to 18 m (about 40 to 60 ft) of water.

Many individuals of both sexes become sexually mature at age 2, although some may not do so until age 3 or 4. Fecundity increases with size, with extremes of 5,210 and 321,000 eggs being reported.

Rising water temperatures in the spring stimulate spawning. New England populations begin reproducing in late March to early April. Within a region, estuarine populations spawn somewhat later in the spring than do freshwater populations. Spawning may last several weeks at specific sites. Individual females may release eggs during two or three different periods within a spawning season. After spawning, adults return to deeper, more open waters of the estuary.

Eggs adhere to the substrate after they are released and fertilized. They hatch in about 30 hours at 20 degrees C (68 F), and as much as 96 to 108 hours at 15 degrees (59 F). Newly hatched larvae remain in the spawning area for up to 13 days, then slowly drift downcurrent into estuaries, where inshore areas serve as nursery grounds. Juveniles may stay in these areas for up to 1 year, although some researchers report large schools of young juveniles venturing into more open water during the day and moving inshore at night. After the first year of life, young perch tend to use the same habitats as adults.

White perch grow slowly and tend to stunt (exhibit notably small sizes for each age) in high-density populations. Most larger adults are 20 to 25 cm (8 to 10 in) in length and weigh less than 0.5 kg (1 lb). Maximum size appears to be about 38 cm (15 in) in length and 1 kg (2 lb). Average lengths at age for white perch from the Parker River estuary, Massachusetts, are presented in Table 21. Maximum age in the Parker River population was 8 years for males and 6 years for females. A 17-year-old has been reported from a landlocked population in Maine.

Table 21. Length at age in cm (inches in parentheses) for white perch from the Parker River estuary, Massachusetts

Age	*Length*
1	9.1 (3.6)
2	16.6 (6.5)
3	20.4 (8.0)
4	22.4 (8.8)
5	24.6 (9.7)
6	26.0 (10.2)
7	26.8 (10.6)
8	27.6 (10.9)

Source: Data from Mosher 1976.

White perch feed upon microscopic zooplankton as fry and upon progressively larger prey as the perch grow. Adult white perch eat shrimp, squid, crabs, fish eggs, and fishes. Individuals 23 cm (9 in) long or larger depend upon small fishes such as mummichogs, pipefish, silversides, smelt, eels, and river herring as the major component of their diet.

White perch are eaten by a variety of fish and bird predators.

Status and Management

Significant commercial harvests occur from Massachusetts to North Carolina, with commercial fishing centering around the Chesapeake Bay. Total East Coast commercial landings ranged from 272 to 1,412 mt (598,000 to 3.1 million lb) between 1960 to 1980, although landings in the 1970s averaged only about two thirds of those in the 1960s. Recreational harvest was an average of 5.6 times greater than commercial harvest between 1960 and 1980 (based upon comparative recreational and commercial data presented in Stanley and Danie 1983). Recreational landings are centered in the Middle Atlantic states, with New England landings composing only a modest proportion of the total harvest.

Various studies have calculated that combined effects of fishing and natural mortality result in the loss of from 49% to 59% of adult males and 53% to 65% of adult females annually. The higher mortality rate for females is probably due to their faster rate of growth; they reach a desirable size for harvest at a somewhat earlier average age than males.

Angling and Handling Tips

This panfish has engendered an avid following among anglers because of its availability and its fine flavor. The white perch can be caught throughout the year, whether in the heat of summer or through winter's ice. This scrappy species is available to most anglers, including youngsters fishing from shore, at very little cost. Anglers can fill a bucket or stringer in a short period of time with little effort. Additionally, coastal and estuarine perch reach excellent size for a panfish species.

Early spring is a fine time to fish for white perch. Large schools gather at this time of year to spawn in estuaries or in upstream fresh water. You should fish in areas of estuaries near outlets from streams or ponds to find these schools. Fishing during rising tides can be particularly productive. Trolling or chumming with crushed clams or mussels in a mesh bag may attract a school of hungry perch.

Once a school is found, the action can be furious, a strike occurring with each cast of lure or bait. Many anglers feel the best perch in a school are found closest to the bottom; thus, the closer the bait is fished to the substrate, the better the catch. The frenzy of strikes in an actively feeding school can frustrate anglers hoping to work their bait down toward the larger fishes.

Anglers prefer either ultralight or light/medium-action spinning outfits with 4–8-pound test line to fish for white perch. If bait-fishing, enough weight must be used to hold the bait near the bottom. In fast currents this may require up to a 1-ounce sinker. Rigs vary from a single- or double-hook setup to a fish-finder. To rig for double or single hooks, tie a bank sinker to the end of the line, add an 8-inch leader with a No. 6 hook about 15 inches from the sinker and a second above the first if desired. The fish-finder employs a No. 6 hook on the end of the line, an egg sinker through which the line passes, and a split shot that prevents the sinker from slipping over the hook. The fish-finder prevents the fish from feeling the sinker initially and allows the angler to feel the slight tugging that indicates a fish has picked up the bait.

Bait is most advantageous early in the year when cold water reduces white perch activity, but it can be effective in all seasons. Sea worms, bloodworms, grass shrimp, clam necks, and night crawlers are effective baits, but live killifish or other bait fishes may be best. Hook live bait fishes through the upper back, behind the dorsal fin.

Jigging and casting lures can also be productive. Small spoons, spinners, miniature Kastmasters, and Hopkins lures up to ½ ounce are favored by anglers. Tipping the lure's hooks with a bit of worm, clam, or squid can increase the catch. Let lures sink to the bottom and retrieve them slowly in short pulls, using a light hand to set the hook. White perch often strike behind the hook on the lure. Placing the eyelet of a second hook over the barb of the lure's hook may increase the catch.

Many anglers rate the white perch among the best-tasting of any catch. Its white, firm flesh is beautifully suited for the pan or grill. Try cooking your cleaned and scaled fish whole on the grill. Apply a light coating of mayonnaise to prevent sticking and grill about 10 minutes per inch of thickness, turning the fish once. A light brushing with barbecue sauce after turning will add spice to the flavor.

White perch are also delicious sautéed. After dressing, scaling, and removing the head, coat with a light layer of commercial Cajun spices, and dust very lightly with flour to limit spattering. Heat an iron skillet; then add some olive oil and swirl it around to coat the pan. Sauté the perch at 350 degrees F for 6 minutes on the first side and 5 minutes on the other. After removing the fish from the skillet, add a little white wine or water and heat it slowly to use as a sauce. Serve with corn bread and salad.

You might also enjoy oven-frying a batch of filleted perch. Dip the fillets in beaten egg or milk and shake in a bag of flavored bread crumbs. Bake for 5 minutes per side at 350 degrees F in a glass dish or pan that has been rubbed with olive oil. Serve with a slice of lime.

References

Moser, T. D. 1976. *Comparison of freshwater pond and estuarine populations of the white perch* Morone americana *(Gmelin), in the Parker River, Massachusetts.* Unpublished master's thesis, University of Massachusetts, Amherst.

Stanley, J. G., and D. S. Danie. 1983. *Species profiles: life histories and environmental requirements of coastal fishes and invertebrates (North Atlantic)—white perch.* U.S. Fish Wildl. Serv. Biol. Rep. 82(11.7). U.S. Army Corps of Engineers, TR EL-82-4. 12 pp.

Black Sea Bass

(*Centropristis striata*)

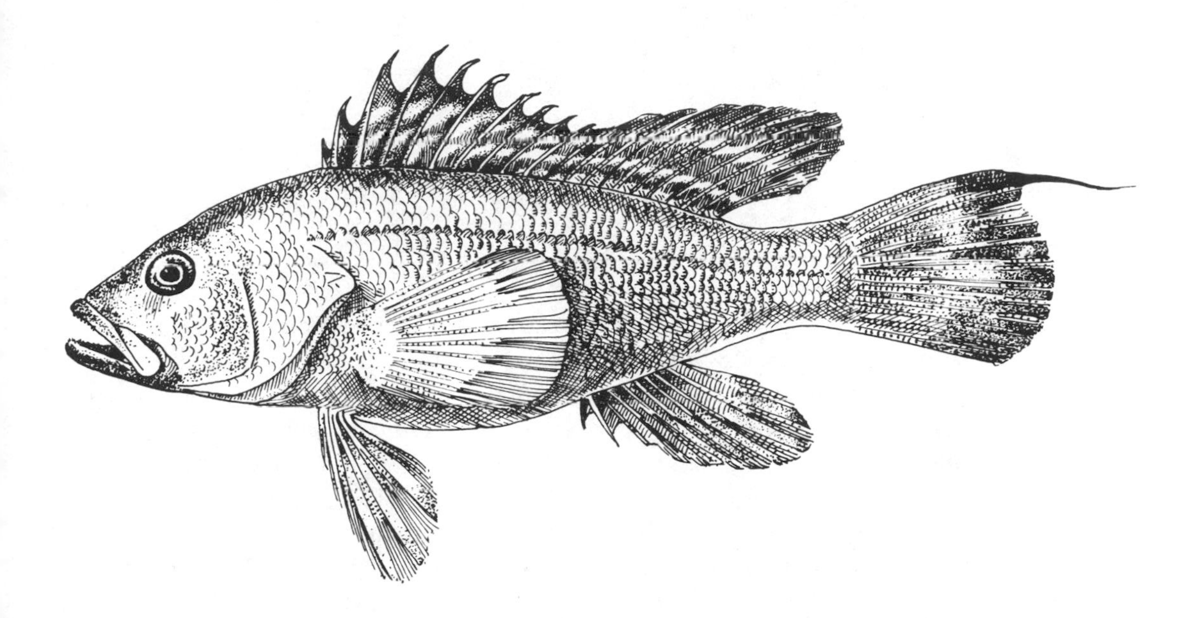

The black sea bass occurs along the Atlantic coast of the United States from Cape Cod to Florida, reaching greatest abundance between the capes of New Jersey and North Carolina. This species forms two distinct stocks, or populations, along the Atlantic coast, one north and one south of Cape Hatteras (NC). The northern stock does not occur in the Gulf of Maine, but it is an important groundfish west and south of Cape Cod.

The black sea bass is a fairly stout-bodied fish, with a long dorsal fin and large pectoral and pelvic fins. The rounded tail sometimes has a long streamer trailing out from the top edge. Each gill cover has a flat spine near the outer edge. Mature males have a fleshy dorsal hump just in front of the dorsal fin.

The background color of the black sea bass (smoky gray, brown, or bluish black) is mottled with darker patches and light speckles. The belly is only slightly lighter than the sides. The dorsal fin is marked with whitish mottling; all the other fins have dark spots. Young sea bass are green or brown with a dark lateral stripe running from the head to the tail.

Biology

Black sea bass generally overwinter at depths of 73 to 183 m (240 to 600 ft), with individuals inhabiting deeper waters in the New Jersey–New York region than in the Middle Atlantic region to the south. Larger adults generally occur in deeper waters than do smaller ones. Few fish are found north of Cape May (NJ) in the winter, although some are known to travel extensively between Nantucket Shoals and Cape Hatteras at depths of nearly 335 m (1,100 ft). In the spring this species displays a general northward and inshore movement, expanding its range as far north as Cape Cod from May to October. Adults migrate to coastal spawning areas and juveniles to estuarine nursery grounds. During the summer, adult sea bass gather around rocky bottoms, sunken wrecks, old pilings, and wharves. At this time of year they are most abundant at depths of less than 36.6 m (120 ft). Young-of-the-year and yearling bass tend to summer in estuaries, which are critically important nursery grounds for this species. In southern New England, young-of-the-year bass start to enter estuaries in August and move offshore to depths of 55 to 110 m (180 to 360 ft) during the fall. The largest adults in southern New England tend to begin their annual offshore and southerly migration as early as August, whereas juveniles and smaller adults migrate later in the fall.

Although the black sea bass can tolerate temperatures as low as 6 degrees C (43 F), it is more abundant at temperatures of 9 degrees C (48 F) or above.

The black sea bass has an unusual life cycle: Most individuals are hermaphroditic, reproducing as both female and male at some time in their lives. Although some are males from the time they reach sexual maturity, most produce eggs when they first mature. At some subsequent point, the ovary tissues in these fish become nonfunctional, and at the same time testes commence production of sperm. The age at which individuals undergo the sex change is variable, although many do so between 2 and 5 years of age, and most by the time they are 8 years old. In heavily exploited populations in which larger, older males are selectively harvested, the resulting dearth of males causes females to change sex at a younger age and smaller size than would be the case in populations less depleted by fishing. The effects of reduced abundance of males and reduced average size of females on the reproductive capacity of sea bass populations are not yet understood.

Table 22. Standard length at age in cm (inches in parentheses) of black sea bass

Age	*Males*	*Females*
1	8.6 (3.4)	8.0 (3.1)
2	17.2 (6.8)	16.7 (6.6)
3	21.7 (8.5)	20.9 (8.2)
4	24.5 (9.6)	23.8 (9.4)
5	28.2 (11.1)	26.4 (10.4)
6	32.0 (12.6)	28.7 (11.3)
7	34.6 (13.6)	28.6 (11.3)
8	39.2 (15.4)	30.9 (12.2)
9	40.5 (15.9)	

Source: Data from Alexander 1981.

Black sea bass reproduce from February to July, with the spawning season starting earliest in the southern portion of their range and progressing northward as spring passes. The center of reproduction for the northern stock is between Chesapeake Bay and Montauk in water 18 to 45 m (59 to 148 ft) deep. Off the southern New England coast, they reproduce from mid May until the end of June. The eggs are buoyant, floating in the water column until they hatch 1½ to 5 days after fertilization. The larvae drift in bays, inlets, and offshore areas; they become bottom dwellers when they have grown to about 1.3 cm (0.5 in) in length.

Most black sea bass do not exceed 0.7 kg (1.5 lb). A 30.5-cm (12-in) fish generally weighs 0.5 kg (about 1 lb); a 46–51-cm (18–20-in) fish weighs about 1.4 kg (3 lb). Males and females of equal age are of similar lengths until about age 4, after which males tend to be progressively larger with increasing age. Evidently, sex reversal from female to male is followed by a period of accelerated growth. Table 22 shows lengths at age for black sea bass captured in the New York Bight.

Juvenile black sea bass feed upon a variety of benthic (bottom-dwelling) invertebrates such as shrimp, isopods, and amphipods, with mysid shrimp commonly constituting more than half their food intake. Adults feed heavily upon rock and hermit crabs, as well as other crustaceans, mollusks, squid, and fishes. Adults also occasionally graze upon attached organisms such as barnacles and colonial tunicates.

Status and Management

The abundance of black sea bass along the East Coast has been declining for more than 3 decades. Furthermore, the average size of fish harvested both commercially and recreationally has been decreasing since about 1950, an indication that larger, older individuals have become increasingly scarce.

Commercial harvests have been based upon otter-trawl and wooden-pot (similar to a lobster trap) fisheries. Annual landings from trawlers are typically greatest from September to March, when black sea bass are distributed more offshore than in summer months. The total annual catch from trawl fisheries peaked in the early 1950s. Yearly landings for the entire Middle Atlantic region peaked at nearly 21 million pounds in 1952 and then plummeted to less than 5% of that level by 1970.

Pots are fished from spring through late fall, enabling fishermen to harvest sea bass in areas where rugged underwater terrain makes trawling ineffective. Pot fisheries developed rapidly along much of the East Coast in the 1950s. After an initial period of annual increases in harvest, pot fisheries have usually declined fairly rapidly even when the number of pots fished continued to increase. Data from the New York State pot fishery clearly exhibit this trend. In 1961, a total of 400 pots harvested 77,000 pounds of black sea bass in New York's waters. In 1965, a fivefold increase in the number of pots fished produced only 2.4 times as great a catch. By 1967, more than 40 times as many pots produced a catch that was 17% smaller than in 1961, and by 1971 catch levels were 37% lower than in 1961. The pot fishery in New England, particularly southern Massachusetts, developed more recently than in more southerly waters but has been expanding rapidly over the last 5 years, especially in Nantucket Sound. This expansion has caused increasing concern about issues of overharvest and gear conflict.

The recreational fishery has added to the decline in population abundance along the Atlantic coast. In 1965, over half of the total catch of black sea bass was credited to recreational fishing. One survey indicated that, by 1970, the recreational catch was at least several times as great as the commercial harvest. Angling pressure has increased markedly in recent years. In the North Atlantic region, including Cape Cod, recreational harvest increased nearly 500% between 1981 and 1986. Over the same time period, recreational harvest increased about 1,400% in the Middle Atlantic region. From Cape Hatteras to Nova Scotia, harvest increased from 8.1 million to 31.2 million bass between 1985 and 1986; in 1986, the black sea bass ranked second to the scup in numbers harvested recreationally in that geographic area. As with the commercial fisheries, the average size of fish caught by anglers has decreased in recent years. The average-size black sea bass caught in 1986 was 22.9 to 25.4 cm (9 to 10 in) in length. Connecticut and Massachusetts managed black sea bass harvest by use of a minimum legal size limit in 1989.

Angling and Handling Tips

When anglers hear someone mention "bass," they often think first of the striper or freshwater black bass. Yet the black sea bass provides a fight and has a flavor that compares favorably to any of the more popular marine or freshwater basses. Long known by commercial fishermen as rock bass, this species also carries the somewhat endearing name of "old humpback," due to the enlarged area above the head of many stocky, old males; this hump yields wonderfully thick fillets.

The best time to fish for black sea bass is from May to July, when they are closest to shore, although they are generally available from spring to autumn around any structures such as boulders, wrecks, jetties, and piers. They can be found from near the shore to depths of about 120 feet, with the largest males more abundant in the deeper waters within this depth range.

The black sea bass is predominantly a bottom feeder, although it will occasionally strike at plugs, jigs, or lures. Thus, bait fishing with strips of squid or fish is the most productive method. The most commonly caught individuals weigh from half a pound to 2 pounds. Bass in this size range will provide a wonderful battle if you use a medium-weight spinning outfit with 8-pound test line. Although a sea bass has a large mouth, use a No. 2 bait-holder hook tied above a small sinker; this fish normally hesitates to grab bait on large cod hooks.

The firm, white flesh of this species is a favorite of many. Bass are easy to fillet, especially when chilled, and yield a thick slice of meat. Some fillets are thick enough to slice lengthwise or to cut into nuggets for frying. Larger bass can be cut into steaks and cooked like striped bass. In restaurants, the black sea bass is often offered as "squirrel fish," and Chinese restaurants serve delicious whole deep-fried bass as "Hunan fish."

Try broiling black sea bass fillets. When broiling, fold under the thin section of the tail area or overlap the tail sections of two fillets, to allow more even cooking. Place the fish in a greased pan, sprinkle with freshly ground pepper and paprika, and dot with butter or olive oil. Broil 5 to 6 minutes on each side, depending upon thickness, until the fillets are golden brown. Be careful not to overcook, as the fillets will dry out. Serve your broiled sea bass with cole slaw.

References

Alexander, M. S. 1981. *Population response of the sequential hermaphrodite black sea bass,* Centropristis striata, *to fishing*. Unpublished master's thesis, SUNY at Stony Brook.

Musick, J. A., and L. P. Mercer. 1977. Seasonal distribution of black sea bass, *Centropristis striata,* in the Mid-Atlantic Bight with comments on the ecology and fisheries of the species. *Trans. Am. Fish. Soc.* 106:12–25.

Bluefish

(*Pomatomus saltatrix*)

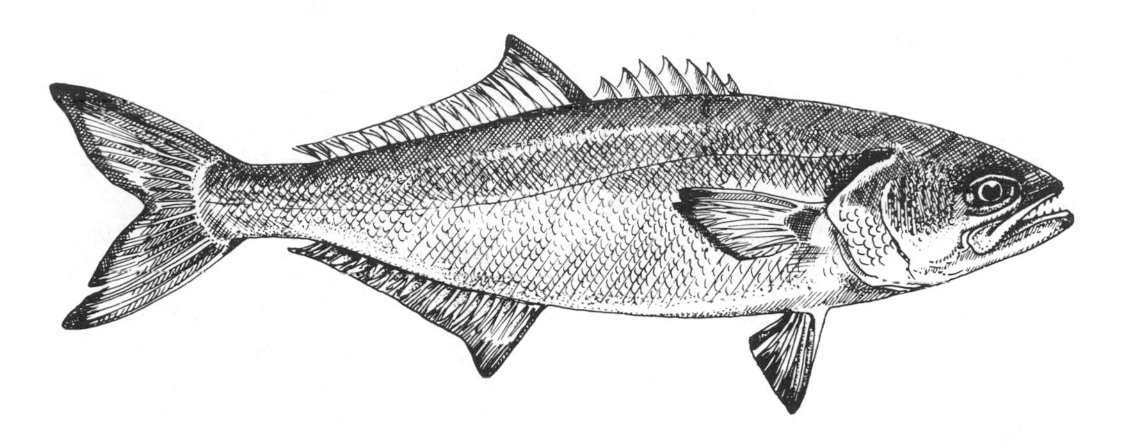

The bluefish (also called the blue, tailor, snap mackerel, chopper, and, as a juvenile, the snapper) is a trophy species hotly pursued by anglers due to its reputation as a champion battler and a voracious predator. This species occurs in temperate-zone continental shelf waters of the Atlantic, Pacific, and Indian oceans. Along the western Atlantic it is abundant from Argentina to Cape Cod, and it occasionally occurs as far north as Nova Scotia. Adults summer throughout coastal waters of New England, whereas juveniles are more abundant south of the Maine coastline.

The name "bluefish" is something of a misnomer, as this species is most commonly a sea-green color above, fading to a silvery shade on its lower sides and belly. The adult bluefish has a stout body and a large mouth that extends posteriorly below and beyond the eye. The lower jaw juts out noticeably. Both the upper and lower jaws are fully armed with large, conically shaped canine teeth. The dorsal fin is divided into two sections. The first section, about half as long and high as the second, has a series of stiff spines supporting the soft tissues of the fin. The second or posterior dorsal fin is equal in length to the anal fin.

Biology

The bluefish inhabits both inshore and offshore areas of coastal regions, with young-of-the-year fish, called snappers, often frequenting estuaries and river mouths. This species normally travels in large schools of up to several thousand similar-sized individuals. One unusually large school sighted in Narragansett Bay in 1901 was estimated to be spread over a 4–5-mile distance.

The bluefish displays an annual migration pattern that is keyed to the seasonal warming and cooling of coastal waters. During summer months, blues concentrate between Cape Cod and Chesapeake Bay and along the northern North Carolina coast. Larger, older individuals tend to migrate farther northward than younger ones at this time. They begin arriving along the southern New England coast during April and May, with substantial numbers typically not arriving before Memorial Day. Smaller bluefish generally arrive first in New England waters, moving into harbors and estuaries. Larger individuals arrive somewhat later in the spring, initially inhabiting deeper waters but moving progressively shoreward into shallow areas as the summer progresses. Adults largely disappear from the coastal waters of southern New England during October as water temperatures cool to 15.6 degrees C (60 F). Many concentrate in deeper waters as far south as the lower Atlantic coast of Florida. Some Atlantic coast individuals may occasionally stray far southward during the winter; one tagged off the coast of New York was recaptured 3 years later in January off the coast of Cuba. Along with a southward movement in the fall, a major migration to offshore deeper waters of the continental shelf is displayed.

Both male and female bluefish reach sexual maturity by the time they are 2 years old. Fecundity ranges from about 900,000 eggs for a 53.3-cm (21-in) female to nearly 1.4 million for a 58-cm (23-in) female.

Bluefish utilize two major spawning areas on the East Coast, the first from southern Florida to North Carolina in the spring, and the second in the Middle Atlantic Bight to southern New England in the summer. Spawning occurs primarily offshore over the continental shelf when water temperatures warm to 18 to 25.5 degrees C (64 to 78 F). Hatching occurs in about 48 hours at 20 to 21 degrees C (68 to 70 F). The greatest abundances of eggs and larvae are found on the outer continental shelf. After hatching, larvae inhabit surface waters and are swept along the continental shelf by prevailing currents. Survival of young blues is highly variable from year to year, being strongly influenced by the circulation patterns of currents on the continental shelf. If they are moved shoreward to suitable habitats, many survive; if they are swept off the continental shelf, higher mortality may result. Young-of-the-year fish produced by spring spawning drift northward along the continental shelf and migrate shoreward by midsummer. Young-of-the-year blues from summer spawning remain offshore on the continental shelf throughout the summer.

Table 23. Mean total length at age in cm (inches in parentheses) of bluefish collected in New England waters

Age	*Connecticut*	*New Hampshire*
1	28.4 (11.2)	26.4 (10.4)
2	45.7 (18.0)	46.7 (18.4)
3	56.6 (22.3)	61.2 (24.1)
4	66.3 (26.1)	70.9 (27.9)
5	74.9 (29.5)	77.7 (30.6)
6	81.0 (31.9)	83.1 (32.7)
7	84.6 (33.3)	86.4 (34.0)
8	87.9 (34.6)	89.4 (35.2)

Source: Data from *Draft Fishery Management Plan,* Mid-Atlantic Fishery Management Council. 1988.

Bluefish rarely exceed 9 kg (20 lb) and 100 cm (about 40 in) in length. Although the North American hook-and-line record bluefish weighed 14.4 kg (31.7 lb), the largest fish caught during a given year generally run between 4.5 and 6.8 kg (10 to 15 lb). A 46-cm (18-in) bluefish weighs about 1.4 kg (3 lb), a 69-cm (27-in) about 4.1 kg (9 lb), and an 81-cm (32-in) about 6.8 kg (15 lb). Table 23 shows lengths at age of bluefish collected in New England waters.

Snappers eat a variety of small-bodied animals, such as polychaetes, shrimp, and other small crustaceans, small mollusks, and small fishes such as the bay anchovy, killifish, and Atlantic silversides. Adult bluefish are opportunistic feeders, commonly focusing upon schooling species such as menhaden, squid, sand eels, herring, mackerel, and alewives, as well as scup, butterfish, and cunners.

Bluefish generally feed in schools, actively pursuing prey in tidal rips or in inshore shallows where food is easier to catch. The feeding behavior of this species is legendary. Bluefish dash wildly about within schools of prey species, biting, crippling, and killing numerous small fishes, most of which subsequently are eaten. They frequently drive schools of prey species into shallow inshore areas where it is easier to cripple or catch individuals trying to escape. Occasionally, during particularly frenzied feeding activity, schooling fishes such as the menhaden will literally be driven to shore, leaving a number of individuals beached along the wave line. Although this occurs relatively infrequently, an occasional beach littered with dead fish has given rise to the bluefish's exaggerated reputation as a vicious predator. This reputation actually has a long history. S. F. Baird, the first U.S. fish commissioner, wrote in 1871 that bluefish have a "voracity and bloodthirstiness which, perhaps, has no parallel in the animal kingdom. . . . The fish seems to live only to destroy, and is constantly employed in pursuing and chopping up whatever it can master. I am quite inclined to assign to the bluefish the very first position among the injurious influences that have affected the supply of fishes on the coast. Yet . . . , it is probable that there would not have been so great a decrease of fish as at present but for the concurrent action of man." This conclusion, certainly curious by today's thinking, reflects the mind-set of Baird's time—that is, the assignment of anthropomorphic characteristics to explain the behavior of animals and an unwillingness to admit that the decline in coastal fisheries resources was a result of human actions rather than natural causes.

Small bluefish are eaten by a variety of predators, including adults of their own species. Adults are eaten by blue sharks, tunas, and billfishes and are considered a major food item of the shortfin mako shark.

Status and Management

In New England waters, the bluefish has a long history of periods of abundance interspersed with periods of scarcity. Records from colonial times indicate that bluefish populations collapsed from high to low densities in New England during the mid 18th century. Similarly, the number of bluefish was greatly reduced along the north shore of Massachusetts Bay twice between the mid 19th and 20th centuries. Bluefish south of Cape Cod Bay showed a pattern of high densities prior to 1930, low densities from that time to the mid 1940s, and a rebound to high densities by 1950. These cycles of abundance and scarcity, typical throughout the East Coast, are greatly influenced by annual reproductive success and the survival of offspring.

Commercial fishing historically composed only a very modest proportion of total annual landings, averaging perhaps 10% of the harvest. North Carolina and Virginia take nearly half of the yearly commercial catch. About 90% of the total annual recreational catch normally is taken from waters reaching from North Carolina to Massachusetts. The bluefish ranked first in weight landed for all Atlantic coast recreational species each year from 1979 to 1987. This species also ranked first in numbers landed from 1979 to 1983 and was in the top three from 1984 to 1987. The total harvest by recreational anglers was stable from 1979 to 1983, after which a nearly 40% drop in landings occurred in 1984 and 1985. Approximately 32 million bluefish at a total weight of 56,364 mt (124 million lb) were landed by anglers in 1987. This harvest, although markedly higher than those in 1984 and 1985, was still 20% lower than the annual average between 1979 to 1983. The number of angling trips taken in pursuit of bluefish reached an all-time high in 1987, but the catch per unit of effort (weight of bluefish caught divided by the total number of fishing trips) in the same year was 35% lower than the 4-year average between 1979 and 1983 and was the second-lowest of any year since 1979.

Snappers and 1-year-old bluefish dominated recreational catches through the 1980s, whereas fishes over 8 years of age were landed only rarely. During this decade, concern grew over the waste of young bluefish, as many anglers typically discard most or all that they catch. Although many anglers enjoy the challenge and thrill of landing this species, its reputation for strongly flavored meat often unjustly lowers its promise as a "keeper" for the table, thereby resulting in many discarded fish and wasted angler harvest. By 1988, the abundance of bluefish along the Atlantic dropped to about 35% and 41% of levels calculated for 1982 and 1985, respectively (the two years of highest abundance between 1982 to 1988). Throughout the 1980s, the fishery was harvested at or slightly above a level that bluefish populations could sustain. Over 70% of all juveniles and adults were estimated to die each year from the combined effects of fishing and natural causes. The Atlantic States Marine Fisheries Commission and the Atlantic coast fisheries management councils developed a draft management plan in 1988 to address conservation of this fishery. This plan was presented at a public hearing in the summer of 1989 to gather public comment (a required step in the process of developing a management plan). Major provisions of the plan included:

1. Limiting possession to 10 bluefish per angler
2. Limiting commercial harvest to 20% or less of the total yearly catch
3. Requiring a permit for commercial hook-and-line anglers to exceed the possession limit

Angling and Handling Tips

Many anglers consider the bluefish to be, pound for pound, one of the finest game fishes in New England waters. As the striped bass fishery has declined, the angling public has increased its pursuit of this premier sport species.

The greatest success in angling for snappers occurs from August through September. Fishing for adult bluefish generally improves through the summer as more adults start moving into inshore areas and extends through October, after which waters cool and the fish migrate offshore and southward. Adult blues are pursued along most of the New England coastline, northward through mid Maine. On a coastwide basis, two to three times more angling effort is directed toward this species from rental and private boats than from the shoreline.

Snappers are caught in estuaries and bays. Many anglers prefer light spinning rods with a less-than-8-pound test line when fishing for snappers. Adults are caught along rocky and sandy shores and from boats. The feeding frenzy of adult bluefish, identified by a flurry of activity by bait fishes at the water's surface, is a signal to the angler that action is near.

When fishing for adults, anglers may use surf rods spooled with 15–40-pound test line and a reel with a smooth drag to tire the blue during its first energetic runs. Adventurous anglers may choose an 8–9-foot graphite rod strung with 10–17-pound test line.

Small poppers are often used for snappers or small adults, and a variety of plugs, large poppers, surface lures, and sand-eel-type jigs for larger adults. Poppers and plugs should be retrieved at an uneven rate, with intermittent pauses and brief snaps of the rod tip. Surface plugs are reeled gently, and the rod tip should be lifted swiftly to simulate a splash after retrieval of every 5 to 6 feet of line. When trolling from boats, the boat speed must allow the lure to express proper action; anglers who want to fish in deeper water should use wire line rather than slow the boat markedly.

Pogies, mackerel, or eels are the preferred live baits. When these are not available, many types of cut bait also do well. Wire leaders are a must when bait-fishing in order to prevent a hooked bluefish from cutting the line with its sharp teeth.

The quality of the flesh, and thus its flavor, will be best if the bluefish is gutted and iced as soon as possible after capture. The soft-textured bluefish flesh has a high oil content. When concentrated, fish oils can create a strong flavor that may not be favored by some people. If you wish to reduce the full flavor of the blue's flesh, remove the outer dark layer of meat from each fillet. Bluefish fillets can also be marinated in vinegar, lemon or lime juices, or wine, or they can be cooked with fresh acidic vegetables such as tomatoes and onions. These methods will lighten the flavor as well as retain the oils that confer the health benefits associated with eating fish.

References

Baird, S. F. 1873. *Report on the condition of the sea fisheries of the south coast of New England in 1871 and 1872.* U.S. Government Printing Office, Washington DC.

Draft fishery management plan for bluefish. 1988. Mid-Atlantic Fishery Management Council, and the Atlantic States Marine Fisheries Commission.

Friedland, K. D., G. C. Garman, A. J. Bejda, A. L. Studholme, and B. Olla. 1988. Interannual variation in diet and condition in juvenile bluefish during estuarine residency. *Trans. Am. Fish. Soc.* 117:474–479.

Pottern, G. B., M. T. Huish, and J. H. Kerby. 1989. *Species profiles: life histories and environmental requirements of coastal fishes and invertebrates (Mid-Atlantic)—bluefish.* U.S. Fish Wildl. Serv. Biol. Rep. 82(11.94). U.S. Army Corps of Engineers, TR EL-82-4. 20 pp.

Scup
(*Stenotomus chrysops*)

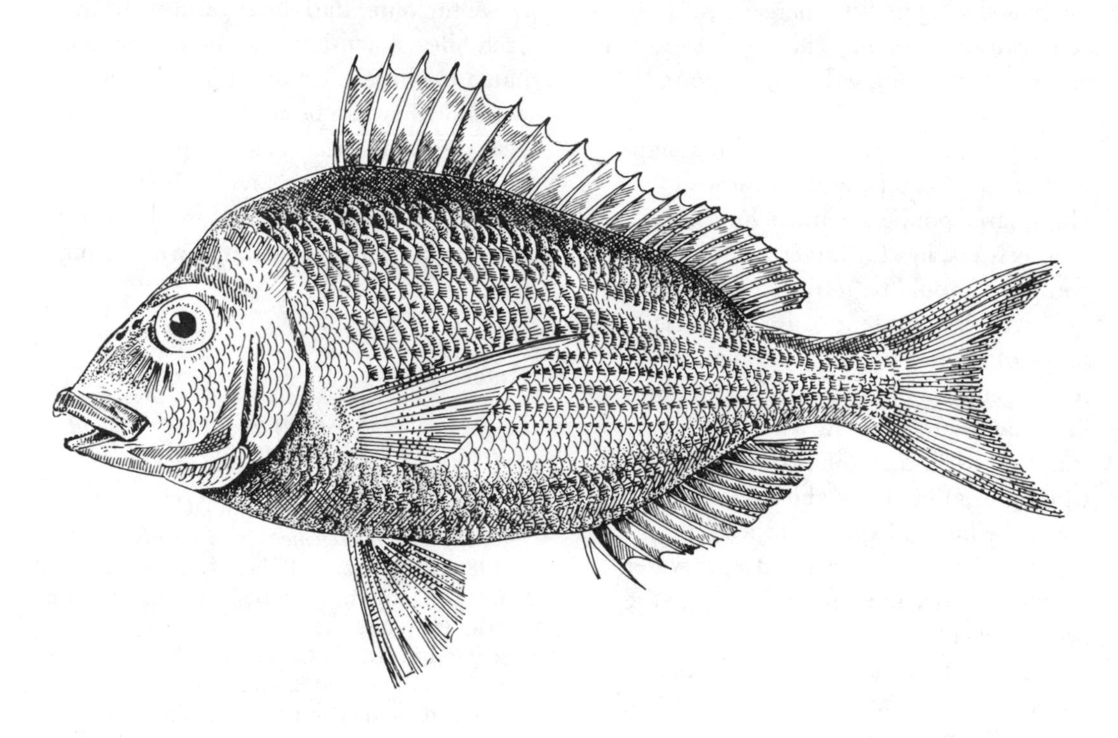

The scup, or porgy, known for its fine flavor and its avaricious pursuit of baited hooks, occurs along the continental shelf of eastern North America. It is most common from Cape Cod to Cape Hatteras, North Carolina, and is encountered occasionally north to Cape Ann.

The scup's laterally flattened body is about twice as long as it is wide. The head, concave dorsally, has a small mouth and high-set eyes. The scup has one long, continuous dorsal fin, which has a series of one short and 11 long spines anteriorly. The anal fin also has one short spine followed by several long ones. The tail is deeply concave and sharply pointed on the corners.

The scup's body is a dull silvery color flecked with light blue and displaying 12 to 15 inconspicuous horizontal stripes. The head is marked with dark patches, and the belly is white.

Biology

Adult scup form into schools of similar-sized individuals in areas with smooth or rocky bottoms. They are particularly plentiful around piers, rocks, offshore ledges, jetties, and mussel beds. Scup make seasonal migrations to inshore summer grounds in the spring and offshore winter grounds in the fall. They move inshore to coastal areas of southern New England by May and linger there until October, when most swim to deeper waters offshore and southward to waters off central and southern New Jersey. Larger fish arrive in coastal waters somewhat earlier in the spring than smaller ones. During the summer, individuals older than 3 years tend to stay in open coastal waters at depths of 2 to 36.6 m (6 to 120 ft); younger scup enter shallower areas of bays. Young larvae live in very shallow estuarine waters. Juvenile and adult scup move into harbors and along sandy beaches during high tides and then into deeper channels as the tides recede. Scup overwinter at depths of 73 to 183 m (240 to 600 ft). Winter distribution appears to be closely associated with water temperature; scup tend to concentrate in offshore areas where the temperature is 7.3 degrees C (45 F) or higher. Water temperatures can also critically affect scup at other times of the year. In New England, water temperatures in the early fall occasionally plunge below the scup's tolerance level, killing large numbers of fish.

Both males and females reach sexual maturity in their second year. In southern New England, scup spawn from May to August, with the peak level of activity typically in June. The buoyant eggs hatch about 40 hours after fertilization.

Although scup may reach 15 years of age and up to nearly 2.7 kg (6 lb), few adults exceed 0.9 kg (2 lb) in weight and about 35.6 cm (14 in) in length. Mortality due to predation and fishing is high for this species; it has been estimated that fishes older than 3 years constitute only about 10% of the total population. Table 24 shows average size at age of scup.

Within several days after hatching, the larvae, having used all yolk reserves, begin to feed upon copepods and other microscopic animals. Adult scup feed upon bottom invertebrates including small crabs, amphipods, gammarids, squid, annelid worms, clams, mussels, snails, jellyfish, and sand dollars. Scup are eaten by a variety of predaceous fishes. Each year as many as 80% of all juvenile scup fall prey to larger predators such as the cod, bluefish, striped bass, and weakfish.

Table 24. Mean length at age in cm (inches in parentheses) and weight in g (pounds in parentheses) of the scup

Age	*Fork length*	*Weight*
1	12.8 (5.1)	60 (0.1)
2	19.6 (7.7)	184 (0.4)
3	22.9 (9.0)	279 (0.6)
4	25.4 (10.0)	369 (0.8)
5	26.7 (10.5)	426 (0.9)
6	28.2 (11.1)	490 (1.1)
7	30.1 (11.9)	590 (1.3)
8	31.8 (12.5)	683 (1.5)
9	33.4 (13.1)	781 (1.7)
10	34.6 (13.6)	862 (1.9)
11	35.5 (14.0)	921 (2.0)
12	36.3 (14.3)	981 (2.2)
13	36.7 (14.4)	1006 (2.2)

Source: Data from Morse 1978.

Status and Management

Recreational fishing constitutes a significant proportion of the total harvest of scup. From 1977 to 1986 an average of 30% (ranging from 15% to 46%) of the harvest along the East Coast was taken by anglers. The proportion of the total harvest that was taken recreationally during this time period was lowest in 1984 at 15%; however, this increased to 33% in 1985, 46% in 1986, and 34% in 1987.

Scup populations on the East Coast have displayed periodic cycles of abundance over the last 20 years, with any change in population density generally lasting for 2 to 4 years before being reversed. Commercial and recreational catches peaked in the 1950s and 1960s, declined markedly by the early 1970s, and recovered to relatively high levels before 1980. The total harvest in 1986, 14,300 mt (31.5 million lb), was the highest since inception of the 200-mile fishery management zone. Much of the increase in harvest in the 1970s is attributed to an increase in fixed gear and otter-trawl activity in the southern New England region. Total harvest remained somewhat steady through the early 1980s, peaking in 1986; however, the proportion of total harvest attributable to commercial vessels generally declined from 1981 to 1986. Commercial landings in 1985 and 1986 were about 20% below the average for the previous 7-year period.

Due to concern over the impact of fishing pressure on scup stocks, by 1989 Connecticut and Massachusetts were managing harvest by establishing minimum legal size limits.

Angling and Handling Tips

The name "scup" has been derived from the Indian *mishcuppauog,* which was shortened to *scuppaug* and, by the 20th century, to *scup.* Scup are also known as porgies, which is commonly mistaken for "pogies," another name for the unrelated menhaden.

Scup feed frantically and fight energetically when hooked, thereby providing angling enjoyment for the entire family. This little scrambler is especially fun for children, as a school of actively feeding scup typically provides nonstop fishing action. If you are fishing with your children, don't expect to have time for anything more than serving as bait master; the action will keep you hopping from child to child with bait.

Scup provides particularly exciting battles when anglers use either a medium-weight spinning or light-weight surf outfit carrying 10–20-pound test line. Some anglers prefer jigging small lures, but the overwhelming majority prefer bait-fishing. In a typical rig, a bank sinker is tied to the end of the line, and one to three snelled hooks (No. 1 to No. 8) are then tied to loops 6 to 10 inches above the sinker. Sea worms, squid strips, and pieces of clam or fish work well as bait. Squid strips are favored because they last well through the frantic attacks a school of scup will make on baited hooks.

Although scup are quick to grab bait, they are difficult to hook. For greatest success, anglers need to become adept at setting the hook as soon as the tip of the fishing rod shows the slightest dipping, or at lifting the baited hooks gently off the bottom to induce the fish to strike sharply rather than allowing them to nibble at the bait. Upon finding a school of actively feeding scup, some anglers lower the bait to the bottom, count to five, and set the hook, rather than risk having the bait stolen while waiting for the subtle nibble of this fish.

Scup do not spoil as quickly as many other species, which must be immediately cleaned or placed in ice. Still, timely icing and cleaning are recommended in order to enjoy the full sweetness of the scup's flavor. Simple methods of preparation are best. Try fried or poached scup, served with melted butter and a slice of lemon or lime. Large scup are delicious when grilled over a charcoal or gas fire. Cut vertical slits into each fish and drip butter or mayonnaise into the slits before grilling. Leaving the scaled skin on the fillets will help hold the flesh together when turning the fish on the grill. Serve this tasty treat with some sliced lemon.

Reference

Morse, W. W. 1978. Biological and fisheries data on scup, *Stenotomus chrysops* (Linneaus). Sandy Hook Lab., Northeast Fisheries Center (NMFS/NOAA). Tech. Ser. Rep. No. 12: 41 pp.

Weakfish
(*Cynoscion regalis*)

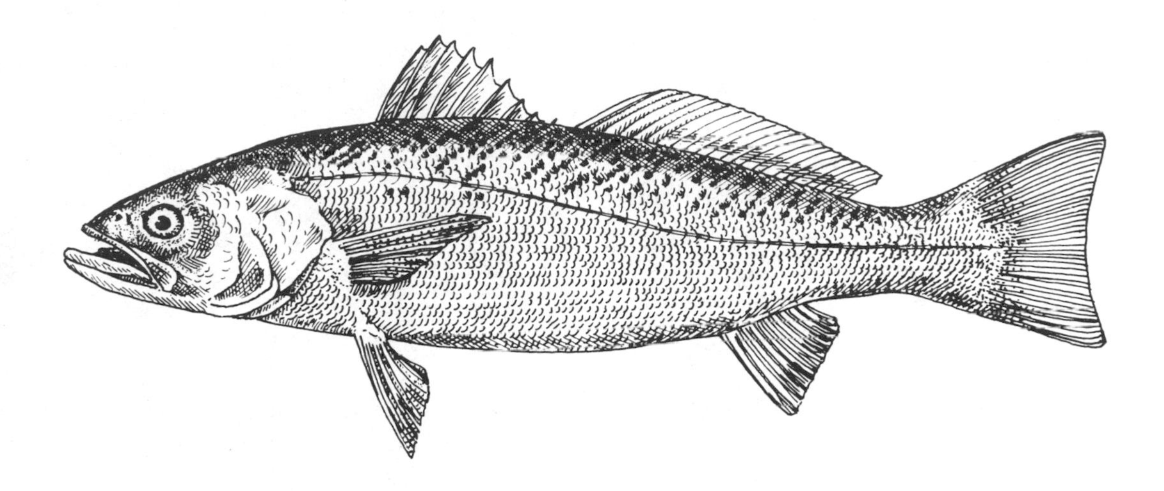

The weakfish, also called the squeteague or gray sea trout, is one of 20 species of the drum family (Scaeinidae) found along the U.S. Atlantic and Gulf coasts. This species ranges from southern Florida north to Massachusetts Bay and occasionally to the waters off Nova Scotia. Abundant from the Virginia coast to the New Jersey shoreline, it occurs only as a summer transient in New England waters. Its occurrence north of Cape Cod is too spotty to sustain a recreational fishery.

The weakfish is a streamlined species, the body being about four times as long as it is deep. It has a pointed snout and large mouth. The upper jaw is equipped with two large canine teeth and is shorter than the lower jaw, which projects noticeably beyond it. There are two separate dorsal fins. The anterior dorsal is triangular in shape and has a series of spines; the posterior dorsal is about three times the length of the anterior. The anal fin is less than half as long as the posterior dorsal. The tail is broad and slightly concave in outline.

Weakfish are dark green or olive above and blazed with purple, blue, and gold on the back and sides. Vertical rows of small dark spots occur on either side of the body. The belly is white. The dorsal fins and tail are dark, the pelvic and anal fins are yellow, and the pectoral fins are olive, marked with yellow near the base.

Biology

The weakfish usually appears in New England in May and leaves by October. Weakfish younger than age 4 generally migrate south along coastal areas in the fall and winter and north in spring and summer. Fish older than age 4 move offshore as well as south in the fall, and inshore as well as north in the spring. Some adults in the New England region may only move offshore to deeper, warmer waters in the winter. Larger fish tend to migrate farther north than smaller ones; thus, the New England weakfish fishery may possess a higher percentage of large individuals than is characteristic of the coastwide fishery. Weakfish tend to move about in schools of similar-sized individuals. Evidence from Delaware Bay suggests that in the summer, at least, some of these schools are made up mostly of individuals of a single sex.

During the summer, weakfish inhabit inshore areas such as open sandy shorelines, bays and estuaries, and even salt-marsh creeks. They typically occur in water less than 9 m (about 30 ft) deep. Weakfish usually move about near the surface, although they also feed off the substrate in estuaries and the surf of the open coast.

About 50% of all individuals of both sexes become sexually mature at age 1, although some fish do not do so until age 3 or 4. The fecundity of females increases with age and size and decreases with geographical latitude. Females in the southern portion of this species' range do not grow as large or live as long as more northerly distributed weakfish, but they produce several times as many eggs per body length as do the northern fish. Thus, even though southern females have a shorter life-span, they produce about as many eggs during their lives as the northern weakfish.

This species reproduces in the Middle Atlantic region from May to October. Spawning takes place at night in estuaries and inshore areas. In a given area, the larger adults arrive at the spawning grounds and reproduce earliest in the spawning season. In years of high population abundance, the weakfish may spawn as far north as Woods Hole or even in Massachusetts Bay.

The buoyant eggs hatch within 48 hours after fertilization at 20 to 21 degrees C (68 to 70 F). Initially, the larvae swim within the water column, feeding upon microscopic plankton. When the larvae reach about 1 cm (0.4 in), they become demersal and move to the protection offered by muddy creeks and coves.

Table 25. Calculated mean length at age in cm (inches in parentheses) of weakfish captured from Cape Cod to Delaware Bay

Age	*Female*	*Male*
1	20 (7.9)	20 (7.9)
2	32 (12.6)	31 (12.2)
3	48 (18.9)	46 (18.1)
4	58 (22.8)	56 (22.0)
5	64 (25.2)	63 (24.8)
6	68 (26.8)	66 (26.0)
7	70 (27.6)	66 (26.0)
8	72 (28.3)	68 (26.8)
9	73 (28.7)	71 (28.0)
10	75 (29.5)	69 (27.2)
11	81 (31.9)	70 (27.6)

Source: Shepherd and Grimes 1983.

Weakfish can grow to around 8 kg (17 to 18 lb). However, fish weighing more than 5.5 kg (12 lb) and measuring longer than 91 cm (about 36 in) are very rare, and most caught by anglers weigh less than 2.7 kg (6 lb). Although avidly pursued as a sport species, the relatively slow-growing weakfish is at least several years old before it is large enough to be viewed as a "keeper" by many anglers. Females are significantly larger than males after age 6. Table 25 shows average lengths for weakfish captured in the Cape Cod to Delaware Bay region. Fish in this region live longer and grow larger than fish farther to the south. There has been a general coastwide trend toward increased size at age between 1929 and 1981. Faster growth rates seen in recent decades are believed to have been stimulated by a major reduction in population abundance caused by fishing.

Juvenile weakfish feed heavily upon small crustaceans, particularly shrimp, crabs, polychaete worms, and clams.

Adults occasionally feed upon invertebrates, but fishes such as sand lances, silversides, scup, mackerel, butterfish, and particularly juvenile menhaden constitute the major portion of their diet.

Status and Management

Historically, the abundance of weakfish along the southern New England coast has fluctuated greatly, and this species occurs north of the southern shore of Cape Cod only when its abundance peaks in the southern New England region. The fluctuations in abundance are probably influenced by variation in reproductive success and recruitment of young fish into the population. There was a general decline in coastwide abundance from 1910 to the late 1960s, followed by a recovery during the 1970s. This decline was probably caused by fishing. Coastwide commercial harvests peaked at over 18,800 mt (41.4 million lb) in 1945. Catches dropped to 12,545 mt (27.6 million lb) in 1947, then steadily declined to the lows seen in the 1960s (1,409 mt, or 3.1 million lb, in 1963). Increased abundance in the 1970s led to increased commercial landings, reaching 13,045 mt (28.7 million lb) in 1979.

In recent years, the levels of commercial and recreational landings have been roughly equal, with recreational catch occasionally exceeding commercial. For example, the commercial catch between New Jersey and North Carolina in 1970 was 3,000 mt (6.6 million lb), and the recreational catch in the same year totaled 7,000 mt (15.4 million lb).

Although supporting only a modest fishery in New England, the weakfish is avidly pursued throughout the Middle Atlantic states. An estimated 11.1 million fish were recreationally harvested in the Middle Atlantic region in 1986, which ranked this species sixth among all recreational species of that region.

Of the New England states, only Rhode Island regulated the harvest of this fishery in 1989. A coastwide management plan was being developed in 1989 by the Atlantic States Marine Fisheries Commission to address future management approaches.

Angling and Handling Tips

Many an angler has been startled by the drumming or croaking sound emanating from a landed weakfish while it is being held in the hand. The weakfish, like all other members of the drum family, produces this sound by using specialized musculature to vibrate its swim bladder.

To catch weakfish, the angler needs to be aware of the bottom structures that provide cover and cause currents to gather food for this species. These sites include docks, breakwaters, rises in the bottom contour (such as sandbars), rocks that produce rips of water, deepwater drop-offs, and channels.

Typical gear includes a light-to-medium spinning outfit with a smooth drag and 8–15-pound test line. Weakfish cannot be "horsed in" with a heavy line because hooks will tear away from their soft mouths; hence the name "weakfish."

Artificial lures are most effective in the spring and fall. Weakfish strike at a variety of lures, including bright plastic shrimp, worms, bucktails, yellow leadheads, twisters, tubes, jigs, metal lures, and sandeel imitations.

Anglers generally find baits more successful than lures in the heat of the summer. Weakfish can be taken on a variety of baits including shrimp, sea worms, bunker, tinker mackerel, eels, strips of squid, and sometimes snapper blues. Live baits on a 6/0 hook with a strong leader often work best for big weakfish. Many anglers will use strips of bait on a standard two-hook bottom rig with 2/0 or 3/0 hooks, up to 2 feet of monofilament leader, and a sinker (flat or round types for bottom-fishing in calmer water, pyramid for fishing in surf).

Anglers who have patience achieve the greatest success in landing the weakfish. The bail of the reel should be open to allow the weakfish to run with the bait after picking it up. When the fish stops running to swallow its catch, count to 10 and set the hook, which by that time should be deep into its throat. Retrieve line slowly to enjoy the battle and to prevent the hook from tearing away if the fish is hooked in the mouth. Be sure to have a gaff or net ready to take the catch out of the water.

The flesh of the weakfish, white with a reddish midstripe, has a mild, almost sweet, flavor. On-the-spot icing is necessary to preserve its delicate flavor and texture.

Weakfish can be prepared in a variety of ways. One popular approach is to dip in milk, dust with flour, and sauté in butter for 5 minutes, turning once. Weakfish can be steamed in broth until the meat flakes; squeeze on some lemon juice and serve with steamed broccoli.

References

Shepherd, G. R., and C. B. Grimes. 1983. Geographic and historic variations in growth of weakfish, *Cynoscion regalis,* in the Middle Atlantic Bight. *Fish. Bull.* 81:803–813.

Shepherd, G. R., and C. B. Grimes. 1984. Reproduction of weakfish, *Cynoscion regalis,* in the New York Bight and evidence for geographically specific life history characteristics. *Fish. Bull.* 82:500–511.

Cunner
(*Tautogolabrus adspersus*)

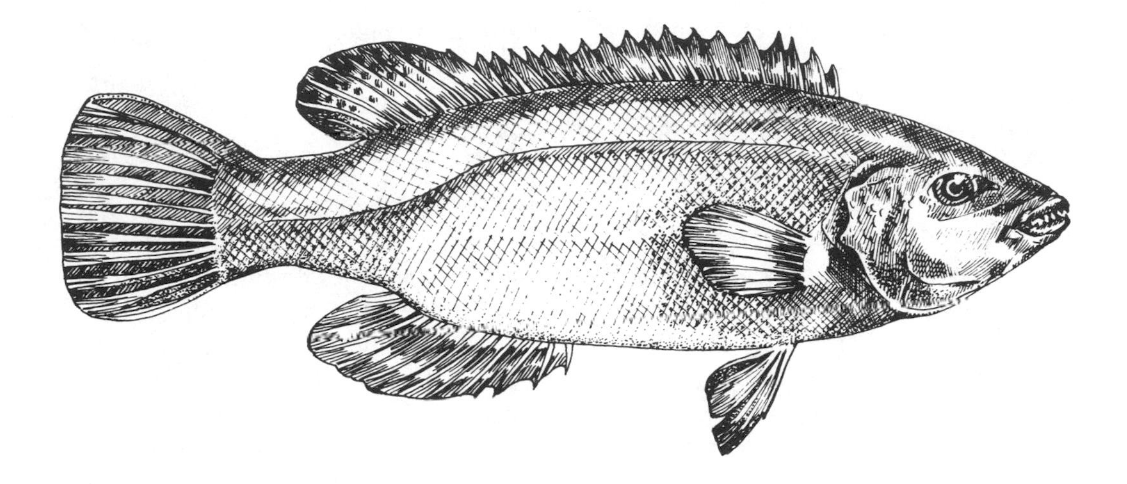

The cunner, a ubiquitous little species occurring throughout the New England coast, is native to the Northwest Atlantic from Newfoundland to the capes of New Jersey and occasionally to the Chesapeake Bay. It is most abundant within the Massachusetts Bay region.

The cunner is distinctive in appearance, differing markedly from most other species of fishes caught by anglers in New England waters. This species has a small mouth containing several rows of jaw teeth, arranged so as to give the fish a "buck-toothed" appearance. The cunner has a single long dorsal fin, with a series of about 18 sharp spines on the anterior three quarters. The pectoral fins, pelvic fins, and tail are rounded in outline. The roof of the mouth and floor of the throat each contain a patch of knoblike teeth that are used for grinding food. The cunner is similar to the closely related tautog but differs from this species in having a slimmer body, a more pointed snout, and noticeably thinner lips.

The color of the cunner is as variable as the wide number of background colors found in the habitats in which it lives. For example, individuals caught among red seaweeds tend to display reddish or rust-colored tones, those found on mud bottoms are often dark sepia, those on sandy substrates are pale brown speckled with black dots, and those taken from deep waters may be a rich red. Any of these colors may have a striking bluish iridescence or may be mingled together in a mottled pattern.

Biology

Cunners inhabit very shallow inshore waters from the tidemark seaward in the northern part of their range but tend to be found at depths of at least 4.6 to 6.1 m (15 to 20 ft) south of New York. Cunners often move up into the larger salt creeks of estuaries. They generally remain within 5 or 6 miles of shore throughout their range, although occasional fish have been collected in the shallow areas of Georges Bank. Cunners do not form into schools; they are found on the bottom, alone or in small groups, resting or swimming slowly among the eelgrass and rocks.

The cunner tends to be a year-round resident of the New England coastline, exhibiting no migrations to other regions regardless of the season. However, this species may move to deeper waters during periods in the summer and winter in order to escape drastic and rapid temperature changes occurring in shoreline areas. The cunner is rarely found in waters over 21 degrees C (about 70 F).

Males and females become sexually mature by the age of 2. Cunners in the Massachusetts Bay region reproduce from late spring through September. Peak spawning activity usually occurs during late May and early June. The buoyant cunner eggs hatch 2.5 to 3 days after fertilization, depending upon water temperature. The tiny larvae begin feeding upon microscopic copepods as soon as their yolk sacs are absorbed and their teeth first develop.

Cunners may reach a maximum size of 1.4 kg (3 lb) and 38 to 43 cm (about 15 to 17 in) in length. However, most are 15.2 to 25.4 cm (6 to 10 in) long and weigh less than 0.5 kg (1 lb), and few individuals exceed 30.5 cm (12 in) in length. One-year-olds are generally 5.1 cm (2 in), 3-year-olds 12.7 to 15.2 cm (5 to 6 in), and 6-year-olds about 20.3 to 25.4 cm (8 to 10 in) in length.

Juvenile and adult cunners feed actively throughout daylight hours, eating many species of invertebrates including amphipods, isopods, shrimp, small lobsters and crabs, worms, mussels, barnacles, and small clams. Hard-shelled foods such as mussels and clams are crushed by the cunner's powerful jaws and sharp teeth. Small fishes such as silversides, sticklebacks, and mummichogs normally form only a modest proportion to the cunner's diet.

Cunners are eaten by a variety of fish species, including sculpin, tomcod, and skates. Night herons and other fish-eating birds catch younger cunners in shallow bays and estuaries.

Status and Management

The cunner was a favored panfish in the 19th century. During the 1870s, small boats fishing out of Boston harvested up to 136.4 mt (300,000 lb) of cunners annually. At the turn of the century, Maine reported annual harvests of 68.2 to 127.3 mt (150,000 to over 280,000 lb). Due in part to a reduction in market interest, commercial harvests dropped dramatically shortly after 1900; only 4.5 mt (10,000 lb) were harvested along the entire Massachusetts coastline by 1919.

Cunners have retained an identity as a recreational species due to their availability and to the fact that they are almost bothersome in striking baits intended for other species. Although most are caught incidentally and released because of their small size, cunners remain a favorite of youthful anglers who are looking for constant action and are unconcerned about the size of the fish that provides it. Due to this species' modest importance as a recreational fishery, no New England state managed harvest of the cunner as of 1989.

Angling and Handling Tips

The cunner, or bergall, is noted for the quickness with which it cleans bait from the hook. This species' protruding teeth make it an ideal bait stealer, so the best way to catch this pesky little fish is to use the toughest-texured bait you can find. Small strips of squid or conch are popular baits. These should be attached to a small hook with minimal weight tied to the line. Large sinkers will mask the tug produced by the cunner grabbing at the baited hook. The line should be held in the fingers in order to feel the light taps produced when the bait is grabbed. Since cunners are small, the hook should be set with moderate force; too great a tug will lose the fish. Anglers catch the most fish and get the best battle out of this feisty species by using an ultralight spinning outfit with 4-pound test line.

Early records indicate that hook-and-line-capture bergalls occasionally weighed several pounds. However, most individuals landed will weigh less than 0.2 kg (0.5 lb) and measure no more than 25.4 cm (10 in) in length. The size of the cunner was not viewed as a drawback by the 19th-century cook, who could fit several cunners into a single pan and cook them simultaneously. Consumer preference today has turned from panfishes to larger-bodied species with thick fillets. However, the cunner offers flavorful, lean, white flesh to those with the patience needed to clean it.

If you plan to prepare whole "dressed" cunners, you will need to scale them. Scaling is easily accomplished with a commercial scaler, a dull knife, or even the edge of a spoon. After scaling, make a cut completely through the fish, starting behind the gills and moving to the vent. This will allow you to remove the head, gills, and most of the gut with a single cut. Use a spoon to remove the entrails that are still in the body cavity. Whole cleaned cunners are best when fried or baked with appropriate seasoning. After cooking, you can easily pull fins and bones away from the flesh.

With patience, the cunner can also be filleted. Fried or baked fillets that have been dipped in a beer batter are a dining favorite. To make a beer batter, mix one or two eggs, ¼ to 1 cup of dark beer or ale (the amount determines the thickness of the batter when cooked), 1 cup of flour (or 1 cup of pancake mix), and 1 teaspoon of baking powder (leave out the baking powder if using pancake mix). Add salt and freshly ground pepper to taste.

Also, try using the heads and frames for a fish stock, or court bouillon. Tie the heads and frames in a piece of cheesecloth and drop into boiling water for 30 minutes, using a quart of water for each pound of fish. Remove the cheesecloth and frames, and add 1 cup of red or white wine, ¼ teaspoon of poultry seasoning, two small whole onions, two sliced carrots, two sliced cloves of garlic, one bay leaf, a little salt, and some freshly ground pepper per quart of stock. Bring to a boil and simmer for 20 minutes. Serve hot, followed by the fried fillets.

Remember, if you are fishing in the future for big fishes and are being harassed by this little thief, you have three choices. You can switch to artificial lures, move to another spot, or catch enough of these avaricious creatures to make an excellent meal.

Reference

Serchuk, F. M. 1972. *The ecology of the cunner,* Tautogolabrus adspersus *(Walbaum) (Pisces: Labridae), in the Weweantic River estuary, Wareham, Massachusetts.* Unpublished master's thesis, University of Massachusetts, Amherst.

Tautog
(*Tautoga onitis*)

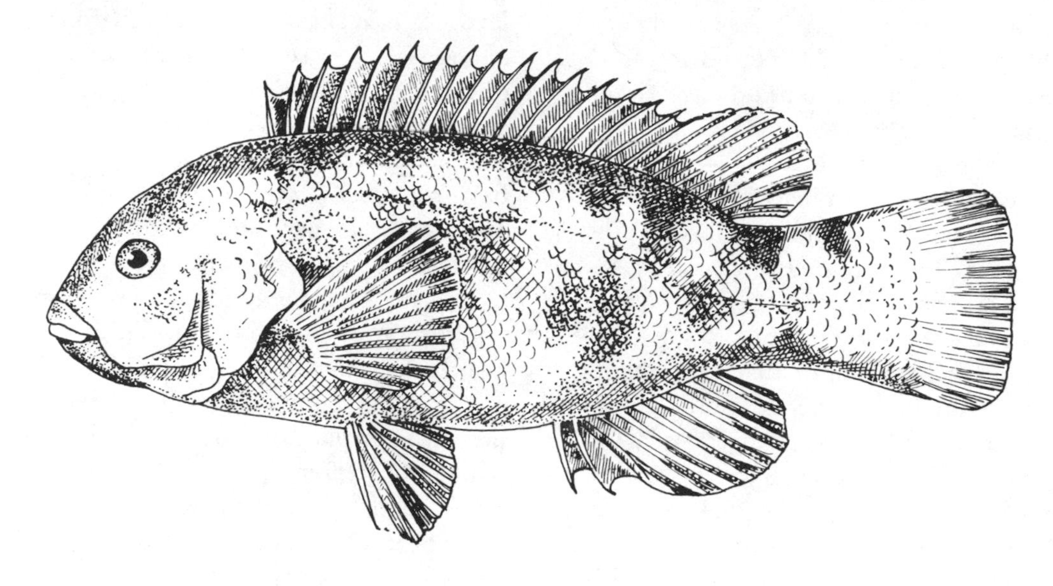

The tautog (also called blackfish and tog), lives along the Atlantic coast from Nova Scotia to South Carolina, reaching greatest abundance along inshore waters from southern Cape Cod to the Delaware capes. It does not sustain a recreational fishery north of Massachusetts.

The tautog is a stout fish with a blunt nose and a thick-lipped mouth with large conical teeth in front and flat crushing teeth in back. The single dorsal fin originates over the gill slit and runs back nearly to the tail. There is a series of stiff, sharp spines on the anterior three quarters of this fin. The anal fin has three spines, and the paired pelvic fins have one spine each.

The tautog ranges from dark green to black dorsally, with these shades mottling to a lighter background color on the sides. The belly is only slightly lighter than the sides. The white chin characteristic of large tautog has led to the name "white chin."

Biology

Tautogs are typically encountered within several miles of shore in water less than 18 m (60 ft) deep north of Cape Cod. More southern populations can be found as far as 10 to 12 miles offshore in depths of 18 to 24 m (about 60 to 80 ft). Tautogs frequently follow flood tides inshore to feed in the intertidal areas and drop back to deeper waters with the following ebb tides.

Tautogs are found in association with cover, hovering around steep, rocky shorelines or hiding near wrecks, wharf pilings, piers, jetties, mussel and oyster beds, and boulder-strewn bottoms. They generally stay within localized home ranges while feeding and resting. While on their summering grounds, tautogs establish a "homesite," a protected spot in which they rest every night. They are so inactive at night that divers can easily catch them by hand as they lie motionless on the bottom. Adults range widely from their homesites when feeding, but young tautogs remain near theirs and are thus protected from predators.

Tautogs do not undertake extensive seasonal migrations but tend to move inshore as water temperatures rise in spring, after overwintering in large groups offshore in waters 15 to 46 m (50 to 150 ft) deep where the bottom is covered with large boulders. Fish less than 25 cm (10 in) long remain in shallow inshore areas throughout the winter. Some tautogs remain offshore all year, exhibiting no movement except when searching for food or cover.

Both sexes mature at 3 or 4 years of age. Fecundity increases with size. Females 30.5 cm (12 in) long and 0.5 kg (1 lb) in weight produce about 30,000 eggs, whereas females 50.8 cm (20 in) long and 2.3 kg (5 lb) produce about 196,000 eggs per season.

Tautogs reproduce from May until August, with peak spawning activity occurring in early summer at water temperatures of 16.7 to 21.1 degrees C (62 to 70 F). Most spawning takes place inshore in areas dominated by eelgrass beds. After reaching sexual maturity, many fish return to the same spawning area each year throughout their lives. Although they intermix in large groups for the rest of the year, tautogs tend to remain in small, discrete groups during the spawning season. Males and females spawn in pairs. Large dominant males may aggressively exclude smaller males at the same site from spawning with females.

The fertilized eggs are buoyant, floating for about 2 days before hatching. Within 4 days after hatching, the larvae begin feeding on microscopic plankton.

Although capable of reaching relatively large sizes, tautogs are very slow-growing. Although they can live up to 34 years and exceed 10 kg (22 lb), the average fish caught by anglers is 6 to 10 years old and weighs 0.9 to 1.8 kg (2 to 4 lb). Males achieve greater lengths at age, and live longer than females. Table 26 shows weights and age ranges for specific lengths of this species. Annual mortality due to natural causes is estimated to average about 50% per year.

Table 26. Average weights in kg (pounds in parentheses) and age ranges corresponding to specific lengths in cm (inches in parentheses) of tautogs

		Age	
Length	*Weight*	*Male*	*Female*
15.2 (6)	0.1 (0.2)	2	3
20.3 (8)	0.2 (0.4)	3	3
25.4 (10)	0.4 (0.9)	4	4
30.5 (12)	0.6 (1.3)	5	5
35.6 (14)	0.9 (2.0)	6	7
40.6 (16)	1.3 (2.9)	8	9–10
45.7 (18)	1.9 (4.2)	9–10	11–13
50.8 (20)	2.5 (5.5)	12–15	13–20
55.9 (22)	3.3 (7.3)	16–18	
61.0 (24)	4.5 (9.9)	18–27	

Source: Data from D. Colek, unpublished manuscript, as modified in Cooper 1967 and Briggs 1969.

Juvenile and adult tautogs feed during the day exclusively, with feeding peaks occurring at dawn and dusk. Tautogs commonly feed in intertidal zones during high tide. Their diet is dominated by invertebrates such as mussels, clams, crabs, sand dollars, amphipods, shrimp, small lobsters, and barnacles. Juveniles and adults living around shoreline ledges feed heavily upon blue mussels. Their large canine jaw teeth are used to tear mussels from the substrate, and their grinding pharyngeal teeth crush the mussels before they are swallowed.

Status and Management

Tautog population levels have been generally stable since colonial times. Although a popular recreational species, the tautog historically had little market value and thus was not commercially exploited. However, both commercial and recreational harvests increased markedly in the 1980s.

Between 1983 and 1986, the poundage of tautogs landed by gear other than hook and line increased nearly three times in Massachusetts alone. This increase in harvest, generally occurring throughout coastal areas of southern New England and New York, is due in part to an increase in the tautog's market value as other commercial species with greater traditional value become less abundant and harder to catch. Coastwide recreational harvest increased nearly fivefold between 1981 and 1986; numbers caught annually in the New England region grew from 767,000 to 4.2 million during this 6-year period.

Slow-growing species such as the tautog can easily be reduced in abundance by exploitation. One of the first signs of overexploitation of such species is a marked reduction in the average size of fish harvested; this phenomenon was clearly being exhibited by tautog fisheries in the 1980s. Because of this, Connecticut and Rhode Island used minimum size limits to regulate the harvest within their territorial waters in 1989.

Angling and Handling Tips

The tautog is hard-fighting, tough on tackle, and excellent on the table. This is one of the first species available to anglers in the spring and one of the last available in the fall. Anglers are particularly successful from April through May and in early fall when tautogs are concentrated in the greatest numbers along shorelines. Tautogs are caught either from a boat at anchor or by casting anywhere along rocky shorelines.

A rod with "backbone" is required to catch this battling fish. Most anglers choose a medium-action spinning or conventional rod with 20–30-pound test line, and use a "no-hardware" two-hook rig with a sinker tied to the bottom. Anglers use bait such as a large piece of sea worm, whole or halved crabs (green, rock, hermit, or fiddler), and pieces of conch, snail, or cracked clam.

It is important to stay alert after casting or lowering the bait into the water, as tautogs often hit the bait as soon as it reaches the bottom. All slack line should be taken in as soon as the bait stops sinking. Once a fish picks up the bait, let it tap once or twice and then set the hook hard, lifting the tautog away from the bottom before the line becomes entangled in rocks.

The fine flavor of the tautog has often been likened to that of the red snapper. Traditionally, it has been considered an ideal chowder fish. Its firm, mild-flavored flesh also lends itself well to baking and broiling when using recipes developed for species such as striped bass.

References

Briggs, P. T. 1969. A length–weight relationship for tautog from the inshore waters of eastern Long Island. *N.Y. Fish Game J.* 16:258–259.

Cooper, R. A. 1966. Migration and population estimation of the tautog, *Tautoga onitis* (Linneaus), from Rhode Island. *Trans. Am. Fish. Soc.* 95:239–247.

Cooper, R. A. 1967. Age and growth of the tautog, *Tautoga onitis* (Linneaus), from Rhode Island. *Trans. Am. Fish. Soc.* 96:134–142.

Olla, B. L., and C. Samet. 1977. Courtship and spawning behavior of the tautog, *Tautoga onitis* (Pisces:Labridae), under laboratory conditions. *Fish. Bull.* 77:585–599.

Olla, B. L., A. J. Bejda, and A. D. Martin. 1974. Daily activity, movements, feeding, and seasonal occurrence in the tautog, *Tautoga onitis*. *Fish. Bull.* 72:27–35.

Atlantic Wolffish

(Anarhichas lupus)

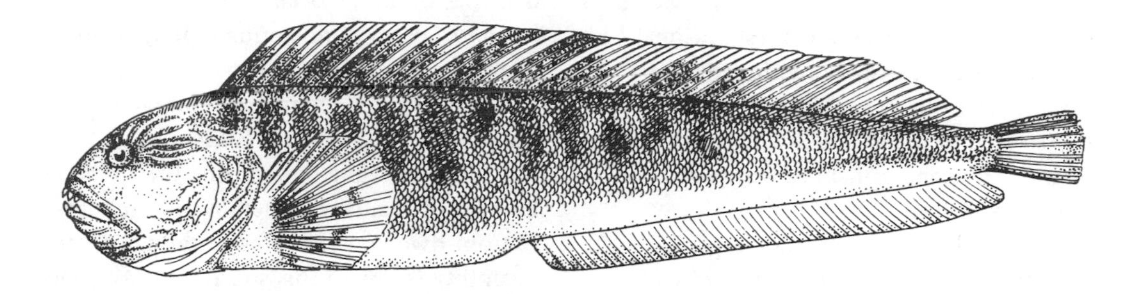

The Atlantic wolffish (or ocean catfish), a deepwater species whose name comes from the large canine teeth protruding from its mouth, occurs on both sides of the North Atlantic. It is common along the North American coast from northern Canada to the Gulf of Maine and northwestern Georges Bank and occasionally is found as far south as Vineyard Sound.

Wolffish species are easily distinguished from other fishes encountered by New England anglers. The wolffish has a large head and an elongate, posteriorly tapering body. The dorsal fin extends from above the opercle to the end of the body. The anal fin is also elongate, and the pectoral fins are large and rounded; wolffishes lack pelvic fins. The head is rounded in cross-sectional perspective, due to the massive musculature associated with these fishes' powerful jaws. The jaws are armed with a row of about six large, stout canine tusks, and the roof of the mouth carries three series of crushing teeth, the middle series consisting of a double row of large, rounded molars united into a solid plate. The spotted wolffish, a northern species that occasionally strays southward into the Gulf of Maine, is easily separated from the Atlantic wolffish by the numerous, variable-sized dark spots on its body and dorsal fin.

The color of the Atlantic wolffish varies from a dark olive green to brown, bluish gray, or slate. It has a series of vertical bars along its sides, and the belly is a dirty white color.

Biology

The Atlantic wolffish is a bottom-dwelling species that reaches greatest abundance on stony substrates at depths of 40 to 350 m (131 to 1,148 ft) and temperatures of 0.4 to 4.0 degrees C (33 to 39 F). It can be found in waters as shallow as 5 to 15 m (16 to 49 ft) in the spring and summer. This is a sedentary species, which spends extended periods of time in very restricted areas. Most individuals tagged in a study in Canadian waters were recaptured, up to 5 to 7 years after tagging, within 2 to 5 miles from the original site of capture. Only 3 of 20 recovered fish were recaptured far from the site of tagging (210 to 530 miles). Atlantic wolffish tend to be solitary, occurring in sparse numbers even in suitable habitat. SCUBA observations indicate that an adult wolffish removed from a site is soon replaced by another, suggesting that wolffish possess a homesite or territory, within which other individuals normally do not occur. Modest seasonal migrations have been noted for this species. Some wolffish move toward the coastline to feed in the spring and summer and migrate to deeper, offshore waters to spawn in the fall.

The smallest mature female captured in an extensive Icelandic study was 25 cm (10 in) long. This fish was about 4 years of age. The size by which 50% of the females in Canadian populations become mature ranges from 50 to nearly 70 cm (about 20 to 27.5 in) in length, depending upon geographic area (southerly distributed fishes mature at significantly greater lengths). The age of these females was not determined. However, wolffish of this size range in the Gulf of Maine region would be about 6 to 10 years of age. Thus, the wolffish spends an extended period of time as a juvenile before reaching sexual maturity.

Fecundity is low compared to many marine species. Females 40 cm (15.8 in) in length produce an average of 2,400 eggs; females 100 cm (39.4 in) in length average nearly 22,700; maximum fecundity is about 35,000 eggs.

Spawning occurs in the fall. Males and females commonly pair at eventual nest sites before spawning commences. Spawning occurs in larger holes or cavities under boulders, after which the male remains in the nest hole to protect the egg mass that has been laid there. Typically, only one egg mass occurs within a nest. Females do not feed during pairing as they prepare to spawn, and males do not during the time they are protecting eggs in the nest. Thus, most adult wolffish are emaciated by the time the spawning season ends. In Newfoundland, most eggs have hatched by the beginning of December. The timing of the spawning season has not been clearly described for the Gulf of Maine, but peak densities of larvae occur between November and January, indicating that spawning probably peaks in midfall.

The eggs are demersal, but larvae initially reside in the water column after hatching. At some time during their first months of life, larvae metamorphose into bottom-dwelling juveniles. Although total fecundity is low in this species, the average size of eggs is quite large because of the great amount of yolk present. This energy reserve allows relatively more extensive embryological development to occur before hatching, thus producing large and presumably well-developed larvae. The protection of eggs by males and the well-developed state of young larvae result in higher survival rates for the offspring of this species than is true for fishes that release tremendous numbers of small, buoyant eggs that are left unprotected. Thus, low egg production is countered by higher survival rates of eggs and larvae.

Growth rates are notably slow for this species (Figure 12). General literature indicates that males grow somewhat faster, and attain slightly larger sizes, than do females. This does not appear to be the case for wolffish from the Gulf of Maine/Georges Bank area. No significant differences occur between growth of males and females from this region.

Wolffish feed mainly upon invertebrates, with starfishes, brittle stars, sea urchins, sand dollars, whelks and other gastropods, and bivalves occurring frequently in their diet. Individual fish often have only one type of food item in their stomachs. This is particularly true of larger adults, which tend to feed more heavily upon large-

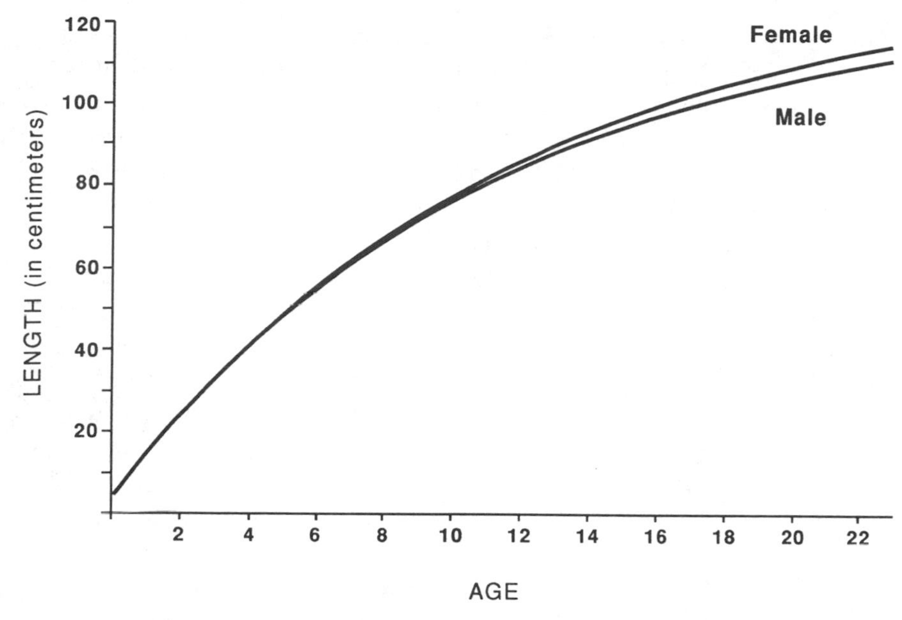

Figure 12. Calculated length at age in cm for male and female wolffish from the Gulf of Maine Georges Bank region (calculations from Ross and G. A. Nelson, manuscript in preparation).

bodied, hard-shelled invertebrates than do smaller wolffish. Large wolffish are often found with intestinal tracts full of shell fragments from numerous quahogs or scallops crushed by the grinding molariform teeth in the fish's mouth. Large adults also feed upon fishes such as the ocean perch; however, fishes never constitute a major part of this species' diet.

Status and Management

Wolffish have historically held minor importance as commercial and recreational fisheries. The combined catch of the Atlantic wolffish and the spotted wolffish in the Northwest Atlantic averaged 14,500 mt (31.9 million lb) from 1975 to 1979. Annual commercial harvests from the Gulf of Maine/Georges Bank region averaged about 1,000 mt (2.2 million lb) from 1979 to 1987, although declines of about 100 mt (220,000 lb) occurred annually after 1982. Recreational harvests were insignificant throughout the 1980s.

Harvests of this species have remained low because of lack of availability, not because it is undesirable as a food species. The wolffish's tendency to be sedentary and solitary and its general unavailability in the fall during reproduction have prevented the development of efficient harvest techniques. Most commercially landed wolffish are by-catch, netted by trawlers focusing on more abundant species. Recreational harvests are similarly limited by the lack of availability. Natural constraints on harvest would therefore seem to protect wolffish populations from depletion caused by exploitation. However, this does not appear to be the case.

Because of the relatively low rate of offspring production and the late age at sexual maturity, wolffish populations cannot rapidly replace individuals taken by fishing. Annual recruitment of new individuals into an unexploited population is stable, and modest in extent over time. Wolffish populations do not appear to experience periodic years of exceptional reproductive success, during which unusually high numbers of surviving offspring are produced. This species combines low natural mortality with consistently modest offspring production. Under these circumstances, even moderate fishing mortality can deplete population abundance.

Declining harvests in the 1980s occurred concurrently with reduced catches noted during the Northeast Fisheries Center's (NMFS, NOAA) spring survey research cruises. Thus, by the late 1980s, evidence strongly suggested that fishing was reducing abundance of the Gulf of Maine/ Georges Bank stock. However, due to the modest nature of the historical harvest, no management strategy had been developed to regulate harvest as of 1989.

Angling and Handling Tips

Due to its bizarre appearance, the wolffish is a storied recreational species with a considerable reputation as a voracious and ferocious predator. Like many species, the wolffish snaps its jaws open and shut when landed by the angler. Although often accused of purposely seeking to bite the hand that caught it, this fish is merely responding with the same reflex that many fishes display in the same circumstances. However, its thick and powerful jaw muscles which it uses to crush large clamshells, and its pointed canine teeth can seriously wound careless fingers and hands. The wise angler uses a pair of long pliers to remove hooks or, better still, cuts the line and worries about retrieving the hook only when the fish is dead and being filleted.

Wolffish, dwellers of deep waters and hard substrates, are usually caught by anglers fishing for cod, cusk, and other bottom dwellers. The only area where nearshore anglers consistently encounter this species is near Eastport, Maine, where a deep onshore drop-off attracts deeper-water species such as the wolffish and redfish.

Wolffish are caught using tackle suitable for large bottom-dwellers, such as the cod. Traditional gear includes a 6–8-foot boat rod and a 3/0 to 6/0 conventional reel spooled with 40-pound test monofilament or Dacron line. A sinker large enough to hold the baited hook firmly on the bottom should also be used. Clams are preferred bait, but starfishes and even sea urchins will attract this species to the hook.

Although they may be landed year-round, wolffish feed less actively during the fall, when they shed their old canine teeth and grow new ones and when they are spawning and guarding nest sites. However, they will still suck up a shucked clam if it is dropped in their own "backyard."

The wolffish, commonly marketed as ocean catfish, is excellent table fare. It is called *loup de mer, lobo, gata, Zeewolf,* and *bavosa lupa* in Europe, where it is a very popular addition to restaurant menus.

Wolffish can be deep-fried or marinated in a barbecue sauce and grilled. For an unusual treat, try Skewered Serpent. To prepare this dish, make a garlic marinade by mixing four to six sliced or crushed cloves of garlic, 1 tablespoon of salt, 1 cup of red or white wine, 1 tablespoon of seafood seasoning (or a combination of a little thyme, rosemary, oregano, and some bay leaves), and 1 or 2 teaspoons of hot pepper sauce. Place wolffish fillets in a glass baking dish and cover with marinade. Store in the refrigerator overnight, turning several times. Drain, reserving the marinade for basting.

After marinating, lay fillets flat and cut

lengthwise into long strips 1 inch wide. Thread the strips of fish onto wire skewers making *s*-shaped loops, so the strips look like snakes or folded ribbons. Use double-pronged skewers, if available, for ease of turning.

Skewered fish can be broiled, but grilling may be preferable. If grilling, brush lightly with olive oil before cooking over high heat for 10 minutes or less, turning once. Use tongs to lift skewers gently every few minutes so the fish won't stick to the grill. When the strips are lightly browned and flaky, remove from the grill and add the "serpent eyes" (stuffed green olives placed at the ends of the skewers with the pimento stuffing showing). Serve one "serpent" to a person on either a large leaf of romaine lettuce, a bed of raw or cooked spinach, or pasta. Skewered Serpent is a dish that can establish your reputation as a chef and one that will be remembered by family and friends alike.

References

Albikovskaya, L. K. 1982. Distribution and abundance of Atlantic wolffish, spotted wolffish and northern wolffish in the Newfoundland area. *NAFO Sci. Coun. Stud.* 3:29–32.

Jonsson, G. 1982. Contribution to the biology of catfish (*Anarhichas lupus*) at Iceland. *Rit Fiskideildar* 4:1–26.

Templeman, W. 1984. Migration of wolffishes, *Anarhichas* sp., from tagging in the Newfoundland area. *J. Northw. Atl. Fish. Sci.* 5:93–97.

Templeman, W. 1985. Stomach contents of Atlantic wolffish (*Anarhichas lupus*) from the Northwest Atlantic. *NAFO Sci. Coun. Stud.* 8:49–51.

Templeman, W. 1986. Some biological aspects of Atlantic wolffish (*Anarhichas lupus*) in the Northwest Atlantic. *J. Northw. Atl. Fish. Sci.* 7:57–65.

Atlantic Mackerel
(*Scomber scombrus*)

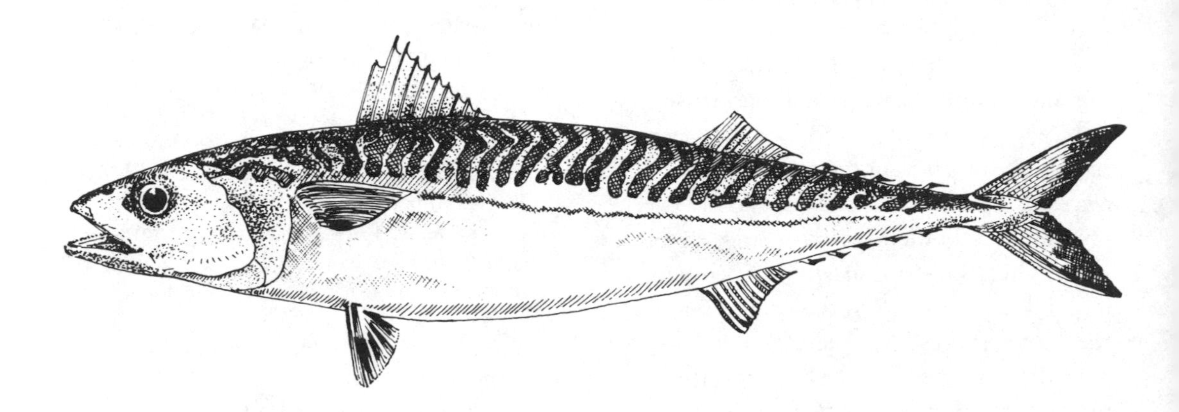

The Atlantic mackerel, a popular recreational species because of its seasonal abundance and voracious feeding habits, is native to both sides of the Atlantic Ocean. On our coast, it ranges along the continental shelf from Labrador south to Beaufort, North Carolina, and is abundant enough to sustain fisheries from the Gulf of St. Lawrence to Cape Hatteras.

The swiftly swimming mackerel has a streamlined body that is about five times as long as it is wide; a long, pointed head; and a mouth with numerous small teeth. It has two large dorsal fins and one anal fin. The tail is broad and deeply forked and is preceded by a series of finlets on the dorsal and ventral sides. The tiny scales of the mackerel make the fish feel velvety when stroked.

The upper sides of the mackerel are dark gray to blue, and the head is almost black. Up to 33 blackish bands run vertically on each side of the body. This bold striped pattern easily distinguishes the mackerel from other recreational species in coastal New England. A dark stripe runs lengthwise along the body below the banded area. The lower sides and the gill covers are silvery, and the belly is silvery white.

Biology

The mackerel is typically an open-ocean fish, but large schools occasionally stray into estuaries and harbors in search of food. Most mackerel inhabit the inner half of the continental shelf, with none straying beyond the shelf's outer edge. Although frequently found near the water's surface, they also can be found as deep as 183 m (600 ft).

The mackerel ranges over extensive areas, traveling in schools that often contain many thousands of fish. All individuals within a specific school tend to be the same size. Mackerel swim at high speeds for extended periods of time while searching for food (for example, small 1-year-old mackerel can swim at an average speed of 13 miles per hour). Because cruising speed increases significantly with age and size,

Table 27. Length at age in cm (inches in parentheses) of Atlantic mackerel from the northern stock

Age	*Length*
1	25.7 (10.1)
2	30.1 (11.9)
3	32.2 (12.7)
4	34.0 (13.4)
5	35.0 (13.8)
6	36.2 (14.3)
7	37.4 (14.7)
8	38.3 (15.1)
9	38.9 (15.3)
10	37.9 (14.9)
11	39.6 (15.6)

Source: Data from MacKay 1979.

conformity of body size within a specific school is necessary to allow all fish to maintain similar swimming speeds.

There appear to be two separate population contingents of Atlantic mackerel that display different spawning and migratory habits. The more southerly of these two groups overwinters between Chesapeake Bay and Long Island. In the spring they move inshore between Cape Hatteras and Delaware Bay, then migrate northward, spawning in April and May off the New Jersey and Long Island coasts. After spawning, these southern mackerel move northward along the New England coast, arriving in the Gulf of Maine by July, where they summer until moving southward once more in October. The northern group of mackerel moves inshore to the southern New England coast by late May, then migrates north and east along the Nova Scotia shore, where it is joined by additional individuals moving inshore from the continental shelf in that region. Mackerel summer in the Gulf of St. Lawrence, where they spawn in June and July. Northern fish leave the Gulf in September, moving south and west along Nova Scotia to the Gulf of Maine. They move off the coast by December to overwinter in deeper continental shelf waters.

Many males and females reach sexual maturity at the age of 2, and all do so by age 4. The fecundity of females increases with age and size, an individual female spawning from 550,000 to 1 million eggs per season.

Mackerel spawn near the surface, and the eggs float in the water column. Hatching occurs within 90 to 102 hours when the water is 13 to 14 degrees C (55 to 57 F). Larval mackerel form into schools about 40 days after hatching, at which time they are approximately 5 cm (2 in) long.

The mortality rates of larval mackerel fluctuate from year to year but are generally very high. Fewer than 10 larvae out of every 1 million eggs produced may survive years when wind-driven currents drive them far offshore over the continental shelf. Predation and starvation are other factors accounting for the high death rate of young fish.

Although potentially reaching sizes in excess of 3.2 kg (7 lb), Atlantic mackerel typically do not exceed about 47 cm (18.5 in) fork length and 1.4 kg (3 lb) in weight. Most caught by anglers are 30.5 to 35.6 cm (12 to 14 in) in length. The maximum age for mackerel is about 20 years. Table 27 lists lengths at age for mackerel from the northern stock.

Young mackerel feed on microscopic copepods. As they grow, they feed on progressively larger prey. Adults will eat any fishes smaller than themselves, feeding heavily upon small herring, sand lances, and young mackerel. They also commonly consume a variety of invertebrates such as copepods, crab larvae, squid, and shrimp. Their voracious feeding behavior leads them to strike at a wide array of baits thrown in their paths by anglers. In the spring, mackerel are thin-bodied because they have eaten very little during the winter; conversely, in late summer and fall they are usually fat from feeding upon abundant inshore foods.

Numerous animals feed upon mackerel, including whales, porpoises, mackerel sharks, thresher sharks, spiny dogfish, cod, tunas, bluefish, and striped bass. Squid feed on small mackerel, and sea birds snatch this species from surface waters.

Status and Management

The Atlantic mackerel has exhibited rapidly fluctuating population densities since colonial times. Fluctuations typically occur more rapidly and to greater extremes than do those of many other species. Peak years of abundance usually have produced record harvests; for example, in 1885, 45,455 mt (100 million lb) were landed in Massachusetts ports alone. Such periods were normally followed by rapid reductions in catch; after 1885, harvest levels dropped to as low as 263.6 mt (580,000 lb) for the entire East Coast in 1910. In 1916, the catch was 25 times greater than that in 1910. Gulf of Maine harvests in peak years were occasionally 50 to 100 times greater than in poor years.

Although the mackerel ranked behind only the haddock, cod, and redfish (ocean perch) in commercial value in the northeastern United States during the 1940s, harvest did not truly intensify until an extensive foreign distant-water fishing fleet entered the fishery in the 1960s. In spite of increasing harvests, mackerel biomass increased from 600,000 mt (1.3 billion lb) in the early 1960s to 2.4 million mt (5.3 billion lb) by 1969; this increase in biomass was felt to be the result of four consecutive years of highly successful reproduction (1966 to 1969). As harvests increased through the early 1970s (peaking in 1973 at 430,400 mt; 946.9 million lb), population biomass began to decline. In response to this, the ICNAF Commission established catch quotas for all nations participating in the commercial fishery. However, biomass continued to decline, dropping to 500,000 mt (1.1 billion lb) by 1978 (Figure 13). This decline was in part due to eight successive years of only modest reproductive success starting in 1970.

Exclusion of the distant-water fleet from the fishery with the passage of the Magnuson Act markedly reduced harvest; commercial landings in 1977 were less than one fifth those of the record catch of 1973. U.S. commercial harvest was low due to the modest value of this species in domestic markets. Population biomass levels stabilized between 1978 and 1981; then, due to modest harvest levels and improved reproductive success in the early 1980s, biomass tripled by 1986. By the mid 1980s, stocks had recovered enough to allow the United States to enter into joint ventures with distant-water vessels, thus markedly increasing the harvest by vessels other than those from America (Figure 14).

Recreational harvests averaged 3,000 mt (6.6 million lb) from 1979 to 1986. This accounted for 2% to 16% of the total domestic and foreign harvest for that time period, an average of about 8% annually. However, recreational landings accounted for 19% to 65% (an average of 42%) of the U.S. harvest. The proportion of the domestic catch landed by anglers declined in the latter half of the 1980s due to an increase in activity by the U.S. commercial fishery.

Since passage of the Fishery Conservation and Management Act, the Mid-Atlantic Fishery Management Council has been responsible for developing Atlantic mackerel management plans. As of 1989,

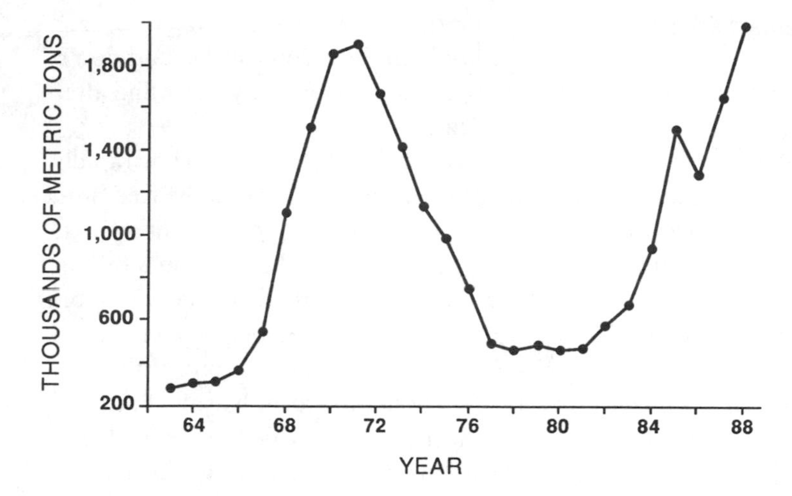

Figure 13. Estimated biomass of age-1 and older Atlantic mackerel from the Labrador North Carolina region (from *Status of the Fishery Resources off the Northeastern U.S. for 1989,* NOAA Mem. NMFS-F/NEC-63).

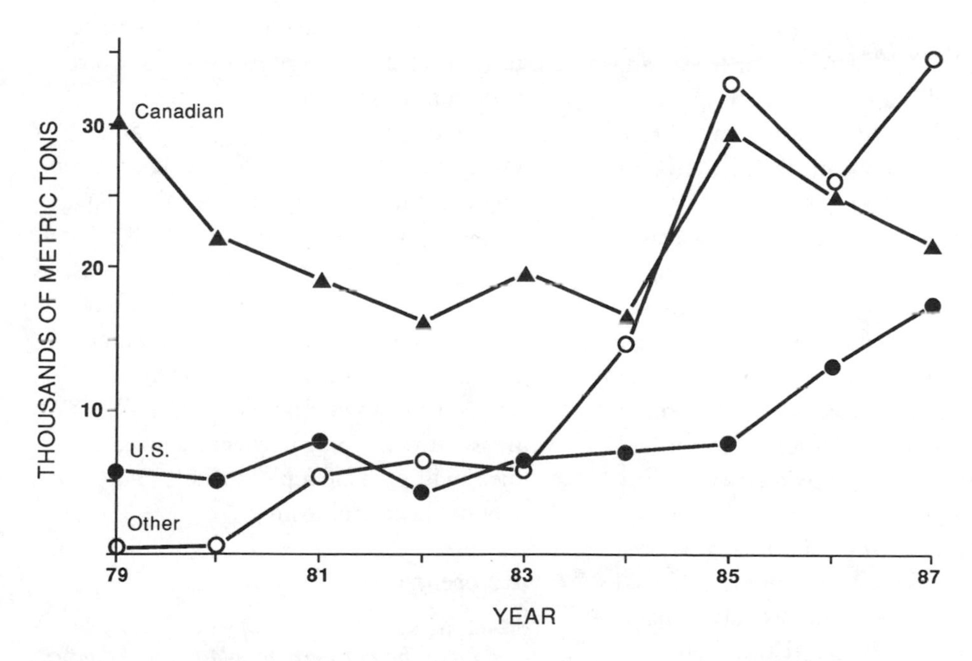

Figure 14. Annual harvest of Atlantic mackerel from Labrador to North Carolina by U.S., Canadian, and all other fleets (from *Status of the Fishery Resources off the Northeastern U.S. for 1988,* NOAA Tech. Mem. NMFS-F/NEC-63).

no New England state regulated the harvest of this species.

Angling and Handling Tips

The Atlantic mackerel can be found anywhere along the New England coast, from deep water to shallow bays, usually inhabiting the upper 10 to 25 feet of the water column. Prime fishing occurs in the fall, when mackerel are fattened from a summer of feeding. Anglers fish for them from boats or shoreline sites such as piers, jetties, bridges, rocky islands, and beaches.

A medium spinning rig spooled with 15-pound test line is best for casting, although adventurous anglers may use medium- or light-action spinning rods with a single 1-ounce mackerel jig or a saltwater fly rod rigged with a streamer. Any metal lure that resembles a sand eel or other bait fish can be used when casting (for example, a miniature Hopkins, Kastmaster, or leadhead, or even small plugs). After casting, lures should be allowed to flutter down through the water for a moment, then jerked and retrieved rapidly.

Many anglers fishing from a boat use the mackerel tree, a small diamond jig preceded by two or more 1/0 surge tube worms. The tree is jigged so that it resembles a larger fish chasing small bait fishes. Typically the jig is dropped to the bottom, lifted with a jerk, and allowed to settle; this action should be repeated at a rapid pace.

Bait-fishing with sand eels, sea worms, squid, or small fishes on long shank hooks with on-line sinkers can also catch mackerel. Mackerel strike hard and then momentarily mouth the bait before attempting to swallow it. Therefore, set the hook on the second strike, and don't allow the fish to have any slack line after hooking.

Since mackerel lose their flavor rapidly if not kept cool, they should be iced immediately upon capture. If ice is not available, they should be kept wet in a burlap bag that is frequently dampened; evaporation from the bag will help keep the mackerel cool. The mackerel is one of the easiest species to clean. Its soft skin doesn't need to be scaled or skinned. To clean, hold the fish's head in your left hand—belly toward you—and make a single diagonal cut from behind the head to just behind the anal opening. Gut and wash out the body cavity, and your fish is cleaned. If you want mackerel fillets, remove each side of the fish, leaving only the center backbone. Lay the fillet flat and make a cut on either side of the center line to remove the fine bones from the rib section in a single ¼-inch-wide strip.

Salted mackerel with potatoes and biscuits is a traditional New England Sunday breakfast. Today's tastes might have changed, but the mackerel can still provide a delightful meal. Its meat is dark, soft-textured, and has a sweet flavor when fresh. Mackerel is excellent when marinated in citrus juices for a half-hour and grilled skin down on a covered grill. Brushed with spicy barbecue sauce, this meal will provide a pleasurable ending to a day of successful fishing.

Reference

MacKay, K. T. 1979. *Synopsis of biological data of the northern population of Atlantic mackerel.* NMFS Tech. Rep. 885: 26 pp.

Atlantic Bonito
(*Sarda sarda*)

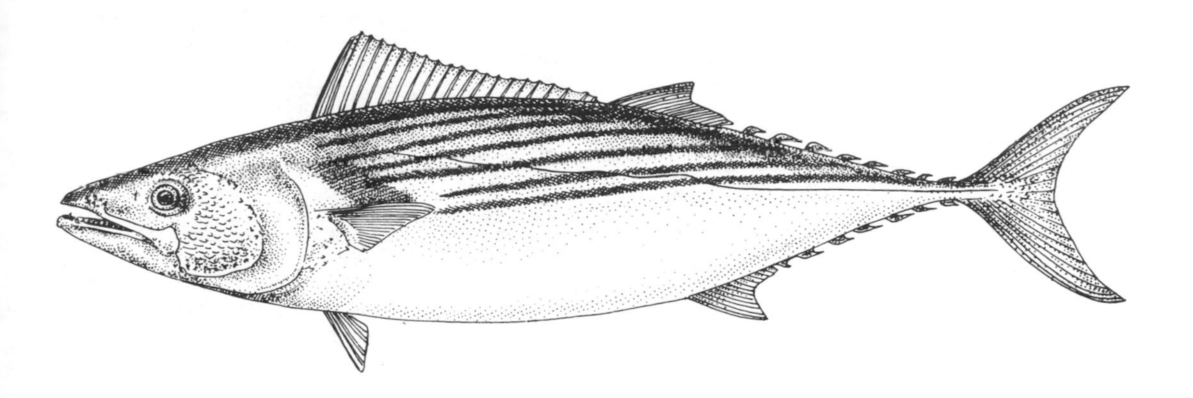

The Atlantic bonito (also called the skipjack and horse mackerel) is an open-ocean species of tropical and temperate coastal areas on both sides of the Atlantic, in the Gulf of Mexico, and in the Mediterranean and Black seas. It is common along the Atlantic coast of North America northward to Cape Cod and is infrequently sighted as far north as Casco Bay, Maine, and the outer coast of Nova Scotia.

The bonito has a series of finlets on the dorsal and ventral surfaces of the caudal peduncle that places it with other tunas and mackerels in the fish family Scombridae. The bonito is shaped like a small tuna, having a stout body and slender caudal peduncle. Unlike the false albacore, its body is fully scaled. It can easily be separated from other tunas caught in New England waters by its dorsal fin, which is markedly longer than it is high.

The bonito is steely blue dorsally and silvery on the lower sides and belly. The upper sides are barred with series of dark bands running obliquely across the lateral line toward the tail.

Biology

The bonito is a highly mobile, schooling species found throughout open waters of the continental shelf but normally restricted to waters less than 200 m (656 ft) deep. It is found in water temperatures from 12 to 25 degrees C (54 to 77 F) but is most abundant where the temperature is between 15 and 22 degrees C (59 to 72 F). It is a summer and fall transient on the New England coast, migrating southward during the rest of the year.

Most individuals reach sexual maturity at the age of 2 years, although some do so at 1 year. Fecundity of females increases with size and age. Studies have reported varying numbers of eggs for specific-size females, but fecundities generally range from 700,000 eggs for small females (about 40 cm, or 18 in, in length) to as many as 3 million to 6 million for large females (longer than 70 cm, or 28 in). Spawning typically occurs in the summer to the south of New England, although fertilized eggs have been collected off the coast of Massachusetts.

Juveniles display fast growth, increasing in length nearly 0.3 cm per day during their first summer of life. Information concerning adult length at age is variable, as shown in Table 28. Maximum size is over 91 cm (36 in) and about 6.8 kg (15 lb).

Larval bonito feed upon copepods and small fish larvae. Juveniles and adults eat squid and a variety of fish species, including mackerel, alewives, menhaden, sand lances, Atlantic silversides, and smaller bonito. Juveniles and adults are daytime feeders, displaying peaks of activity in the morning and evening. Bonito are often sighted leaping from the water in large numbers as they chase schools of smaller fishes near the water's surface.

Table 28. Calculated length at age in cm (inches in parentheses) of the Atlantic bonito

Age	Black Sea	Sea of Marmara 1957	Sea of Marmara 1958
1	31.5 (12.4)	47.6 (18.7)	41.1 (16.2)
2	41.5 (16.3)	59.2 (23.3)	52.9 (20.8)
3	48.8 (19.2)	64.5 (25.4)	59.2 (23.3)
4	56.2 (22.1)	68.7 (27.0)	63.8 (25.1)
5	61.2 (24.1)		
6	66.5 (26.2)		
7	69.5 (27.4)		

Source: Data selected from summaries in Yoshida 1980.

Status and Management

The Atlantic bonito supports active commercial fisheries in the eastern North Atlantic and the Mediterranean Sea but does not do so in the northwestern Atlantic off the New England coast. Management of this species, as for all Atlantic tunas, is largely the responsibility of the International Commission for the Conservation of Atlantic Tunas (ICCAT). Due in part to the lack of intensive commercial fishing, harvest was unregulated as of 1989.

Angling and Handling Tips

The streamlined and powerful bonito, one of the swiftest of game fishes, is avidly pursued by many anglers who have experienced the battle provided by this little "torpedo." Bonito feed in schools, driving bait-fish species to the surface, often clearing the water in spectacular leaps while in pursuit of their next meal. Although a "blue-water" species, the bonito can be found cruising close enough to the coastline or islands of southern New England in summer and early fall that even anglers with small boats can enjoy fishing for this compact version of the mighty tuna.

Swimming at speeds up to 30 or 40 miles per hour, the bonito strikes hard and runs deep and fast. Landing one or two of these battlers is a taxing but satisfying and exciting experience.

Use a light saltwater or a medium/heavy spinning outfit to get the greatest battle out of this scrapper. Trolling rods should be light with a fast-tapering tip to absorb the bonito's powerful strike. For trolling or jigging, use a smaller conventional reel that has a fast retrieve and a good drag that is lightly set. When a bonito strikes and starts to run, it can peel up to 100 yards of line from the reel in very little time. When it reverses direction and heads back toward the boat, line must be retrieved as quickly as possible. That's when a 5:1 ratio spinning reel will prove worth its weight in gold!

Although some anglers use lines as light as 4–8-pound test, 12–15-pound test is appropriate. When fishing with bait, use No. 1 to 2/0 short-shanked hooks. Bonito strike baits such as squid, mullet, or other small, whole bait fishes and a variety of lures such as small Kastmasters, Hopkins, Rebels, spoons, feathered jigs, and tailless poppers. Troll bait or lure at speeds in excess of 8 miles per hour, keeping rigs near the wake of the boat; the noise and turbulence tends to attract this species.

Chumming is also useful. Butterfish, spearing, sand eels, or other fishes can be used as chum. Lay out a chum slick and drift or anchor, depending upon the current. Bait with a fresh fish big enough to hide the hook in the head. Use a small weight attached 2 to 3 feet above the hook to keep the bait down in the water.

The bonito is a light-fleshed tuna that is wonderful either grilled or broiled. It needs less special handling than larger tunas, because its small size allows it to be bled easily and iced quickly. Bleeding the fish will produce a lighter flavor. To bleed, make one cut under the head and gills, a second behind the pectoral fin, and a third into the area in front of the tail. Next, fillet the fish and cut off the darker meat near the backbone; if left, this meat can impart a bitter flavor to your bonito. Rinse as soon as possible in water and ice, continuing until the water running off the meat is clear.

For a lighter color, soak the fillets in a solution of 2 cups of salt to 1 gallon of water in a noncorrosive container for at least 30 minutes. Remove, rinse with fresh water, and store chilled. Canning in jars works well if you have a pressure cooker. Follow canning directions for fish. Like most tuna meat, fresh bonito doesn't store frozen for more than 3 months. For increased storage life, cook in a canner for 20 minutes and freeze.

Mediterranean ethnic groups such as the Basques love the full flavor of the bonito when cooked in simple casseroles. This species can be cooked as you might a bluefish or mackerel, but its firmer texture and lighter color beg to be marinated and grilled. It can be marinated in wine vinegar, orange juice, and even lemon-lime soda. For a tasty treat, try marinating in barbecue sauce for an hour, then grilling for 10 minutes per inch of thickness, turning once. Its firm flesh will hold together like a piece of red meat.

Teriyaki Tuna is a special treat that will be enjoyed by everyone. This recipe works wonderfully with any type of fresh tuna. Cut 1-pound pieces of tuna steak into ¼-inch-thick slices about 2 inches long and place in a bowl. Add three large cloves of chopped garlic, 2 tablespoons of soy sauce, 1 tablespoon of teriyaki sauce, ½ teaspoon of seafood seasoning (such as Old Bay), 1 tablespoon of olive oil, and ⅔ of an inch of freshly shredded ginger. Mix and allow to marinate in the refrigerator for 1 hour. Meanwhile, thinly slice ½ pound of mushrooms and some sweet red, green, or yellow peppers (all three together make an attractive dish) and add to 2 tablespoons of olive oil in a hot wok or iron skillet. Heat and stir until mushrooms and peppers are wilted; then remove. Put the marinade and fish into the hot pan and sear at high heat for 2 to 3 minutes. Add mushrooms and peppers and ½ cup of white wine, stirring constantly for 8 minutes. Remove the tuna and vegetables and reduce sauce by one third. Serve over angel-hair spaghetti or rice.

Reference

Yoshida, H. O. 1980. *Synopsis of biological data on bonitos of the genus* Sarda. NOAA Tech. Rep. NMFS Circ. 432: 50 pp.

False Albacore
(*Euthynnus alletteratus*)

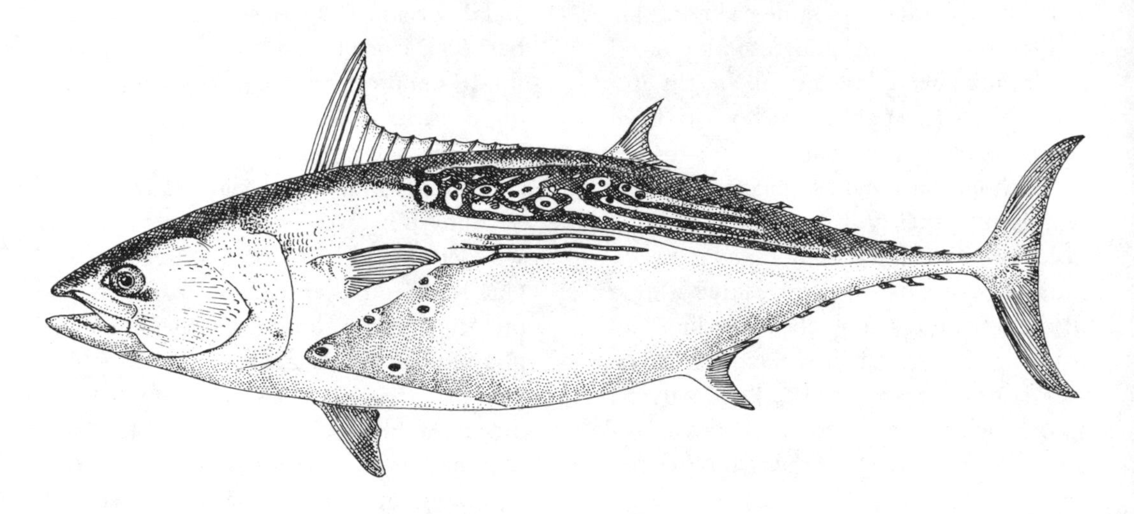

The false albacore (or little tunny), noted for its battling skill, is found along both coasts of the Atlantic Ocean and in the Mediterranean Sea, the Sea of Marmara, and the Black Sea. In the western Atlantic, it occurs from the Gulf of Maine to the coast of Brazil, as well as in the Gulf of Mexico. It is only occasionally seen north of Cape Cod.

Unlike other New England tunas which have minute scales over most of their bodies, the false albacore is scaleless, except on the anterior upper part of the trunk, around the pectoral fin, and along the lateral line. Also, unlike the bonito, lateral stripes occur only above the lateral line. Its second dorsal fin is noticeably shorter than the first, or anterior, one, which separates it from the bluefin and yellowfin tunas. The little tunny is steel-blue dorsally, fading to white on the lower sides and belly. It has several blackish spots between the bases of the pectoral and pelvic fins.

Biology

The little tunny is usually found in near-coastal waters or, if away from continental coastlines, around offshore shoals or islands. Along the Atlantic coast of the United States it is restricted to the continental shelf, rarely occurring in slope waters or seaward. It usually frequents areas with strong currents. This species ranges through such habitats in large, mobile schools of equal-sized individuals, often swimming in mixed schools with other species such as the Atlantic bonito. The false albacore makes seasonal migrations along the Atlantic coast of the United States, moving northward as waters warm in spring and summer. It occurs along the South and North Carolina coasts in May and June of each year, usually appearing in southern New England by August and September.

Sexual maturity has been recorded at a variety of sizes among the geographic regions where this species occurs, with females maturing when as small as 27.2 cm (10.7 in) off the coast of Florida and as large as 57.0 cm (22.4 in) in other regions. The spawning season is prolonged, generally lasting from March to November. A 75.0-cm (29.5-in) female may produce as many as 1.8 million eggs, which will be released in several large batches during the spawning season.

Maximum size for this species has been reported as 122 cm (48 in) in length. Length and weight at age for the little tunny are listed in Table 29.

The false albacore feeds during daylight hours, actively attacking schools of prey species such as the sand lance, herring, mackerel, and young false albacore. Crustaceans, particularly euphausiid shrimp, and squid are also important foods. Crustaceans have dominated the diet of the false albacore in specific studies. This species probably competes with other fishes such as the Atlantic bonito, and marine mammals such as dolphin species, for these foods. This species is eaten by various species of sharks and yellowfin tuna.

Status and Management

The little tunny has historically carried only limited commercial value throughout its range. This species is harvested along the coast of Africa in beach seines, sardine purse seines, traps, and encircling gill nets. Little tunny are taken in the western Atlantic and Gulf of Mexico, but harvest occurs largely in a mixed-species fishery and only at modest levels. In the mid 1970s to early 1980s, the largest catches were taken from the Caribbean and along the northern coast of South America near Venezuela. Like other tunas, management of this species would be accomplished under international agreement. However, because commercial catch levels are so modest, harvest in the Northwest Atlantic was not managed as of 1989.

Table 29. Average length in cm (inches in parentheses) and weight in kg (pounds in parentheses) of false albacore from the Mediterranean Sea

Age	*Length*	*Weight*
1	35.8 (14.1)	0.8 (1.8)
2	53.9 (21.2)	2.8 (6.2)
3	63.7 (25.1)	4.5 (9.9)
4	70.2 (27.6)	6.0 (13.2)
5	75.5 (29.7)	7.5 (16.5)
6	80.2 (31.6)	8.5 (18.7)
7	81.0 (31.9)	9.0 (19.8)

Source: Yoshida 1979.

Angling and Handling Tips

As summer fades to September and coastal waters reach their greatest warmth, the false albacore ("albi") enters the southern New England angling scene. This species, larger than the bonito, usually ranges from 6 to 12 pounds in weight, although individuals exceeding 15 pounds are occasionally landed. Albacore are powerful battlers that can strip line down to the backing or swim toward a boat so rapidly that the fastest retrieval cannot keep up with the slackening line.

Although most abundant offshore, false albacore occasionally drive schools of bait fishes toward the shoreline, bringing the albis well within range of shore anglers. Shoreline fishing is best from rocks or jetties on either side of a high tide. If in a boat, fish around underwater structures or ridges that attract bait species.

For the most exciting battle, use a lightweight surf or boat outfit, such as a 6½-foot saltwater spinning rod with a large-capacity reel that has a good drag. A fast-tapered boat rod with a flexible shock tip will act like a shock absorber, preventing the loss of this fish during its frantic, powerful runs. A conventional reel, such as a Jigmaster, which has large line capacity and a 4:1 retrieval ratio, is well suited to albacore fishing. The drag should be set lightly, since drag is increased as the reel empties and the spool turns faster. You'll get more hits with lighter line; 10–12-pound test monofilament or lighter is best.

You can catch albis trolling, casting, jigging, or chumming. They will strike lures such as the Hopkins, Kastmaster, and various leadheads, and surface lures such as 5-inch Rebels, Rapalas, and pencil poppers. All lures need to be retrieved or trolled at high speed in order to attract this species.

Bait-fishing is also common. A variety of fresh baits or frozen sand eels are successful. When bait-fishing, use as little terminal tackle as possible. Hooks and lures must be tied directly to the line without any hardware. This species has keen eyesight and will strike less often if wire leaders are used. Use short-shanked 2/0 to 3/0 hooks that are well hidden in the bait. If you are fishing from shore, let the bait float out with the tide. If on a boat, use ground pogies and chopped pieces of butterfish mixed in salt water as a chum. When fishing in a chum slick, add a lead shot 3 or more feet above the hook. This will allow you to test different water depths. Drop the bait into the chum near the boat and let it drift with the movement of the water. After the bait drifts 75 to 100 feet, bring it in slowly and repeat the drift. Don't allow the bait to sit motionless in the water at the end of a taut line; keep it moving constantly and experiment with depth.

For an exciting fishing experience, try fly-fishing with white bucktails and saltwater fly rods, as some of the old salts do on Martha's Vineyard. Then the next time someone brags about the large steelhead he caught, tell him about your experience landing a "silver bullet" on a fly rod.

Americans do not favor the albacore for eating as much as the somewhat whiter-fleshed and milder-tasting bonito. As our palates become more worldly and our cooking more sophisticated, our taste for the tunny may change, as it has for other fresh tunas. As with all tunas, the tunny should be bled as soon as it is landed. To bleed, simply hold the fish by the tail and cut behind the gills. Draining the blood over the side of the boat has the added advantage of enhancing the chum slick. The flavor of albacore steaks and fillets can also be lightened by marinating. Sprinkle with salt and soak in orange juice for several hours prior to cooking, though this will also remove some healthful omega-3 oils from the tunny meat.

The false albacore can be dressed and cooked whole, filleted, or steaked. This species is baked, steamed, sautéed, broiled, grilled, smoked, or blackened Cajun-style.

To hold the meat together, leave the skin intact when grilling. To grill, make an Asian marinade of ½ cup of soy sauce, ½ cup of dry sherry or white wine, 2 tablespoons of lemon juice, two cloves of minced or crushed garlic, 4 teaspoons of freshly grated ginger root, and 1 teaspoon of sugar. Soak tunny steaks in this marinade for several hours, dry, and brush with a little olive oil to prevent sticking. Place on a hot grill for 6 minutes per side, turning once and basting with the marinade.

A Mediterranean-style tuna dish that is sautéed/steamed is a delightful change of pace. Prepare a marinade of ¼ cup of olive oil, ½ cup of red wine, 2 tablespoons of lemon juice, freshly ground black pepper, and a light sprinkling of dried basil. Place an albi fillet in the mixture, turn several times to coat, and marinate 3 hours. Next, pour marinade into a deep iron skillet. Add two minced onions and several minced cloves of garlic and lightly brown. Next, add ½ cup of tomato paste, ½ cup of white wine, ½ teaspoon of oregano, and ½ teaspoon of basil. Stir while heating, and add the tuna. Cover and simmer for 10 minutes. Uncover, sprinkle with parsley, and cook 5 minutes more, spooning the sauce over the tuna. Serve on a bed of rice or risotto.

To bake steaks, prepare a marinade of 2 cups of red wine and four cloves of freshly minced garlic. Next, slice several large stalks of celery, two carrots, and an onion in ¼-inch pieces. Dust the fillets or steaks with salt and a little dried basil; then add to the marinade along with the vegetables. Refrigerate for 6 hours to overnight, turning several times.

After refrigerating, remove fish and strain out the vegetables. Put 1 tablespoon of olive oil or butter into a skillet, add the vegetables, and sauté until tender. Oil a baking dish and place an even layer of sautéed vegetables on the bottom. Lay the fillets on top and tuck the thin tail ends under so they won't overcook. Place in a 400-degree oven for 12 to 15 minutes, testing with a fork to see if the fillets will flake. Remove the fillets and keep warm. Return the dish and vegetables to the stove top. Mix 1 tablespoon of cornstarch with ¼ cup of water and add to the sauce in the dish. Heat and stir until the sauce thickens. Serve the sauce and fillets over thin spaghetti seasoned with garlic powder and olive oil. Top with plenty of grated cheese, freshly ground black pepper, and a sprig of fresh parsley.

References

Collette, B. B., and C. E. Nauen. 1983. *FAO species catalogue.* Vol. 2: *Scombrids of the world.* FAO Fisheries Synopsis No. 125: 137 pp.

Yoshida, H. O. 1979. *Synopsis of biological data on tunas of the genus* Euthynnus. NOAA Tech. Rep. NMFS Circular 429: 57 pp.

Yellowfin Tuna
(*Thunnus albacares*)

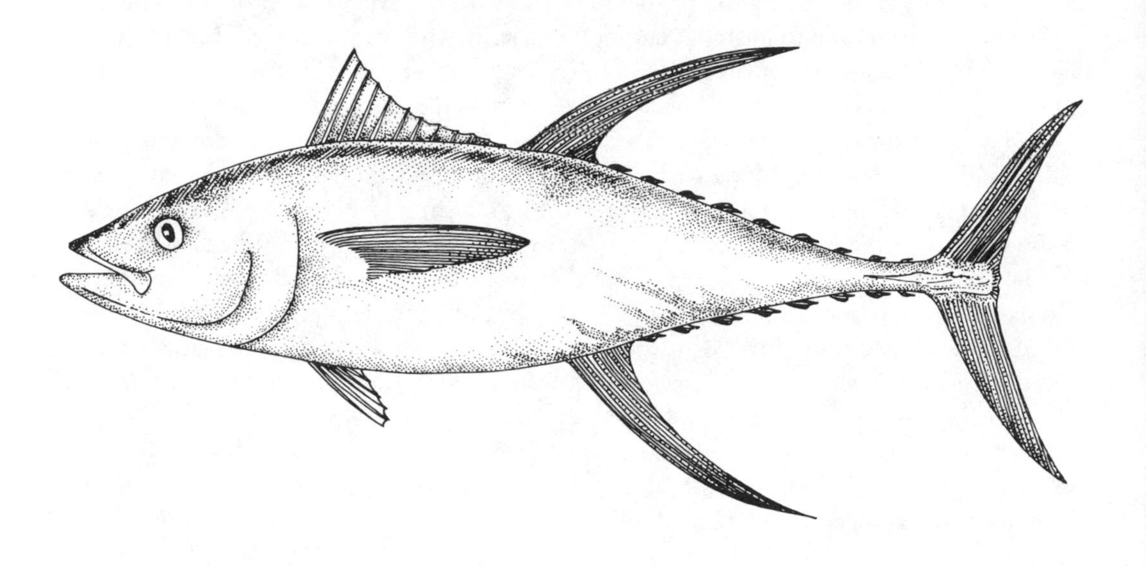

The yellowfin tuna is circumtropical in distribution, occurring from about 40° north latitude to 35° to 40° south latitude throughout the Atlantic, Pacific, and Indian oceans but is absent from the Mediterranean Sea. In the Northwest Atlantic, it occurs along the edge of the continental shelf off Nova Scotia southward.

This species can be distinguished from the bluefin tuna by its relatively longer and sickle-shaped second dorsal and anal fins. Its pectoral fins reach posteriorly to below the second dorsal fin, whereas the bluefin's shorter pectorals do not extend beyond the first dorsal.

The yellowfin is metallic dark blue dorsally, with yellowish sides and a silvery belly. The belly often has about 20 faint and broken vertical lines. The dorsal and anal fins and finlets are bright yellow, and the finlets have a narrow black border.

Biology

This open-ocean, schooling species occurs throughout the upper water column in temperatures between 18 and 31 degrees C (65 to 88 F) and is normally most abundant at 20 degrees C (68 F) or higher. These temperature ranges determine not only where this species is located geographically but also where it occurs vertically within the water column.

Like other tuna species, the yellowfin tuna is warm-blooded, individuals maintaining an internal body temperature that can be significantly higher than the water they are swimming in. Increased body temperatures are believed to increase the power of swimming muscles, which promotes faster speeds and allows for more prolonged periods of rapid swimming while on feeding forays.

Schooling usually occurs near the surface, with most individuals in particular schools being of nearly identical size. Yellowfins can be found in mixed schools with similar-sized individuals of other tuna species. Schools of larger-bodied yellowfins in the eastern Pacific are often associated with schools of porpoises. Large schools of yellowfins frequently concentrate in areas of the ocean that have major upwellings of water, due to the food found in such areas.

Spawning occurs throughout the year, with peaks in the summer months in more northerly and southerly areas of this species' range. Age and size at maturity vary greatly among individuals, with some fish maturing at 12 to 15 months of age and 50 to 60 cm (20 to 24 in) in length, others not until they are up to 120 cm (47 in) long. After hatching, larvae remain in the upper 50 to 60 m (165 to nearly 200 ft) of the water column.

Growth is rapid, with individuals reaching substantial sizes by the end of the first year of life. Table 30 lists calculated lengths at age for yellowfin tuna from the Pacific Ocean. Maximum length appears to be over 200 cm (78.7 in).

Yellowfin tuna feed most actively during daylight hours in near-surface waters. This species feeds upon a wide variety of fishes, cephalopods, and crustaceans, with squid being a particularly prominent part of the diet. One study done on yellowfins captured in the Pacific indicated their diet included at least 37 families of fishes and 8 orders of invertebrates. Smaller yellowfins are eaten by adult tunas, shortbill spearfish, white and black marlins, and sharks.

Table 30. Calculated length at age in cm (inches in parentheses) of yellowfin tuna from the Pacific Ocean

Age	*Length*
1	53.4 (21.0)
2	91.8 (36.1)
3	119.4 (47.0)
4	139.3 (54.8)
5	153.5 (60.4)
6	163.8 (64.5)
7	171.1 (67.4)

Source: von Bertalanffy growth equations, as presented in Cole 1980.

Status and Management

There are important yellowfin tuna fisheries throughout the world's oceans. Historically, up to 35 nations have commercially harvested this species, with Japan and the United States landing the greatest biomass (up to 100,000 mt, or 220 million lb, annually in the early 1980s). Although the greatest harvest levels of the yellowfin have occurred in the Pacific Ocean, the biomass landed in the Atlantic was greater than that of any other tuna species in that ocean in the early 1980s, averaging 154,000 mt (338.8 million lb) per year between 1982 and 1985. Nearly 90% of the harvest in the Atlantic occurred along its eastern coast, particularly offshore from Africa. Purse seine fisheries represented the major means of landing yellowfins in that region. Harvest levels were so high in the early 1980s that concern grew that the Atlantic stock was being overfished. However, in 1984 nearly half of the fishing effort in the eastern Atlantic shifted to the Indian Ocean, where stocks had experienced markedly lower rates of exploitation up to that time. This reduction in fishing effort reduced harvest to levels that would allow desirable stock abundance to be maintained.

Harvest has been much lower in the western Atlantic, with longline and recreational fisheries taking only modest numbers of yellowfins throughout the 1980s. As it does for other Atlantic tunas, the ICCAT recommends harvest regulation to member nations participating in the Atlantic fishery. As of 1989, due to the reduction in catch levels of previous years, the yellowfin stock was deemed in good condition, so minimum size limits were the only means recommended to regulate harvest.

Angling and Handling Tips

In the spring and fall, anglers pursue the yellowfin tuna well out onto the continental shelf in areas with complex bottom topography, including canyons, ridges, and banks. Many anglers take party boats or charter experienced captains and mates to make these offshore ventures. In late summer, when coastal waters are warmest, yellowfins may be found closer to the coastline, well within range of boats owned by many anglers.

Although large 130-pound-class tackle has traditionally been used to fish for the yellowfin, new lighter, stand-up tackle in the 50-pound range is less tiring to the angler and will usually subdue all but the canyon giants. You'll need a quality 4/0 to 6/0 reel with a good drag that is capable of holding at least 300 to 500 feet of 30–50-pound test monofilament line.

The type, size, and color of lure, as well as the speed used to tow it, all need to be considered when trolling. Lures include the traditional Japanese Feathers as well as Smoker baits, Big Eye and Tuna Clones, Silver Bullets, Soft Heads, and Green Machines. Daisy chains (multiple lures tied together in some kind of pattern) are popular with some tuna anglers. When trolling, combine a selection of lures of different sizes on your daisy chain. Use a large lure as an attractor, with midsize (8 to 10 in) lures around it. Bullet-shaped lures should be splashed along the surface, and flat-faced ones should dive with a stream of trailing bubbles.

Bait-fishing is also popular, particularly at night. Hook sizes range from 4/0 to 8/0, with the smaller ones being more easily hidden in the bait and larger ones being strong enough to hold the biggest fish. A 3–4-ounce sinker should be looped with an elastic band that is tied 6 to 10 feet above the hook. If a fish strikes, the elastic band will break, releasing the sinker and preventing it from interfering with reeling in the yellowfin. Use a whole butterfish for bait, hiding one hook inside, and drifting it out from the boat into a chum slick. Add chunks of butterfish to ground menhaden, mackerel, smelt or sand eels to make the chum. Tie a chum box with a bucket of frozen chum to the side of the boat, allowing it to melt as you fish. To drift your bait, pull out a yard or more of line and allow the bait to sink, repeating slowly until up to several hundred feet of line have been peeled and drifted.

Special care is necessary if a yellowfin is to retain the flavorful quality of fresh tuna. To land your fish, gaff it in the gills or lower jaw and use a tail rope to bring it on board; a gaff in the belly will ruin its meat. Once landed, stop the fish from thrashing as soon as possible by covering its eyes and dispatching it with a billy. Because the yellowfin is warm-blooded, elevated body temperatures created during the battle on the line increase the rate of spoilage if the fish is not properly handled immediately. Thus, it should be bled and iced as soon as possible. To bleed, lift the pectoral fin and insert a knife several inches deep into the fin recess, twisting the knife after insertion. Repeat on the other side of the fish. Also cut through the flesh to the backbone on each side between the second and third finlets. After gutting and gilling, chill in a large chest or insulated fish bag. To freeze for storage, cut into steaks and rinse in nonchlorinated water. Before wrapping, place steaks on a tray and freeze overnight. Next, dip the steaks in cool water and put back on the tray; then dip a second time. This will coat the steaks with a thin layer of ice, which prevents freezer burn and the development of off flavors. When thawing, defrost in a colander so that water drains from the steaks as they thaw.

Fresh yellowfin is very popular in the Mediterranean and Asia, but Americans are just learning about the culinary delights associated with eating fresh tuna. We eat nearly one third of all the tuna consumed in the world, but most of this is canned albacore and yellowfin. Fresh yellowfin steaks have a firm texture like red meat and a wonderfully distinctive flavor. All tuna meat has a dark red strip running along its length where fat deposits are stored. This strip, which has a strong flavor like that of the bluefish, can be removed. Also, skin your tuna before cooking.

Fresh tuna steaks can be baked, broiled, fried, grilled, steamed, kabobed, sautéed, or poached to make a fresh tuna salad. Try making appetizers for an enjoyable way of introducing friends to fresh tuna. To make one such appetizer, Tuna Dipping Sticks, cut steaks that are about ¾ inch thick into strips about ¾ inch wide and 3 to 4 inches long. Place in a glass dish and pour enough homemade or commercial Italian salad dressing to coat each piece. Squeeze one or two limes over the strips and marinate for 30 minutes, stirring gently several times. Mix ½ cup of flour, ½ cup of cornmeal, ½ cup of flavored bread crumbs, ¼ cup of grated Parmesan cheese, 1 teaspoon of garlic powder, and 1 teaspoon of Cajun blackened-fish spices (or some ground black and hot red pepper) in a plastic bag. Dip the strips in a beaten egg and shake in the bag to coat. Deep-fry in oil for 2 to 3 minutes, drain in a basket or on paper towels, and keep warm. Serve on a platter as an appetizer, arranged around dipping bowls containing tartar sauce, barbecue sauce, and light soy or Worcestershire sauce.

Like the mako shark, the firm-fleshed tuna can be substituted for veal in such recipes as Tuna Scallopini. For this dish, chill a 3–4-inch-diameter chunk of tuna, and with a sharp knife thinly slice ¼-inch scallopini cutlets. Flatten these cutlets by pressing with the palm of your hand. Place in a bowl and pour ¼ cup of lemon juice over the slices. Allow to sit for several minutes and drain. Place the cutlets in a plastic bag and lightly coat with flour. Meanwhile, heat 2 tablespoons of butter and 2 tablespoons of olive oil in a skillet. Add 1 teaspoon of rosemary, stir several times, and add the scallopini. The pan should be hot but not smoking, and the heat should be kept constant. Sauté, shaking the cutlets in the skillet for about 3 minutes per side, or until browned. Remove the tuna and keep warm in an oven while you deglaze the skillet with 3 tablespoons of Marsala wine (light white wine can be used). Dissolve a chicken bouillon cube in ¼ cup of hot water and add it to the pan along with a teaspoon of butter. Heat, stir, and reduce by one half. Place a little sauce on a plate, and put the tuna scallopini on top of the sauce. Place a sprig of parsley or a scallion on top for a garnish.

References

Cole, J. S. 1980. Synopsis of biological data on the yellowfin tuna, *Thunnus albacares* (Bonaterre 1788), in the Pacific Ocean. In W. H. Bayliff (ed.), *Synopsis of biological data on eight species of scombrids.* Spec. Rep. 2, Inter-American Tropical Tuna Comm., La Jolla, CA.

Collette, B. B., and C. E. Nauen. 1983. *FAO species catalogue.* Vol. 2: *Scombrids of the world.* FAO Fisheries Synopsis No. 125: 137 pp.

Joseph, J. W. Klawe, and P. Murphy. 1988. *Tuna and billfish: Fish without a country.* Inter-American Tropical Tuna Comm., La Jolla, CA.

Bluefin Tuna
(*Thunnus thynnus*)

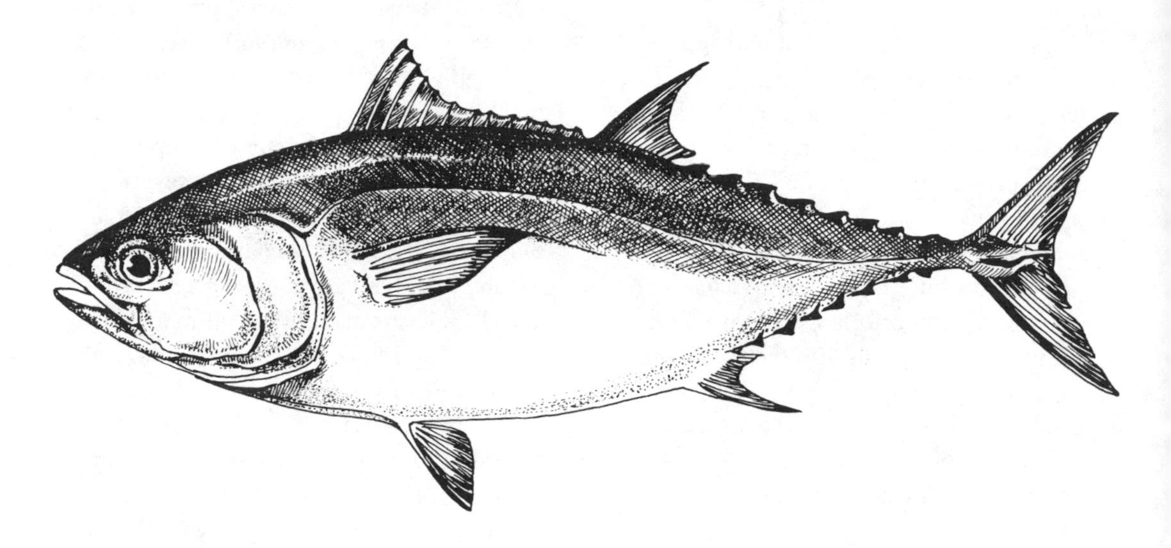

The bluefin tuna, the largest living species of bony fish, is a revered recreational and important commercial species on both the Atlantic and Pacific coasts of North America. In the western Atlantic, it ranges from Newfoundland to the southern West Indies. It is a summer visitor to the coast of Massachusetts.

Although a member of the mackerel family, fishes noted for their streamlined appearance, the bluefin is a robust species. The body tapers anteriorly to a pointed nose and posteriorly to a slim caudal peduncle and a deeply concave tail. Like other members of the mackerel family, the bluefin tuna has a series of small finlets posterior to the second dorsal fin and the anal fin. It differs from the Atlantic mackerel in having the two dorsal fins nearly continuous, and from the bonito in having a tall, narrow-based second dorsal fin. Its second dorsal, anal, and pectoral fins are noticeably shorter than those of the yellowfin tuna.

The bluefin tuna is dark blue to almost black with a gray or green iridescence on the back and upper sides. The lower sides are silvery and are marked with iridescent pink or gray spots and bands. All the fins are dark; the finlets are bright yellow, tipped with black.

Biology

Bluefin tuna travel both in schools and individually within inshore and offshore areas. They rarely occur more than 91 m (300 ft) below the surface and are occasionally sighted swimming at the surface with their dorsal fins exposed above the water.

Bluefin tuna are often categorized into three size groups: juvenile or school tuna, 3 to 32 kg (5 to 70 lb); medium tuna, 32 to 123 kg (70 to 270 lb); and giant tuna, more than 123 kg (270 lb). Bluefins on the Atlantic coast run considerably larger than those on the Pacific coast or in the Medi-

Table 31. Average weight at age in kg (pounds in parentheses) for bluefin tuna

Age	*Weight*
1	4 (8.8)
2	10 (22)
3	18 (40)
4	31 (68)
5	45 (99)
6	67 (148)
7	85 (187)
8	110 (242)
9	130 (286)
10	150 (330)
11	170 (374)
12	200 (440)
13	235 (517)
14	290 (638)

Source: Data from Grosslein and Azarovitz 1982.

terranean Sea. Atlantic giants may reach lengths exceeding 3 m (10 ft) and can weigh well over 455 kg (1,000 lb). Average weights of bluefins up to 14 years of age are listed in Table 31. Maximum age is about 38 years.

Bluefins follow extensive migratory routes along the Atlantic coast, moving northward and inshore in the spring and summer and southward and offshore in the fall. Larger individuals can tolerate waters as cold as 10 to 12 degrees C (50 to 54 F), but smaller ones stay in areas where temperatures do not fall below 15.6 degrees C (60 F). Thus, giant bluefins tend to migrate farther northward and appear in New England waters earlier in the summer than do smaller ones. Giants normally swim through the Straits of Florida in May or June, just after spawning. They follow the Gulf Stream northward, appear in coastal waters of southern New England and Cape Cod Bay in June and July, and arrive off the Maine, Nova Scotia, and Newfoundland coasts shortly thereafter. Medium-size bluefins swim northward into the New York Bight area in June and move inshore. The small, young school tuna initiate their northward migration last, arriving in southern New England in July.

Bluefin tuna reach sexual maturity when they are about 6 years of age. A 130-cm (51-in) fish will produce about 5.2 million eggs, and a 230-cm (91-in) one about 32 million eggs.

Giant bluefins spawn from late April to June in the Caribbean, Gulf of Mexico, and Straits of Florida. Medium tuna may spawn somewhat later in the year as far north as the New York Bight.

Unlike most fishes whose body temperatures match the temperatures of the surrounding water, tuna species have the ability to regulate their internal body temperatures. This remarkable adaptation allows the bluefin tuna to withstand a change of up to 17 degrees C (31 F) in water temperature before its internal systems change more than 2.8 degrees C (5 F). This control of internal body temperature aids the bluefin in maintaining very high swimming speeds over prolonged periods of time.

The bluefin tuna is a noted open-water predator, chasing and feeding upon schooling species such as the herring, mackerel, silver hake, and squid. They also occasionally feed near the substrate. Their chief predators are midsized cetaceans, such as the killer and pilot whales, and mako sharks.

Status and Management

Although historically a valued food fish in the Mediterranean and on the West Coast of the United States, until this century bluefins were considered little more than a nuisance along the East Coast of North America. Landings along the East Coast have increased dramatically as markets have developed since the turn of the century. For example, annual landings for the coasts of Maine and Massachusetts increased from less than 45.5 mt (about 100,000 lb) to over 900 mt (nearly 2 million lb) between 1919 and 1948. Consistently increasing fishing pressure along the entire East Coast during the past several decades has significantly reduced the bluefin tuna resource.

The bluefin tuna is a highly migratory species whose range extends well beyond U.S. territorial waters. The Fishery Conservation and Management Act of 1976 specifically precludes the United States from sole management responsibility for tuna resources. The International Commission for the Conservation of Atlantic Tunas (ICCAT) recommends catch quotas for the bluefin tuna, which are intended to maintain the bluefin resource at levels that will sustain future maximum harvests. Quotas are then allocated to nations participating in this fishery (the United States, Canada, and Japan). Under this allocation system, the United States has received approximately 52% of the total allowable annual harvest from 1982 to 1987. After allocations are established, the National Oceanic and Atmospheric Administration (NOAA) is responsible for establishing regulations restricting harvest levels of U.S. citizens and their vessels. The allocation to the United States in the late 1980s was divided among five categories of fisheries:

1. A limited-entry purse seine fishery
2. A harpoon fishery
3. A general fishery including hook-and-line, handline, and harpoon vessels
4. An incidental-catch fishery where bluefins are landed by vessels fishing for other species or other size classes of bluefins
5. An angling fishery for small fish

Under this system, no one can participate in more than one category. Fishing within any category may be regulated by closed seasons, size and number limitations, and/or gear restrictions. Once allocations intended for a particular category are met, the fishery for that category is closed for the year. Within-season adjustments can be allocated by the assistant administrator for fisheries (NOAA) from a portion of the total allocation held in reserve.

Angling and Handling Tips

The bluefin tuna is the ultimate big-game species. This is one of the most powerful fighters in the ocean. Beyond its allure as a trophy fish, it is also revered as a food fish. The Japanese so prize Northwest Atlantic bluefins that they sell for over 20 dollars a pound in Tokyo sushi bars.

Bluefins are pursued largely from private and charter boats. Bluefins can be landed within 20 miles of the coast, well within reach of medium to large fishing boats. To find bluefins, fish in areas with pronounced drop-offs, deep trenches adjacent to shallower waters, and old wrecks or other bottom structures.

Use a stand-up tuna rod and a 4/0 to 6/0 reel spooled with 30–80-pound test Dacron line. These rods are shorter and easier to use than are trolling rods. A variety of artificial lures including diamond jigs, spoons, leadhead tuna feathers, 6-inch plastic squids, multisquid rigs, daisy chains, and Kona or jet heads can be used.

Chunking enhances your success when bait-fishing. Drop chum overboard at frequent intervals, occasionally adding chunks of quartered butterfish or sand eels, to form a chum slick. Then simply hide a 4/0 to 8/0 hook inside of a whole butterfish, and feed your bait into the current with the slick.

Tuna needs immediate attention if its light flavor is to be enjoyed. Whenever possible, a tuna should be gaffed in the gills or the very anterior part of the body to prevent damage to the meat. Don't let the fish thrash around the deck when landed, or the meat will become discolored. After landing, the tuna should be bled. Bleed as described for the yellowfin tuna. Gut and chill the fish as soon as possible. Later, the tuna can be filleted, cut into steaks, and rinsed to flush out any remaining blood.

Most Americans are accustomed to eating tuna only if it comes from a can. If you are one of these, preparing and eating fresh tuna will be a surprising treat. Its texture is firm, resembling red meat more than fish; when cooked, it is similar to steak. Tuna can be pressure-cooked, broiled, baked, or grilled as a kabob. Try braised tuna. Soak tuna steaks in cold water for an hour, pat dry, dust with flour, and sauté for 10 minutes in butter or olive oil. Add freshly ground pepper, parsley, seafood seasoning, and 1 cup of red or white wine. Turn and cook for an additional 10 minutes. Baste with the liquid until it is reduced by half. Serve with rice and peas, using the remaining liquid as a sauce.

Reference

Grosslein, M. D., and T. R. Azarovitz, 1982. *Fish distribution*. Mesa NY Bight Atlas Monograph 15. NY Sea Grant, Albany.

Swordfish
(*Xiphias gladius*)

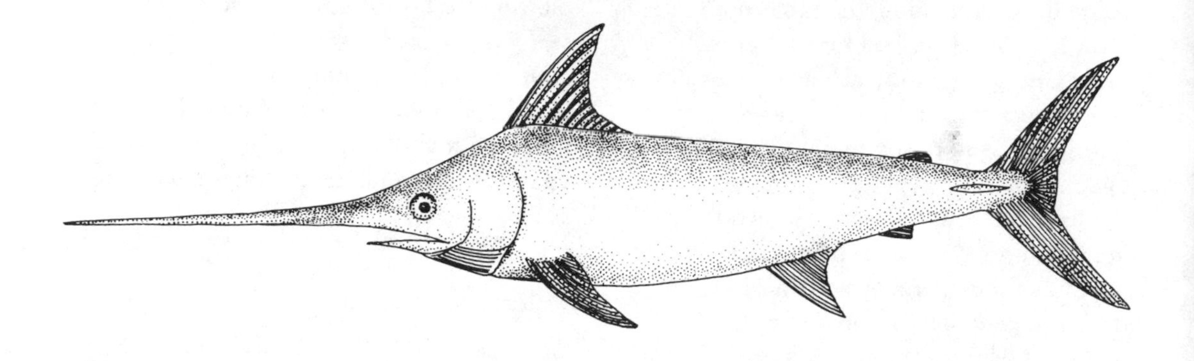

The swordfish is a cosmopolitan species, occurring throughout the world's tropical and temperate oceans, including both sides of the North Atlantic. In the northwestern Atlantic, the swordfish inhabits the Caribbean Sea and the Gulf of Mexico and is found from Florida to as far north as Newfoundland.

The swordfish has a stout body and a slender, tapering caudal peduncle that has a distinctive lateral keel. Its name derives from the upper jaw, which is greatly elongated and narrow, projecting forward from the fish's head like a sword. This species can be easily separated from the white marlin, the only other New England recreational species with a similarly projecting jaw, by two characteristics: Unlike the white marlin, the swordfish lacks pelvic fins; and the first (or anterior) dorsal fin of the swordfish is taller than broad, whereas the reverse is true of the white marlin.

The swordfish is a dark, metallic shade of purple dorsally and laterally, fading to a dusky gray on its lower sides and belly. Its fins are dark with a silvery sheen.

Biology

The swordfish moves through a great range of depths, commonly being captured from the surface to nearly 500 m (1,640 ft) down. Individuals have been recorded as deep as 1,000 m (3,280 ft). Swordfish are summer and fall visitors to New England waters, entering the warming Northwest Atlantic coastal waters from far offshore in the Gulf Stream. Evidence suggests that such onshore–offshore seasonal migrations are much more prevalent than are north–south migrations along the coastline.

Temperature determines the northern limit of distribution along the coast, as the swordfish typically does not occur in water temperatures below 13 degrees C (55 F). Individuals less than 160 cm (63 in) in length are not normally found in temperatures less than 18 degrees C (64 F); thus, a high percentage of swordfish landed in New England are large adults. While in New England the swordfish is most abundant on the outer half of the continental shelf, on the offshore areas of Nantucket

Shoals and Georges Bank, and between Georges Bank and Browns Bank.

Swordfish migrate daily as well as seasonally. Most individuals are found at the surface at night and migrate to deeper waters during daylight. However, large females often bask at the surface on sunny days, their quiet basking making them easy targets for the active harpoon fishery directed at them in the past.

This species' behavior adds to its popularity as a recreational fishery. Swordfish are often seen leaping out of the water, a spectacular sight, especially when displayed by large individuals. The cause of this behavior is not known, although it may aid in ridding the fish of remoras. The pugnacious behavior of the swordfish is well storied. Reports of attacks on whales are probably highly exaggerated; however, there are numerous accounts of a harpooned or hooked swordfish driving its sword completely through the planking of wooden-hulled boats. Bigelow and Schroeder (1953) recount observations of a harpooned swordfish diving with such force that it buried itself in mud beyond its eyes in over 102 m (336 ft) of water. The *Alvin,* a two-man submarine used for oceanographic research, was once attacked by a swordfish at a depth of 700 m (nearly 2,300 ft).

Swordfish spawning, which occurs only in water temperatures higher than 24 degrees C (75 F), peaks in April in the Northwest Atlantic, although some winter spawning occurs in the Sargasso Sea. Spawning has been recorded in the Caribbean, the Gulf of Mexico, and waters off of Florida. Eggs are buoyant, and larvae average about 0.4 cm at hatching.

Swordfish grow rapidly (Table 32). Females grow faster, live longer, and weigh more at a given length than males. Maximum size of females is about 445 cm (14.6 ft) and 540 kg (1,188 lb).

Table 32. Length at age in cm (inches in parentheses) and weight at age in kg (pounds in parentheses) of female swordfish

Age	*Length**	*Weight*
1	100 (39.4)	4 (8.8)
2	120 (47.2)	15 (33)
3	140 (55.1)	40 (88)
4	160 (63.0)	70 (154)
5	175 (68.9)	110 (242)
6	185 (72.8)	
7	200 (78.7)	
8	215 (84.6)	

*Fork length, from the tip of the snout to the midpoint of the fork in the tail.

Source: Scott and Scott 1988.

Squid and octopods are the most important foods in the swordfish's diet. Fishes such as the Atlantic mackerel, silver hake, redfish, Atlantic herring, haddock, bluefish, cod, and butterfish are also important. The swordfish uses its sword to slash prey, which it then captures and swallows whole. Smaller squid and fishes may be slashed indiscriminately as the swordfish feeds through a school of prey. However, reports of indiscriminate slashing of great numbers of individuals that are not eaten are probably inaccurate.

Young swordfish are eaten by a variety of fishes, including blue sharks, tunas, marlins, mako sharks, and other swordfish. Sperm whales, killer whales, and sharks are listed as the only natural enemies of adults.

Status and Management

The swordfish was not a favored food species until the end of the 19th century, bringing such low prices that only a very modest fishing effort was directed toward it. However, by the turn of the century, market prices increased as it gained popularity as a table food. Throughout the 1900s the demand has consistently exceeded the harvest to a level that persistently maintained its high market value.

The swordfish catch in New England ports averaged about 909 mt (2.0 million lb) per year during the first 3 decades of the century. Prior to World War II, the highest yearly catch was slightly over 2,300 mt (5.1 million lb). Catches rapidly returned to over 900 mt annually after the war ended.

During this entire time, nearly all swordfish harvested commercially were harpooned. So few were caught with handlines or longlines that Bigelow and Schroeder (1953) felt such methods did not "figure to any extent in the total catches and are not likely to." Catches in the New England domestic fleet historically have constituted only a moderate proportion of the total harvest from the western Atlantic.

In the last several decades, harpooning remained a major means of commercial harvest, and directed longline fisheries increased in importance. Both domestic and foreign catches of this species increased during this time period. After passage of the Magnuson Act, an Inter-Council Swordfish Committee was established among the several councils (including the New England) overseeing regions possessing active swordfish fisheries. In 1985, responsibility for development of a management plan was shifted to the South Atlantic Fishery Management Council (SAFMC), with the other original participants, including the New England Council, serving in an advisory role.

Swordfish spawning biomass in the western Atlantic declined continuously after 1979, by 1989 dropping to about 40% of the estimated 1978 level. This reduction in adult biomass, a clear indication that quotas established for domestic harvest were not protecting the viability of the resource, led the South Atlantic Council to propose a management plan (to go to public hearings early in 1990), which included the following:

1. A total allowable catch (TAC) for the domestic fishery
2. Severe controls on, or prohibition of, directed fishing, depending upon incidental-catch levels occurring in tuna fisheries
3. Prohibition of nighttime longline fishing after TACs are reached
4. Prohibition of imported fish from the same stock after TACs are reached
5. Prohibition of the use of drift-entanglement gill nets in the swordfish fishery
6. A recreational fishing allocation that would include a minimum legal size limit, the prohibition of sale of recreationally landed fish, and the use of rod and reel only

Although able to regulate domestic harvest within territorial waters, the United States had little leverage in the foreign harvest, except as a member of the ICCAT, under whose jurisdiction international management agreements concerning conservation of swordfish are accomplished. The inability to control foreign harvest at one point led the New England Council to differ publicly with the management recommendations of the South Atlantic Council (*New England Council Newline,* 1989, Vol. 2, No. 2), noting that "international action is required to ensure the success of any swordfish management program. With the development of foreign cooperation, this Council . . . will be prepared to restrict U.S. fishermen." Thus, the two councils were torn between creating conservation measures that would be imperfect without international support or taking no action in conservation but allowing the domestic fleet to share in the economic benefits accrued from overharvesting the depleted stock.

Angling and Handling Tips

From many anglers' point of view, the swordfish is one of the greatest fighting machines in the Northeast. This species starts slowly, then builds its battle to the very end, providing more muscle-draining sport than many might wish to experience! Swordfish are somewhat solitary travelers who roam the open seas, keeping a distance from each other. This species scientific name (*X. gladius*) references the short sword carried by the ancient Roman foot soldier. The swordfish uses its bill to slash food, not to spear it, swallowing its prey after stunning or killing it.

The best time of year for swordfishing is from June or July through early fall, when waters warm to 60 degrees F or higher, and during full and new moons. Fishing is most productive over drop-offs and steep ledges of offshore canyons, particularly at night. Calm nights are not only better for drifting but safer for the angler.

When swordfishing, use a stand-up tuna rod with backbone and a large conventional reel that can hold hundreds of yards of 50–80-pound test Dacron line. Use 12/0 to 14/0 hooks on a very heavy monofilament leader of 10 feet or more, attached to the Dacron line by a heavy swivel. Swordfish do not strike artificial lures but will take a variety of baits. The best bait is squid, although mackerel and bonito are also used. After capture, large squid should be placed in a plastic bag and put on ice to keep them fresh. Some swordfish anglers peel the skin off of the squid, claiming the white flesh can be seen better by swordfish. Light sticks attached about 12 feet above the bait can greatly improve luck when angling at night, as they attract both bait fishes and swordfish. When drifting, use multiple baits set at 50 to 300 feet. Two to 5 ounces of lead sinker may be required to drop the bait to the greater depths.

Swordfish can also be pursued during daylight hours. Large females sometimes bask at the surface at midday. They can be sighted by their dorsal and tail fins, which stick well out of the water. After spotting a basking individual, approach it very quietly, trolling the bait to within 20 to 30 feet, avoiding undue noise or a wake. Since a basking swordfish is not feeding, you may need to present the bait to it many times before it will strike. It will mouth bait softly before taking it; thus, the drag should be set to free-spool, since the swordfish gulps rather than strikes. Lightly thumb the reel to avoid backlash, and don't react to the first, second, or even third tugs. Count to 10 before setting the hook, because the deeper into the throat the hook is set, the better the chance of landing the fish. Recover all slack line, flip the reel's lever, and set the hook hard at least three times. Swordfish usually start the fight by diving to deep waters. When they rise to battle on the surface, their aerial displays can provide an exciting show.

After landing, clean the fish as soon as possible. Steaks can be cut using a mallet and cleaver, heavy knife, or saw. An easier approach is to cut off a fillet at the thickest part of the fish, then take steaks from it while the boned side of the fillet lies downward. Save the thinner part near the tail for double-sided steaks. Steaks should be at least three quarters of an inch thick, since swordfish dries out rapidly when cooked.

The swordfish is one of the most popular restaurant seafoods worldwide. Its white flesh is firm, fine-grained, and dense.

It is neither fishy nor bland in flavor, and its texture is closer to tender beef than to most other fishes. Its quality actually improves after freezing. It is excellent grilled, broiled, blackened Cajun-style, smoked, baked, used for kabobs, and as a replacement for veal in many Italian dishes.

Marinate the steaks before grilling, using a marinade consisting of 1 cup of olive oil, a shake of oregano, four finely chopped cloves of garlic, some freshly ground pepper, and some sprigs of rosemary. Press both sides of each steak into the marinade and place in a glass baking dish. Pour the marinade over the steaks and refrigerate for several hours. Place the steaks on a hot grill, letting them sear on one side for 6 to 7 minutes, as you would a piece of beef. Add some of the marinade to softened butter and blend with a fork. After turning the steaks, spread some of this mixture over the meat and cook for an additional 4 to 5 minutes. Steaks can also be broiled using this same method.

French Bread Kabobs are a tasty dish friends and family will long remember. Simply use chunks of fish left over after filleting, or cut 1½-inch cubes from steaks. Make the marinade described above, and add: the juice of one lemon, ½ cup of finely chopped onion, ½ inch of freshly grated ginger, and additional rosemary. Coat the cubes with the mixture, pressing the flavorings into the flesh. Cover the fish with olive oil and marinate in the refrigerator, turning several times. Next, cut a loaf of thin French bread into chunks, leaving enough crust on each piece so that it will hold on a skewer. Remove the fish from the marinade and drain. Dip a piece of the bread into the marinade and put it on the skewer, followed by a cube of fish. Alternate bread and fish in this way until the desired number of skewers are filled; then sprinkle very lightly with salt. Grill over a hot fire or cook in a preheated oven or broiler. Turn and baste skewers often, cooking 10 to 12 minutes, or until the bread and fish are lightly browned. Serve on a platter surrounded by wedges of lemon and topped with fresh parsley.

References

Scott, W. B., and M. G. Scott. 1988. Atlantic Fishes of Canada. *Can. J. Fish. Aquat. Sci.* 219.

Stillwell, C. E., and N. E. Kohler. 1985. Food and feeding ecology of the swordfish *Xiphias gladius* in the western North Atlantic Ocean with estimates of daily ration. *Mar. Ecol. Prog. Ser.* 22:239–247.

White Marlin

(*Tetrapturus albidus*)

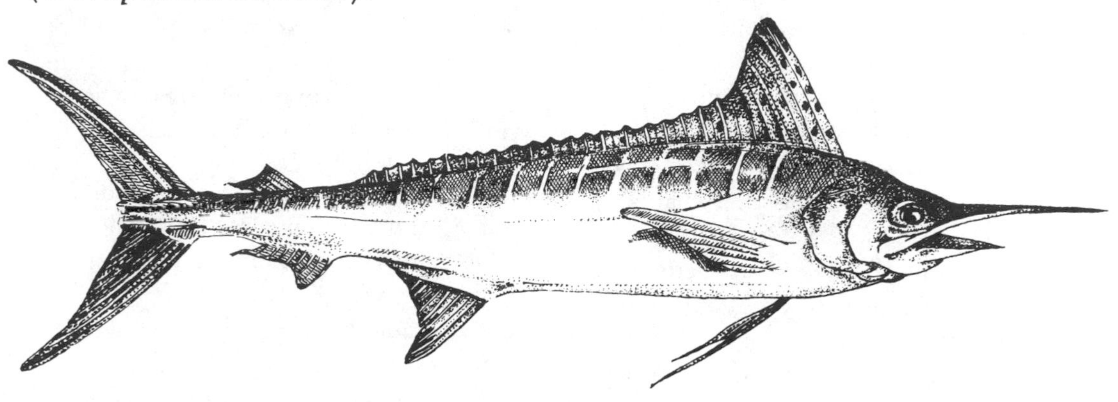

The white marlin, a highly prized billfish species of the eastern coast of the United States, is native to the western Atlantic Ocean from the northern coast of Brazil to Cape Cod. It is abundant throughout the year from the Caribbean to southern Florida and is seasonally abundant from Cape Hatteras, North Carolina, to Cape Cod.

The white marlin resembles the larger and more southerly distributed blue marlin but differs in having a rounded instead of a pointed anterior dorsal fin and lighter coloration on its sides and belly. The white marlin has a much shorter "sword" (elongate projection of the snout and upper jaw) and a much longer dorsal fin than the swordfish; it also has scales, whereas the adult swordfish is scaleless.

The white marlin is bright greenish blue dorsally, fading to pale blue laterally and to white on its belly. The sides are striped with as many as a dozen light blue or purple vertical bands. The dorsal fin is blue with dark spots.

Biology

White marlins are oceanic, ranging from coastal waters to well beyond the continental shelf of the United States. They are transient in Massachusetts, moving northward to the southern New England region in the summer and migrating southward by the beginning of fall. The marlin occasionally occurs as far north as Nova Scotia, but the waters off the southern and eastern shores of Cape Cod are its normal northern limit. Major currents appear to play an important role in migratory and distribution patterns of this species, which is frequently found in greatest abundance within or near the Gulf Stream when along our eastern coast. White marlins are often found in shoals, submarine canyons, or steep drop-offs of the ocean floor where current boundaries, upwellings, and thermal fronts tend to concentrate food resources.

Female white marlins reach sexual maturity at a weight of about 20 kg (44 lb). A 27.3-kg (60-lb) female produces about 5.1 million eggs and a 36.4-kg (80-lb) female about 10.5 million eggs in a year.

White marlins spawn during April and May throughout the Caribbean and Gulf of Mexico and in the Straits of Florida. Little is known about the survival of larvae and juveniles. Each year up to 40% of all adult white marlins die from either fishing or natural causes.

Although white marlins up to nearly 60 kg (over 130 lb) have been landed by New England anglers, most weigh less than 27 kg (60 lb). Little is known about the growth rates of this species, but females tend to grow larger than males. The maximum age is thought to be somewhat greater than 10 years.

White marlin larvae feed on microscopic organisms until they are large enough to pursue larger prey. Adults feed mainly on small schooling species, including herring, squid, and anchovies, but the composition of their diet can differ markedly from one region to another.

Status and Management

A recreational fishery for white marlins has existed off the coast of the United States since the early 1900s. However, prior to 1950, this fishery was limited to a few areas. Since then, a coastwide fishery has developed for this species.

In recent years, competition for the white marlin has intensified among the recreational fishery, the yellowfin tuna and swordfish longline fisheries that capture it as by-catch in more southerly waters, and a directed commercial fishery. Mortality due to commercial harvest is felt to be reducing this resource to potentially critical levels. The average total harvest in the years 1977 to 1979 was about half the average for the previous 10-year period, which was approximately 1,000 mt (2.2 million lb), or 44,000 fish. Total harvest in the 1980s has increased somewhat but remains below that of the 1966–1976 period.

An Atlantic Billfish Fishery Management Plan, prepared jointly by the New England, Mid-Atlantic, South Atlantic, Caribbean, and Gulf of Mexico regional fishery management councils and approved in 1988, addresses the conservation of the white marlin. This plan contained regulations meant to protect the white marlin resource for recreational harvest. Regulations in the plan include:

1. Prohibiting the sale of any landed marlins, in order to prevent further development of a commercial market for this species
2. Requiring the release of all fish weighing less than 22.7 kg (50 lb), in order to reduce significantly mortality caused by the recreational fishery
3. Prohibiting longline and drift-net commercial vessels from keeping the white marlin as by-catch, in order to maximize the release of live fish by the commercial boats most likely to catch them.

Angling and Handling Tips

In recent years, an increase in the number of offshore private and charter boats that fish for billfish species has led to greater numbers of marlin being sighted and caught in our waters. Anglers usually call upon an array of indicators to locate this pelagic species. In order to have a successful trip, the angler must be able to interpret correctly a number of factors, such as: water temperature changes, weed lines, water color changes, fathom curves, and the presence of schools of bait fishes or of sea birds. Therefore, many prefer to charter an experienced boat captain when fishing for the white marlin.

Unlike the closely related blue marlin, which tends to travel singly, white marlins occasionally gather in small groups while feeding. White marlins can be spotted as they swim at the surface with their dorsal fins breaking the water. Once sighted, careful presentation of bait or lures can often lead to a strike.

Those anglers who use the more traditional method of trolling prefer stout rods and reels that can handle 450 yards of 50-pound test line. Anglers experiment with sizes, shapes, styles, and colors of lures to achieve the greatest success. Hex heads, daisy chains, Green Machines, and Smokers are all favored lures.

Many anglers prefer to present live or frozen bait to fish seen finning the surface waters. Spinning outfits fitted with 20-pound test line are typically used when casting. The line is doubled and tied directly to a size 6/0 offset "sailfish" hook. Live tinker mackerel, scup, and northern ballyhoos or frozen Spanish minnows or mullet are commonly used baits.

In the United States, marlins historically have been considered a nonfood game fish; thus, anglers in the past typically released their catch. Catch-and-release is still being promoted as one means of conserving the marlin resource. However, as consumers become more willing to experiment with fish that have traditionally not been eaten, more white marlins are being kept as seafood. The marlin is esteemed in the Caribbean and Latin America, where it traditionally has been grilled, broiled, or baked. Smoked marlin is also considered a delicacy. Marlins can be cooked using any recipe written for swordfish, mako shark, or even fresh tuna.

Marlin kabobs offer an interesting dining change. Marinate chunks of marlin in olive oil and lemon juice for at least 1 hour. Alternate the meat on the skewer with vegetables such as zucchini, yellow squash, onions, green peppers, tomatoes, and broccoli; then, brush with the marinade. Season with garlic powder and freshly ground pepper and grill on a hot fire or broil, turning and basting, for 15 to 20 minutes. One skewer per diner served on a bed of rice makes a flavorful meal.

Reference

Draft fishery management plan for the Atlantic billfishes. 1987. South Atlantic Fishery Management Council.

Redfish
(*Sebastes fasciatus*)

The redfishes (also commonly called rosefishes or ocean perches) form an assemblage of several closely related species that inhabit deeper waters of the coastal Northwest Atlantic. Although considered a deepwater species, *Sebastes fasciatus* occurs in shallower waters than its sister species, and thus it is the redfish captured by anglers fishing offshore for deepwater groundfish. This species of redfish occurs from Canadian waters to as far south as New Jersey. However, no sizable concentrations occur south of the Gulf of Maine, except in the Great South Channel that separates Cape Cod and Nantucket Shoals from Georges Bank.

S. fasciatus has one dorsal fin, the anterior two thirds of which is supported by a series of stiff spines. The anal fin has three spines of increasing length from most anterior to most posterior. This species has a small tail and large, oval pectoral fins. The redfish has large eyes and a lower jaw that projects in front of the upper. The lower jaw is tipped with a large bony knob. A series of five spines occurs on the cheek, with two near the dorsal fin and two more on the posterior edge of the gill cover.

Redfish vary from an orange or flame red to a darker grayish or brownish red. The belly is usually a paler color than the sides and back. Many individuals have several dusky patches along the back and one on the gill cover.

Biology

Redfish dwell on rocky or hard-bottomed substrates, occasionally in shallow coastal waters to depths as great as 640 m (2,100 ft). They exhibit a shift in depth with seasonal changes in water temperature, moving up from the bottom in the winter and back to the bottom waters in the summer. In the Gulf of St. Lawrence, redfish inhabit the warmer bottom waters of deep channels and do not enter the colder waters above.

Redfish tend to be segregated by size, with juveniles and smaller adults most abundant in shallower depths. The proportion of mature, larger-bodied fish increases with depth.

Table 33. Length at age in cm (inches in parentheses) of selected ages of male and female redfish

Age	*Male*	*Female*
1	6.8 (2.7)	7.3 (2.9)
3	12.3 (4.8)	11.8 (4.6)
6	18.3 (7.2)	18.7 (7.4)
10	25.9 (10.2)	28.1 (11.1)
14	28.1 (11.1)	32.7 (12.9)
17	30.2 (11.9)	34.5 (13.6)
20	29.3 (11.5)	37.0 (14.6)

Source: Data from Kelly and Wolf 1959. Although data are for *Sebastes marinus,* all samples were taken from the Gulf of Maine, which is where the recently identified *S. fasciatus* typically occurs as the only redfish species.

Fifty percent of male redfish mature at 18.5 cm (7.3 in) in length; 50% of the females do not do so until they are about 29.5 cm (11.6 in) in length. The redfish is ovoviviparous, accomplishing fertilization internally and giving birth to posthatching larvae that were nourished by yolk reserves while developing inside the female's oviduct. Insemination of females by males may take place in the autumn, although actual fertilization of eggs and subsequent embryo development occur from February through May. All females have released larvae before the beginning of July.

Although internal egg incubation affords eggs tremendous survival advantages, redfish larvae are quite small at birth (6 mm, or 0.2 in, long) and drift with currents in near-surface waters until mid autumn, when they are about 5 cm (2 in) long. During this pelagic stage, larvae are estimated to suffer a 99% mortality rate, which is typical of many species that have pelagic larvae. Thus, the fecundity of females is intermediate, being much lower than many species that release eggs into the water column but markedly higher than live-bearers such as sharks, which give birth to very large young that suffer notably low rates of mortality. A 30-cm (12-in) female may produce about 50,000 eggs and may release 15,000 to 20,000 living young.

Redfish are long-lived and extremely slow-growing. Table 33 lists lengths at age for selected ages of redfish taken from the Gulf of Maine. Although the largest redfish may approach 8.7 kg (19 lb), most individuals weigh between 0.1 and 0.5 kg (0.3 to 1 lb). Maximum age appears to be about 50 years.

Juvenile redfish eat euphausiid shrimp and planktonic copepods until they are about 5 years of age. Larger juveniles and adults eat euphausiids and other shrimp, with small fishes becoming moderately important seasonally. Adults feed nocturnally, migrating upward from the substrate to feed actively in midwater areas.

Larvae and young juvenile redfish are eaten by cod, halibut, adult redfish, and other fish species.

Status and Management

The redfish has been commercially exploited since the mid 1930s, when freezing techniques that allowed it to be shipped to distant markets were developed. An extensive market for frozen redfish developed in the midwestern United States after retailers started selling it under the name of "ocean perch." The public in that region of the country identified well with this new name, since the freshwater yellow perch had long been a midwestern favorite.

Exploitation grew rapidly, from 500 mt (1.1 million lb) landed in U.S. ports in 1934 to 60,000 mt (132 million lb) in 1941. This level of exploitation was followed by a gradual decline in harvest levels. Declines in stock abundance led to the establishment of harvest restrictions by ICNAF in the early 1970s. Countries participating in the commercial fisheries were assigned total-catch allocations within which they were to restrict their fleet activities. Further declines led to the establishment of lower allocations (from a total of 30,000 mt in 1973 to a proposed 9,000 in 1977). The United States withdrew from ICNAF in March 1977 after passage of the Magnuson Act (FCMA). The redfish resource was not managed under FCMA until 1986. Continued population declines caused consistent reductions in landings from the Gulf of Maine/Georges Bank region (declines in commercial landings are presented in Figure 15). The total commercial catch dropped from 14,700 mt (32.3 million lb) in 1979 to 2,000 mt (4.4 million lb) in 1987. Recreational landings have historically been so modest they are considered an insignificant portion of the total harvest.

Commercial harvest is currently regulated under a fishery management plan of the New England Fishery Management Council.

Angling and Handling Tips

The distinctive red and orange hues and spiny head of this species make the ocean perch one of the easier New England fishes to identify. Although occasionally reaching nearly 2 feet and several pounds, most individuals will measure 8 to 12 inches in length and about half a pound in weight.

Although the ocean perch is commonly harvested by commercial fishermen, anglers rarely encounter it because of the great depths this species inhabits. Ocean perch have been recorded occasionally in Gloucester Harbor in the winter, but most are landed recreationally near the mouth of Passamaquoddy Bay, Maine, in the Eastport area, where a deepwater drop-off close to shore brings this species into closer possible contact with anglers than anywhere else on the New England coast.

The ocean perch is an adept bait stealer when the angler fishes with cod gear and hooks. The best equipment must be solid enough to allow you to fish at 100–200-foot depths but light enough to let you feel, hook, and enjoy landing this smaller-bodied fish. A conventional outfit spooled with 25-pound test line is well suited for perch; heavier line, which drags more in currents, is not necessary for this small species. A 4–8-ounce sinker might be needed to hold the baited hook on the bottom in strong currents, although the lighter the weight, the better. Use a whiting or ling rig. This consists of several small, long-shanked No. 1 or 2 hooks attached by dropper loops above a sinker tied to the bottom of the line. Snelled fluke and flounder hooks also work well. Some anglers like to tie one leader off of the sinker and another on a dropper or swivel 16 inches above the first. Others tie three in succession on droppers, so that if one or two pieces of bait are stolen they still have another chance at a fish. Ocean perch will take almost any bait, including clams, shrimp, worms, or fishes. Whichever you try, be sure to use only a small piece.

The ocean perch is a good table fish, excellent for filleting and quick-freezing. The meat, similar to freshwater perch, is a creamy light tan in color, firm and slightly coarse in texture. The perch has a delicate flavor that lends itself to a variety of sauces and cooking methods. It can be baked, broiled, deep-fried, sautéed, poached, and steamed. In addition, it can be flaked after poaching, boiling, or steaming and used in molded salads, casseroles, or stuffing.

Traditionally, the ocean perch is fried. To fry, roll whole gutted, headless, and scaled perch or fillets in flour, dip in a beaten egg, and shake in a plastic bag with flavored bread crumbs. Sauté or fry in butter or olive oil and serve with quartered lemons.

For a different approach to perch eating, try Yogurt Perch or Perch Volcanoes. For Yogurt Perch, place skinned fillets in a buttered baking dish. Sprinkle ½ cup of Italian dressing and 3 tablespoons of wine

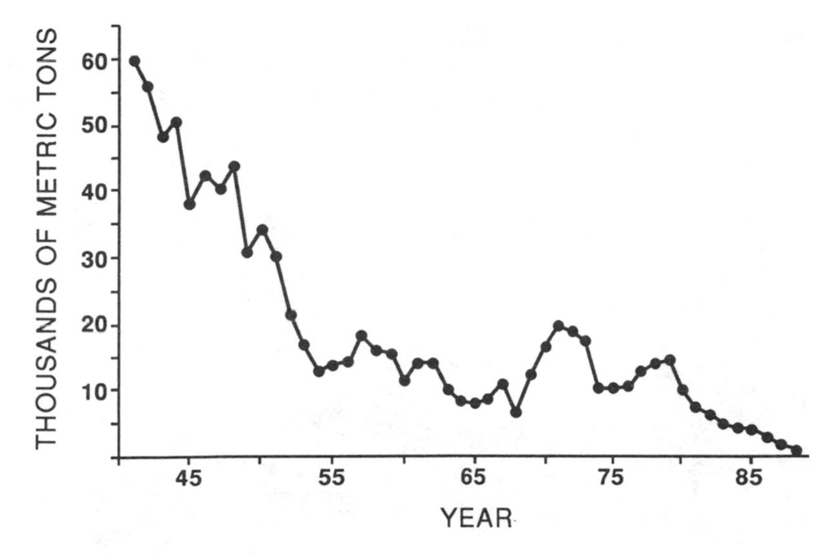

Figure 15. Declines of annual landings of redfish by United States vessels, from the Gulf of Maine-Georges Bank region, 1941 to 1988 (from *Status of the Fishery Resources off the Northeastern U.S. for 1989,* NOAA Mem. NMFS-F/NEC-63).

vinegar on the fillets. Marinate, turning several times, for about 15 minutes. Next, dust the fillets with flavored bread crumbs in a plastic bag and arrange in a single layer in the baking dish. Mix 1 cup of plain yogurt and ½ cup of Parmesan cheese, stirring until the mixture becomes liquefied. Drip the sauce over the fillets, forming a topping. Sprinkle several teaspoons of toasted sesame seeds and some shredded cheddar cheese on the top and add slices of sweet red pepper or some other vegetable for color. Place in an oven preheated to 450 degrees F for 10 minutes, or until the fillets just begin to flake; then, serve from the baking dish before the fish cools. This can be an excellent low-fat dish if you leave out the cheese and use low-fat yogurt.

Perch Volcanoes use flaked perch or almost any other white-fleshed species. Place 2 cups of water in a pan, add a tablespoon of Old Bay or other seasoning, and bring to a boil. Cut up several pounds of skinless fillets into 1-inch pieces and simmer in the boiling water for 8 minutes or until the fish begins to flake. Remove from the heat and drain. Next, place 2 teaspoons of butter or olive oil in a skillet and sauté ½ cup of chopped onion, ½ cup of chopped sweet red pepper, ¼ cup of thinly sliced celery, and 2 teaspoons of chopped parsley until tender. Add ½ teaspoon of salt, three grinds of black pepper, and ¼ teaspoon of seafood seasoning such as Old Bay. Blend 3 tablespoons of cornstarch with 1 cup of regular or low-fat milk and add this and the fish to the pan, stirring gently while heating (do not boil). Place "hills" of cooked rice on individual plates and hollow the centers (like tiny volcanoes); brown rice can provide a nice texture and color contrast. Fill the hollow with the fish mixture. You can top this dish with chopped toasted almonds, bacon bits, or roasted sunflower seeds, and add a dash of paprika for color.

Reference

Kelly, G. F., and R. S. Wolf. 1959. Age and growth of redfish, *Sebastes marinus* (Linnaeus) in the Gulf of Maine. *Fish. Bull.* 60:1–31.

Northern Searobin (*Prionotus carolinus*)
and
Striped Searobin (*P. evolans*)

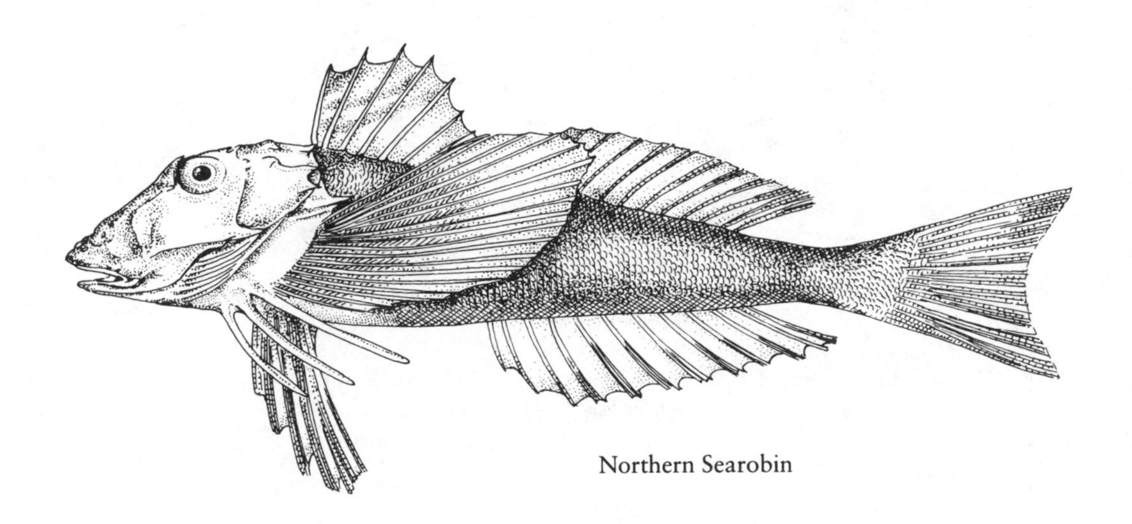
Northern Searobin

The northern and striped searobins are among the most frequently caught species in New England. These strange-looking fishes occur in coastal North American waters from the Bay of Fundy to South Carolina and are most abundant from Cape Cod southward.

Searobins have large, elevated heads with relatively small, slender bodies that taper toward the tail. All fins, other than the tail, are large in relation to the size of the body. There are two dorsal fins, the first (or anterior) one with spines, and one anal fin. The three lower support rays of the pectoral fins are separated from the rest of the fin and are modified into elongate feelers with slightly enlarged tips. The head, covered with heavy, bony plates, has one sharp spine on each cheek, two over each eye pointing backward, one in front of the dorsal fin, and one above the base of the pectoral fin. The striped searobin's mouth is larger than the northern's, extending back almost under the front edge of the eye, and the posterior edge of its tail is straight in outline, whereas the northern searobin's is concave.

Searobins generally are gray or reddish brown dorsally, with several dark blotches along the back, and dirty white or pale ventrally. The dorsal fin is mottled gray, the tail and anal fins are gray or brown, the pelvics are yellow, and the pectorals and modified rays are yellow or orange.

Biology

The bottom-dwelling searobins are found more commonly on smooth and hard substrates than on mud or among rocks. They usually lie on the bottom, with their fanlike pectoral fins spread widely when resting. If disturbed, searobins often burrow into sand, with only the top of the head and eyes exposed above the substrate. These fishes are quite active when feeding and may be caught striking baits near the surface in shallow waters. They use their elongate pectoral rays to stir up the substrate while searching for food. This behavior makes them appear to be scurrying along the bottom, using their pectoral rays as if they were stiff legs; this curious sight attracts many bemused onlookers at public aquariums.

Searobins overwinter well out on the continental shelf in waters up to 170 m (558 ft) deep. They move shoreward in the spring, first arriving in coastal waters of southern New England in late April or May. They remain inshore throughout the summer, inhabiting waters from the tide line to about 73 m (240 ft) deep.

Most searobins reach sexual maturity at age 2. Spawning has been noted from June to early August in Long Island Sound and may occur as early as May in southern New England. The buoyant eggs hatch in about 60 hours at 22 degrees C (72 F).

The striped searobin grows faster and lives longer than the northern searobin. Striped searobins may reach a maximum length of 46 cm (18 in) and 9 years of age, northern searobins nearly 41 cm (16 in) and 6 years of age, although few northern searobins caught by anglers are more than 28 cm (11 in) long. Mean lengths at age of striped and northern searobins collected from Long Island Sound are presented in Table 34.

Young-of-the-year searobins eat copepods. Adults feed heavily upon crustaceans such as amphipods, crabs, and mysid and sand shrimp. Mollusks, squid, annelid worms, and fishes such as small herring, menhaden, and winter flounder are also eaten. Fishes typically constitute only a minor portion of a searobin's diet.

Status and Management

Searobins were commercially harvested in pound-net fisheries through the 1940s, but harvest levels were modest. More recently, searobins are harvested incidentally by vessels fishing for other inshore species. They also are frequently caught by anglers who are fishing with bait for preferred species. As many as several hundred thousand per year may be landed by anglers in the three southernmost New England states. Many are likely to be returned to the water upon capture rather than kept for eating. Since the harvest of these species is largely incidental, and apparently modest in extent, no New England state managed these fisheries through the 1980s.

Table 34. Length at age in cm (inches in parentheses) of northern and striped searobins captured from Long Island Sound

Age	*Northern searobin*	*Striped searobin*
1	12.6 (5.0)	12.8 (5.0)
2	18.2 (7.2)	17.9 (7.0)
3	22.0 (8.7)	22.0 (8.7)
4	24.6 (9.7)	25.3 (10.0)
5	26.3 (10.4)	27.9 (11.0)
6	27.5 (10.8)	30.0 (11.8)
7		31.7 (12.5)

Source: Calculated from formulas presented in Richards et al. 1979, assuming a 22-cm length for age 3 fishes.

Angling and Handling Tips

Although searobins have traditionally had a small demand as a specialty item in New York City, markets have never developed in New England for these fishes. The extra work required to remove searobins from nets has earned these species the reputation of a nuisance "trash fish." Searobins are also more a product of by-catch than a target of recreational anglers. Although the young angler fishing from a jetty in the middle of the summer may be ecstatic about the fish she or he has just caught on a bit of sea worm, most adult anglers grimace and discard this bizarre-looking fish upon removing it from the hook. Searobins are considered a nuisance by many anglers due to their skill as bait stealers.

These species, like so many other fishes, are normally enjoyed only by certain ethnic and nontraditional anglers. Although they do not meet most anglers' idea of an edible species, searobins are excellent table fare.

Searobins are most frequently caught on baited hooks but may also leave their bottom haunt to attack a small spinner. Almost any kind of bait used for bottom-fishing will attract searobins, including shrimp, squid, worms, and clams. A small, long-shank hook is desirable, since it can be more easily removed from these fishes' bony mouths. A fish-finder or egg-shaped sinker with a hole in the center will allow you to feel the light tapping of searobins as they mouth your hook. When you catch one of these little fishes, be sure to keep it cool in order to retain its mild flavor.

Searobins are related to the gurnards of northern Europe, which are marketed on that continent as a desirable food species. The gurnard is one of the fishes used in the traditional French bouillabaisse, or fish soup. Searobins are equally tasty. Although searobin fillets are small, they are mild-tasting, dry, white, and firm, making them useful in many dishes. They can be broiled or added to chowder and are similar to flounder when sautéed or fried.

You'll need a bunch of fillets to make a family meal of these little fishes. To fillet, start behind the large, armored head and the base of the pectoral fins and make a diagonal cut to the front of the dorsal fin. Then, finish as you would any other round fish, cutting down the backbone and removing the meat from above the ribs and along the tail. Searobins have no lateral bones, so your end product will be a small, triangular-shaped white, boneless fillet.

Try broiling the fillets with a little butter or margarine and seasoning. Searobins should be treated like a thin fillet of sole when broiling. Five minutes of cooking time may be enough to allow the fillets to flake when prodded with a fork.

The following dish will also please the entire family: Put ½ cup of flour and ½ cup of flavored bread crumbs into a plastic bag. Add 1 teaspoon of commercial seafood or turkey seasoning and shake to mix. Dip each fillet into beaten egg or milk, place in the bag, and shake until the fillets are coated. Sauté the fillets in olive oil or butter until lightly browned. These can be served as is, with a lemon wedge, or with a cheese sauce.

If you want to use a cheese sauce, dissolve 1 teaspoon of cornstarch in ½ cup of white wine and add to the pan drippings. Stir over low heat, sprinkling lightly with salt and freshly ground pepper. Add more wine if you need to thin the sauce. Place the fillets in a greased baking dish in one layer and pour the sauce over them. Cover with shredded cheddar or other cheese, sprinkle with Parmesan cheese, and broil until the cheese melts and just begins to brown. When serving, use a large spatula so as not to break the tender fillets. Eat as is or try serving them on split French rolls for a delightful meal.

Reference

Richards, S. W., J. M. Mann, and J. A. Walker. 1979. Comparison of spawning seasons, age, growth rates, and food of two sympatric species of searobins, *Prionotus carolinus* and *Prionotus evolans,* from Long Island Sound. *Estuaries* 2:255–268.

Sea Raven (*Hemitripterus americanus*)
and
Shorthorn Sculpin (*Myoxocephalus scorpius*)

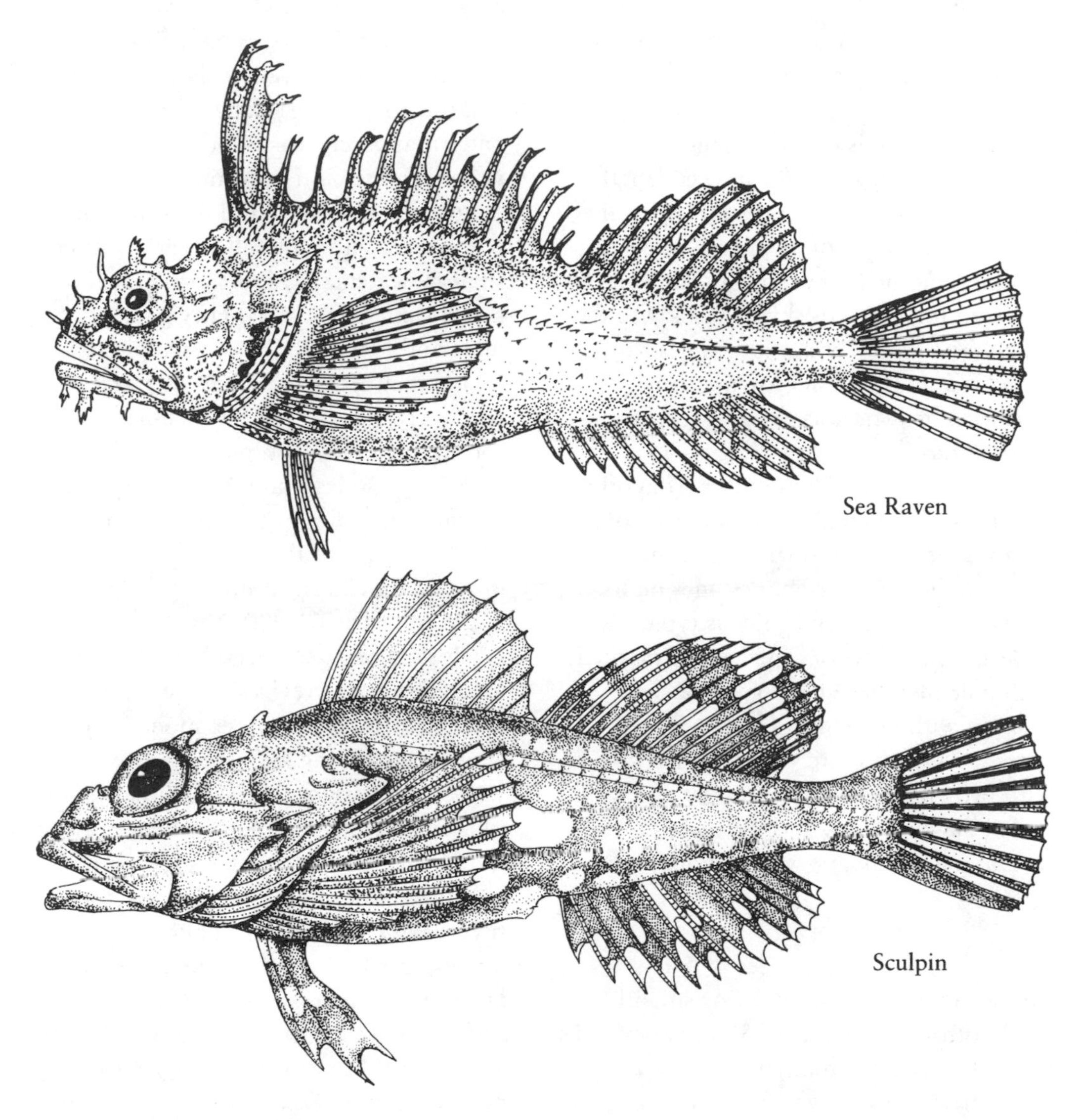

Sea Raven

Sculpin

The sea raven and shorthorn sculpin are members of the sculpin family, a group of fishes represented by more than a half-dozen species in New England coastal waters. These two species are the largest of the New England sculpins, and both are frequently caught by anglers bait-fishing in shallow waters. The sea raven occurs in the Northwest Atlantic from Labrador to the Chesapeake Bay and is common throughout the coastal belt of New England; the shorthorn sculpin occurs from arctic waters southward to New York.

Members of the sculpin family have

large, flattened heads, wide mouths, and slender, tapering bodies. They have unusually large eyes and typically have some arrangement of spines on the head. All sculpins have two large dorsal fins, the first with a series of spines, and large anal and pectoral fins. The tail is small, and the pelvic fins are narrow, with only three long support rays.

The sea raven is easily distinguished from other sculpins by the uneven lengths of the spines in the anterior dorsal fin, the ragged appearance of the membranes of that fin, and the fleshy tabs of skin projecting from the head. Its skin is prickly to the touch. The sea raven's color ranges from a blood red to a reddish brown, purple, or yellowish brown. It has a yellow belly and mottled fins.

The shorthorn sculpin differs from other members of the family in having the combination of three short spikes around its eyes and five to seven short spines on its cheeks. Its background color is typically some shade of brown. The head is marked with pale blotches, the lower sides are mottled with large, distinctive yellow spots, and the fins are variously mottled with brown and yellow patterns. The female's belly is white, and the male's is reddish orange with large white spots.

Biology

The sea raven lives on rocky or smooth hard bottoms in depths of 3 m to about 91 m (10 to 300 ft), although individuals are also infrequently caught from greater depths. Its preferred temperature range is approximately 6 to 9 degrees C (43 to 48 F). The shorthorn is more restricted to shoreline waters, occurring in greatest abundance in depths less than 37 m (121 ft). It occurs on smooth substrates, including mud, sand, and pebbles, and is frequently found among weeds. The shorthorn is often landed by anglers fishing around pilings of wharves, around which this species commonly feeds. All sculpins generally remain within limited, localized areas and do not conduct any extensive migrations. In southern New England, the sea raven does migrate inshore in the fall and offshore in the spring; however, this pattern apparently is not displayed by individuals in the Gulf of Maine.

Both the sea raven and the shorthorn sculpin spawn in the fall to early winter in shallow waters, the shorthorn in depths of 6 to 11 m (20 to 36 ft). Fifty percent of male and female shorthorn sculpin reach sexual maturity by the ages of 5 and 6, respectively; age at maturity is not known for the sea raven. Sculpins generally produce only modest numbers of eggs. The average fecundity of the sea raven is about 15,000 eggs, and the maximum is about 40,000. Shorthorn fecundities range from 4,205 to nearly 61,000 eggs.

Clusters of up to several hundred eggs are laid in rock crevices or the branches of the finger sponge or other sponges by the shorthorn and the sea raven, respectively. Male shorthorn sculpin guard the eggs until they hatch. Male egg guarding is a common behavior within the sculpin family but has not been observed for the sea raven. Hatching occurs about 3 months after eggs are laid and fertilized. Larvae are large at hatching, measuring about 7 to 8 mm in length for the shorthorn and 10 to 14 mm for the sea raven. Larvae are benthic, and individuals remain near the substrate throughout the rest of their lives. The combination of laying eggs in protected sites, parental guarding, and large, well-developed larvae hatched from large eggs provides high rates of survival for offspring of sculpin species.

Sea ravens may attain lengths up to 64

Table 35. Length at age in cm (inches in parentheses) of male and female shorthorn sculpin

Age	*Males*	*Females*
2	17.7 (7.0)	14.9 (5.9)
3	21.9 (8.6)	20.8 (8.2)
4	25.4 (10.0)	25.7 (10.1)
5	28.3 (11.1)	29.9 (11.8)
6	30.8 (12.1)	33.5 (13.2)
7	32.8 (12.9)	36.5 (14.4)
8	34.5 (13.6)	39.0 (15.4)
9	35.9 (14.1)	41.1 (16.2)
10	37.0 (14.6)	42.9 (16.9)

Source: Calculated from growth equations presented in Clayton et al. 1978.

cm (25 in) and weights of 3.2 kg (7 lb), although the average-size individual caught by anglers is usually less than 30 cm (12 in) and 2.3 kg (5 lb). Female shorthorn sculpin may reach 50.6 cm (20 in); males do not exceed 42.2 cm (16.6 in) in length. Lengths at age for male and female shorthorns are presented in Table 35. Maximum age for this species is about 15 years.

Both species eat a variety of bottom organisms. The sea raven feeds heavily upon bivalves, snails, various crustaceans, and sea urchins. Fishes such as the herring, sand lance, tautog, and sculpin species are also eaten. The shorthorn sculpin eats crabs, shrimp, amphipods, bivalves, snails, polychaete worms, and small fishes including herring, gobies, and small cod.

Two behavioral characteristics displayed by these species, along with their appearance, generate great curiosity among anglers who land them. Both species startle anglers by emitting croaking or grunting sounds when being handled. In addition, the sea raven is capable of swallowing such large amounts of water and air that its belly becomes markedly distended, causing it to bob belly-up in the water as it is being brought to boat or shore.

Status and Management

The sea raven and shorthorn sculpin have been harvested for use as bait in New England lobster fisheries. However, this bait fishery was always modest, never having any apparent effect upon population levels. The closely related longhorn sculpin supported an active "trash-fish" fishery in the 1940s, when it was processed into commercial fish meal. However, the menhaden fishery was meeting the demands of the fish meal industry by 1950, and the sculpin fishery declined in activity thereafter.

Both the sea raven and the shorthorn sculpin are frequently landed by anglers fishing for other, preferred species. Although both of these sculpins are good table fish, many anglers discard them after landing.

Due to the limited level of harvest by commercial and recreational activities, the fisheries of both species remained unregulated through the 1980s.

Angling and Handling Tips

Although never targeted species, shorthorns and sea ravens are frequently caught in estuaries or from docks, wharves, and piers. These fishes strike at almost any bait used by anglers, especially clams, squid, snails, and conches, and provide an enjoyable struggle on light spinning tackle with 8-pound test or lighter line. When removing these species from the hook, anglers are well advised to use pliers to avoid their numerous and sharp spines.

Sculpin species have lean, light-textured, delicate flesh that has a light flavor. Smaller individuals can be skinned and gutted, floured, and slowly sautéed whole in garlic and butter. To skin whole small fish, gut and remove the head. Next, hold the front, exposed end of the backbone with a pair of pliers, grip the cut edge of the skin with another pair, and pull in one quick movement. The fins can be removed or left in place until after cooking, when they easily pull free from the flesh.

Larger fishes are well suited for filleting. Fillets can be baked, broiled, or fried gently. They are well known in ethnic markets as excellent soup fishes and work well in cold fish salads.

To broil, dip the fillets in lemon juice, place on a broiler pan, dot with butter or olive oil, and season with salt, pepper, and paprika. Broil for 5 minutes (basting if needed) or until the meat barely flakes. Sculpin species are delicate, so broil on foil for easy removal to a platter. Lightly oil the foil before broiling to prevent sticking.

For a quick and easy meal after a day's fishing, try Sculpin Pasta. Make your usual spaghetti sauce (marinara or without meat) and let it thicken a little by simmering uncovered; the sauce needs to be somewhat thick because the fish adds liquid, thus thinning it. Commercial chunky or pizza sauces also work well and are quicker than working from scratch. Add three or more drops of hot pepper sauce and some fresh basil to liven the dish. Cut fillets into 1-inch pieces, coat lightly with flour, and add to the simmering sauce. Simmer for 5 to 8 minutes and serve over cooked pasta such as linguini or small shells. Sprinkle lavishly with freshly grated Italian hard cheese and serve with French bread.

To prepare a cold fish salad, cut ½ pound or more of fillets into 1-inch pieces and place in a pan with several slices of orange. Add equal amounts of water and white wine to cover, and bring to a boil. Reduce the heat and simmer slowly for 3 to 5 minutes or until the fish begins to flake. After simmering, drain. In a bowl, mix 1 cup of very finely sliced or processed celery and one small onion, a small can of cooked baby peas, several chopped hard-boiled eggs, ½ cup of low-cholesterol mayonnaise, several teaspoons of red wine vinegar, pepper, and salt. Add the fish and mix gently. Serve on lettuce leaves and top with sliced boiled eggs if desired.

Reference

Clayton, G., C. Cole, S. Murawski, and J. Parrish. 1978. *Common marine fishes of coastal Massachusetts.* Mass. Coop. Ext. Serv., Univ. of Massachusetts, Amherst. C-132.

Atlantic Halibut

(*Hippoglossus hippoglossus*)

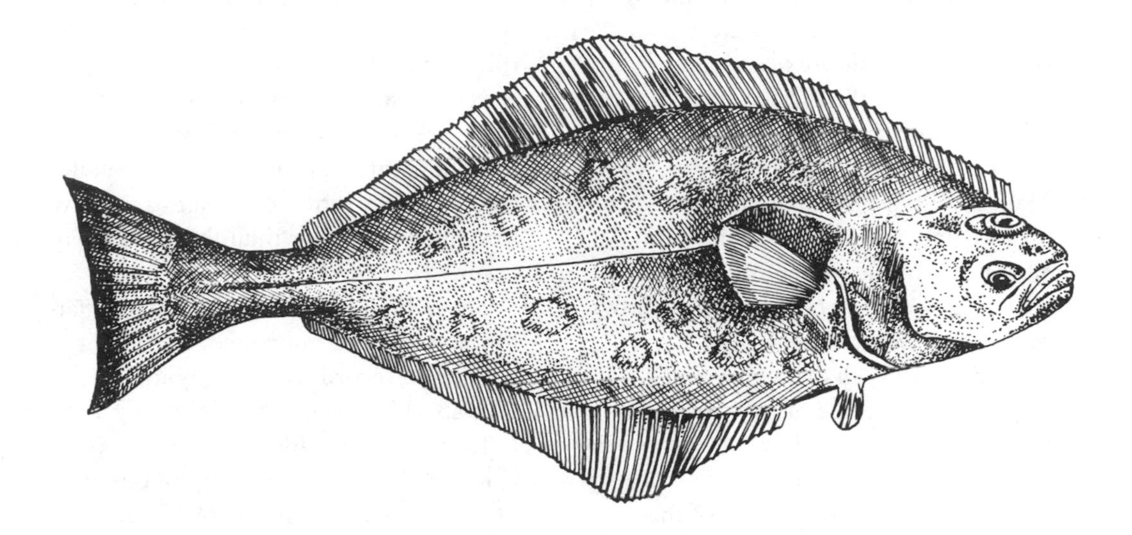

The Atlantic halibut, a giant among the flatfishes, is native to both sides of the Atlantic Ocean. In the western Atlantic it ranges from the coast of Greenland to New Jersey. It is most abundant from Nantucket Shoals to the northern Gulf of St. Lawrence, the southern portion of the Grand Bank, and the deep waters of the outer continental shelf off Labrador.

Like other flatfishes, the halibut has both eyes on one side of the head. A newly hatched larva has one eye on each side of its head but within months accomplishes a major metamorphosis that adapts it to bottom-dwelling. During this process, one eye and the surrounding bony orbit "migrate" to the other side of the head. Unlike most other bottom-dwelling fishes that rest by lying on their bellies, a flatfish rests and swims on its side. Having both eyes on the same side of the head enables a flatfish to rest on the ocean's floor while directing both eyes upward. The halibut is called a "right-handed" species because both eyes are on the right side of its head. When resting on the ocean floor, it lies on its "blind" left side.

The halibut differs markedly from other right-handed flatfishes of the Northwest Atlantic. Its body is less broadly flattened and thicker. It has a large mouth that extends posteriorly to below the eyes and contains sharp, curved teeth. The dorsal and anal fins are decidedly highest at their middle. The posterior edge of the tail is concave, and the lateral line is noticeably arched at its anterior end.

The halibut ranges in color from olive brown to chocolate or almost black on the upper side. Younger halibut are usually paler and blotched. The blind or lower side is pure white in smaller fish or mottled with gray in larger ones.

Biology

Halibut are found in subarctic waters; they prefer temperatures from 2.2 to 8.3 degrees C (36 to 47 F) and depths from 61 to 915 m (200 to 3,000 ft). As young fish, they inhabit shallower areas but move into progressively deeper waters as they grow. Thus, the largest, oldest specimens may be found in the deeper waters inhabited by this species. Although they are bottom dwellers, individuals are occasionally sighted feeding at the surface. They live on sand, gravel, or clay bottoms, tending to avoid mud and rock substrates.

Halibut display seasonal movement patterns associated with changes in water temperature. In the southern portion of their range, they may be found in shallow areas such as Georges Bank and Nantucket Shoals only in the winter and spring; as water temperatures rise in the summer above about 7.2 or 7.8 degrees C (45 or 46 F), these fish withdraw to deeper waters. In the northern portion of their range, halibut tend to move to deeper waters with more stable water temperatures in the fall as shallower areas cool to below 2.2 degrees C (36 F).

Females become sexually mature at about 9 or 10 years of age, males somewhat earlier. The fecundity (number of eggs produced in a given spawning season) of females increases with size and weight; a female Atlantic halibut of about 91 kg (200 lb) produces as many as 2.2 million eggs in a season. Recent studies indicate that individuals in the severely depleted halibut stocks of recent years are displaying a significant decrease in both age and size at sexual maturity.

Halibut reproduce from December to February. In the northwestern Atlantic, peak spawning activity occurs in January. Halibut reproduce on the ocean bottom at depths of from 274 to over 610 m (900 to 2,000 ft) and are believed to congregate at particular spawning sites every year.

Although laid on the bottom, the eggs are subsequently suspended in the water column at depths of 55 to 91 m (180 to 300 ft) and at temperatures between 2.8 and 3.9 degrees C (37 and 39 F). After hatching, larval halibut remain suspended in the water column, feeding upon plankton and drifting in currents. During this time, halibut tend to rise slowly in the water column and drift toward more inshore areas. Thus, by the time they metamorphose into bottom-dwelling juveniles, young halibut settle to the substrate in waters much shallower than those they may occupy later in life.

The halibut is the largest of all the flatfishes. The largest northwestern Atlantic halibut on record weighed about 318 kg (700 lb). However, fishes exceeding 136 kg (300 lb) were considered rarities even before this species was depleted due to fishing pressure. Most "large" halibut commercially landed weigh from 23 to 91 kg (50 to 200 lb). Females are somewhat longer and heavier than males of the same age. The average lengths of selected ages of male and female halibut are presented in Table 36.

Halibut feed mainly on groundfishes but occasionally eat a variety of shellfishes as well. Their diet includes whatever species happen to be most abundant at a particular time, including cod, cusk, haddock, sculpin, silver hakes, herring, sand lances, flounders, crabs, lobsters, shrimp, clams, and mussels. Even sea birds have been found in the occasional specimen. In turn, adult halibut are eaten by seals and Greenland sharks. They serve as a major dietary item for the latter, when available.

Status and Management

In colonial times the halibut was abundant from the Massachusetts coast northward but held little value as a food fish. However, by 1820 a demand for this species had developed in the Boston market; this

Table 36. Mean lengths at age in cm (inches in parentheses) of selected ages of Atlantic halibut from the Gulf of St. Lawrence

Age	*Male*	*Female*
5	61.0 (24)	58.4 (23)
10	91.4 (36)	104.1 (41)
15	129.5 (51)	145.1 (57)
20	147.6 (58)	170.5 (67)
25	154.9 (61)	185.7 (73)

Source: Data taken from Scott and Scott 1988.

demand has strengthened ever since, even as halibut populations have been progressively depleted. The numbers of fish landed annually from Cape Cod, Massachusetts Bay, the coastal Gulf of Maine, Georges Bank, and Nantucket Shoals had dropped significantly by 1850. Halibut populating the deepwater slopes off Georges Bank were also markedly depleted by 1870, even though the difficulty associated with fishing off these areas had largely prevented their exploitation until the middle of the 19th century. Thus, the Atlantic halibut showed the effects of fishing pressure earlier than nearly any other species in the northwestern Atlantic. In recent years, Atlantic halibut have been harvested largely as by-catch, although a modest baited line fishery operating in coastal waters off the Gulf of Maine is partially directed toward halibut. Halibut populations have shown no tendency to recover, even though fishing pressure directed toward them has lessened in the 20th century.

Slow-growing, long-lived species that are slow to reach sexual maturity are particularly susceptible to population collapse in the face of human exploitation, since harvested fish are not replaced very rapidly. Because it has such an extended life cycle, the halibut would require long-term restrictions on harvest in order to recover. Such restrictions would be confounded by the fact that this species is usually landed as by-catch by boats fishing for other, more abundant species. Thus, long-term recovery of the halibut would require prolonged restrictions on fishing for all groundfish species living in similar habitats. For this reason, regulations that specifically restrict commercial fishing have not been established for this species within U.S. territorial waters. Since this species is largely restricted to more offshore waters, no New England state regulated either its commercial or its recreational harvest as of 1989.

Angling and Handling Tips

Few fishes reach the size and provide the battle of the halibut. This prized species is difficult to find due to its scarcity and its preference for relatively deep waters. Party boat captains can find areas with good halibut habitat, but the major element necessary to encounter a halibut is luck. Landing a large individual is dependent upon gear and skill. If you think this species fights like a large flounder, your first halibut will prove to be a surprising experience. An unsuspecting angler can be overwhelmed by the power of a halibut that suddenly bends his or her rod nearly in half.

Halibut can be caught while fishing a wide variety of baits used for other species, or while deepwater jigging for cod. Use medium to heavy cod fishing rods and 6/0 reels spooled with 60–80-pound test Dacron line. When fishing with bait such as clams, add 4 to 5 feet of monofilament leader with large 8/0 cod hooks. To hold the bait on the bottom in greater depths, 10–20-ounce Swedish jigs or 16–20-ounce bank sinkers may be needed.

Halibut that are hooked are occasionally lost due to stripped or broken lines, broken reels, or even twisted gaffs. A halibut frequently makes only a short run with the bait immediately after being hooked. If this happens, the angler should retrieve as much line as possible and prepare for a prolonged battle with a powerful fish.

The thick, firm, and sweet-flavored steak of the Atlantic halibut is a popular, versatile meat that can be baked, broiled, grilled, sautéed, or steamed. For a summer speciality, flour halibut steaks that are about 1½ inches thick and baste with a mixture of melted butter or mayonnaise, lime juice, and white wine. Place the steaks on a hot, oiled grill for a total of 5 to 10 minutes per inch of thickness, turning once. The steaks are done when they flake. Sprinkle with paprika before serving.

Pan-baked halibut is also a favorite. Marinate 1-inch-thick steaks for several hours in ½ cup of olive oil, freshly ground pepper, and seafood seasoning such as Old Bay. Next, drain the marinade and heat it in an iron skillet. Flour the steaks and lightly brown them in the skillet. Place the skillet in an oven at 375 degrees F for 10 to 15 minutes, basting with the marinade. After cooking, remove the fish, add white wine to make a light sauce, and serve the halibut with a quartered lemon.

References

Beacham, T. D. 1982. Median length at sexual maturity of halibut, cusk, longhorn sculpin, ocean pout and sea raven in the Maritimes areas of the Northwest Atlantic. *Can. J. Zool.* 60:1326–1330.

Scott, W. B., and M. G. Scott. 1988. Atlantic Fishes of Canada. *Can. Bull. Fish. Aquat. Sci.* 219.

Winter Flounder
(*Pseudopleuronectes americanus*)

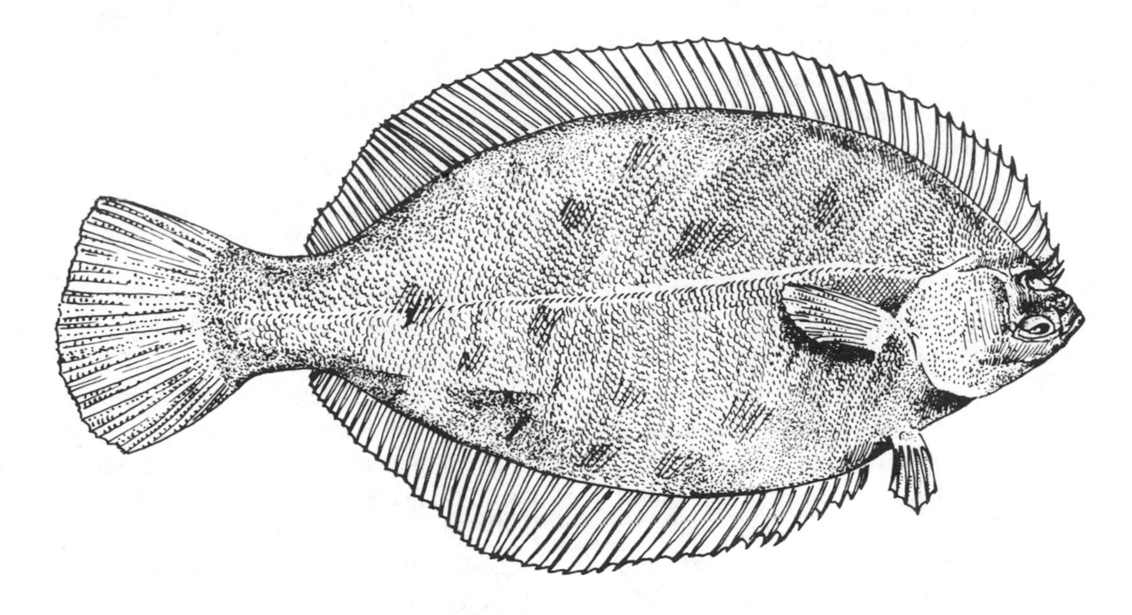

The winter flounder (also called the blackback and lemon sole), one of the most common and popular recreational species in New England, is highly favored because of its fine flavor and thick fillets. It ranges from southern Labrador to the waters of South Carolina and Georgia and is most abundant from the Gulf of St. Lawrence to the Chesapeake Bay. This species' name comes from its tendency to move to shallower inshore waters during the winter months, where it is easily caught. It is frequently called a blackback when it is smaller than 3 pounds and a lemon sole when it is larger.

The winter flounder is referred to as a right-handed flounder because the eyes are located on its right side. This species is oval in shape, with a body that is about twice as long as it is wide. Its small mouth does not extend backward to below the eye. The lateral line runs in a nearly straight line from the head to the base of the tail. The dorsal fin originates opposite the forward edges of the eyes and follows the length of the body at a uniform height. The anal fin also extends the length of the body but is noticeably highest near its midpoint.

The color of this species is highly variable, since the winter flounder can change color to mimic the bottom on which it rests. However, the scaled upper side of the body of most individuals ranges from a muddy or reddish brown to black, and the scaleless underside is white. Smaller fish are generally paler and have a less uniform color pattern than larger, older fish.

Biology

Winter flounder are caught in almost any shallow bay or estuary where the bottom is sandy or silty. This species frequently moves into the brackish water of river mouths and also ranges into the deeper waters of the coast and the shallower areas of Georges Bank. Winter flounder are uncommon at depths greater than 55 m (180 ft), although they have been found as deep as 128 m (420 ft) on Georges Bank. When they are on soft bottoms, they lie buried in the mud, dashing out occasionally to feed on invertebrates moving close by.

This is one of the most stationary of flatfishes; individuals are typically grouped into small, local subpopulations that display limited seasonal migration. Fish overwinter in inshore areas. As summer approaches in areas south of Cape Cod, large adults move offshore to deeper water, followed later in the season by smaller adults. This movement is in response to warm inshore waters; winter flounder prefer temperatures below 15 degrees C (58 F), although they have been found in temperatures as high as 24 degrees C (75 F). Juveniles remain in estuaries for up to 3 years, moving offshore as they grow older. Winter flounder north of Cape Cod tend to remain in bays and harbors all summer, moving into holes or deep channels in the warmest weather.

Both male and female winter flounder normally reach sexual maturity by age 3, although in the northern reaches of this species' range females may not do so until they are 5 to 7 years of age. Fecundity increases with body size and differs somewhat by region (Table 37). In New England, reproduction occurs in estuaries from January to May, with peak activity during February and March when water temperatures are the coldest of the year, ranging from 0 to 4 degrees (32 to 39 F). Evidence suggests that specific individuals return for many years to the same site to spawn.

Table 37. Average fecundity of specific lengths of winter flounder from Rhode Island and Newfoundland

Rhode Island		*Newfoundland*	
Length in cm (in)	*Number of eggs*	*Length in cm (in)*	*Number of eggs*
22 (9)	99,000	25 (10)	193,000
33 (13)	610,000	34 (13)	590,000
43 (17)	1,340,000	44 (17)	604,000

Source: Data from Kennedy and Steele 1971.

Unlike the floating eggs of all other local flatfishes, eggs of the winter flounder clump together in masses on the bottom. Eggs, usually laid on clean sand, hatch 15 to 18 days after being released. By the time the larvae are 9 mm long, they have undergone complete metamorphosis, the left eye having migrated to the right side of the body. Mortality is highest during larval stages, partly due to predators such as the striped killifish and jellyfish.

Growth rates are variable; individuals on Georges Bank, for example, typically grow markedly larger than those living in inshore areas. Female winter flounder generally grow faster than males and attain larger maximum sizes. Table 38 lists average lengths at age for winter flounder from three New England regions. The maximum age appears to be 12 years.

Winter flounder feed mostly during daylight hours. Juveniles and adults feed most actively in intertidal zones during flood or ebb tides, moving to deeper waters at low tide. Larval winter flounder eat microscopic crustaceans, polychaete worms, and protozoans. Juveniles feed heavily upon copepods, amphipods, and polychaetes. Adults feed upon a great variety of organisms including shrimp, clams, polychaete worms, coelenterates, other invertebrates, fish eggs, fish fry, and vegetation. Plant material may compose as much as 40% of the diet. Winter flounder often congregate at spawning sites of species

Table 38. Length at age in cm (inches in parentheses) of the winter flounder from three New England regions

	South of Cape Cod		*Georges Bank*		*North of Cape Cod*
Age	*Male*	*Female*	*Male*	*Female*	*Female*
2	18.8 (7.4)	21.5 (8.5)	25.5 (10.0)	21.5 (8.5)	21.5 (8.5)
4	30.1 (11.9)	35.0 (13.8)	40.2 (15.8)	47.0 (18.5)	34.1 (13.4)
6	37.0 (14.6)	41.8 (16.5)	47.1 (18.5)	56.0 (22.0)	40.1 (15.8)
8	41.1 (16.2)	45.2 (17.8)	50.4 (19.8)	59.6 (23.5)	43.0 (16.9)
10	43.6 (17.2)	46.9 (18.5)	52.0 (20.5)	61.2 (24.1)	44.3 (17.4)
12	45.2 (17.8)	48.7 (19.2)	52.7 (20.7)	61.8 (24.3)	45.0 (17.7)

Source: Clayton et al. 1978.

that lay demersal eggs. Such eggs can be a major source of food for blackbacks. Capelin eggs provided enough energy to account for 23% of the total annual body growth of winter flounder off the coast of Newfoundland.

Predation is a major cause of mortality for larval and juvenile fish. Winter flounder are eaten by fishes such as the toadfish, striped bass, bluefish, monkfish, spiny dogfish, sea raven, and summer flounder, and by birds such as cormorants, blue herons, and ospreys.

Status and Management

Each year anglers catch more winter flounder than any other species of fish in New England coastal waters. Presently, only the bluefish attracts a greater number of anglers. Recreational angling is a major component of the total annual harvest of this species, accounting for 30% to 46% of the yearly poundage harvested coastwide between 1979 and 1985. Recreational harvest may exceed commercial landings in a given year in New England waters. Commercial harvest is accomplished primarily by means of spring and autumn otter-trawling activities. The bulk of commercial landings in New England generally come from Georges Bank.

Although winter flounder populations historically showed less tendency to fluctuate in abundance than some other New England groundfish species, harvest levels had dropped by the mid 1930s. Improved fishing technology and the introduction of the foreign distant-water fishing fleets into the New England region increased fishing pressure throughout the 1950s and 1960s. Passage of the Magnuson Act in 1976 eliminated most fishing by foreign vessels within 200 miles of the U.S. coastline. However, since winter flounder fisheries include an important inshore component that was never exploited by foreign vessels, the exclusion of the distant-water fishing fleets in 1977 did not markedly reduce total harvests. Winter flounder fisheries were most heavily exploited in the late 1970s and early 1980s. Catches declined almost yearly throughout the 1980s. The frequency of winter flounder less than 30 cm (12 in) in length caught on southern New England grounds increased from 3% of the total catch in 1979 to 23% in 1982 and 35% by 1985. This strongly suggests that fewer fish were surviving to the older age groups during that time period.

Commercial harvests are regulated under the Northeast Multispecies Fishery Management Plan of the New England Fishery Management Council and by state regulations in inshore waters. Regulations included minimum legal size limits and area closures to protect juvenile fish. All New England states regulated winter flounder fisheries by means of minimum legal size limits in 1989.

Angling and Handling Tips

Angling starts for this species each year in March and lasts until the end of May. Blackbacks leave warm inshore waters in the summer and don't return until September, when they once more provide fishing action until the weather becomes too unpleasant for fishing. Anglers can fish for winter flounder from docks, jetties, and party and private boats in nearly all New England bays and estuary mouths. Deeper holes at the mouths of estuaries or bays, and the mouths of rivers and streams, are particularly productive fishing sites; effort should be concentrated in places that are open to currents and tides, unchoked by numerous boulders or large patches of vegetation.

Winter flounder provide the most enjoyable action when caught on light tackle. Use 10–15-pound test line on a 6½-foot medium-action spinning rod or a small boat rod. Many anglers use freshwater outfits when fishing from docks or the shoreline, but heavier equipment will be needed when tidal currents require sinkers of more than 1 or 2 ounces to hold bait to the bottom—a must when fishing for this bottom fish. Use a small, long-shanked flounder hook with a split shot fastened several inches up the line. Add a slip sinker or a bank sinker about 12 to 18 inches above the hook. Many anglers believe that attaching yellow beads above the hook or using yellow sinkers will attract greater numbers of blackbacks.

Many recommend chumming. Use ground or crushed clams or mussels in a mesh bag or "chum pot," or toss into the water occasional handfuls of corn from a freshly opened can. Sandworms are considered the best bait for winter flounder, but bloodworms, clams, and mussels are also commonly used. Some anglers prefer strips of squid and even night crawlers. Use very little bait; an inch of worm will work best. Blackbacks can quickly and quietly sneak in and take bait; thus, unattended rods lose fish. The rod should be raised often to check for fish as well as to attract them.

No fish lends itself to more imaginative dishes than does the winter flounder. Its texture and delicate flavor are well suited to sauces, spices, fruits, vegetables, and other seafoods. Winter flounder can be fried, steamed, baked, microwaved, or broiled and can be substituted for other species in most fish recipes.

References

Clayton, G., C. Cole, S. Murawski, and J. Parrish. 1978. *Common marine fishes of coastal Massachusetts*. Mass. Coop. Ext. Serv., Univ. of Massachusetts, Amherst. C-132.

Frank, K. T., and W. C. Leggett. 1984. Selective exploitation of capelin (*Mallotus villosus*) eggs by winter flounder (*Pseudopleuronectes americanus*): Capelin egg mortality rates, and contribution of egg energy to the annual growth of flounder. *Can. J. Fish. Aquat. Sci.* 41:1294–1302.

Kennedy, V. S., and D. H. Steele. 1971. The winter flounder (*Pseudopleuronectes americanus*) in Long Pond, Conception Bay, Newfoundland. *J. Fish. Res. Board Can.* 28:1153–1165.

Summer Flounder

(*Paralichthys dentatus*)

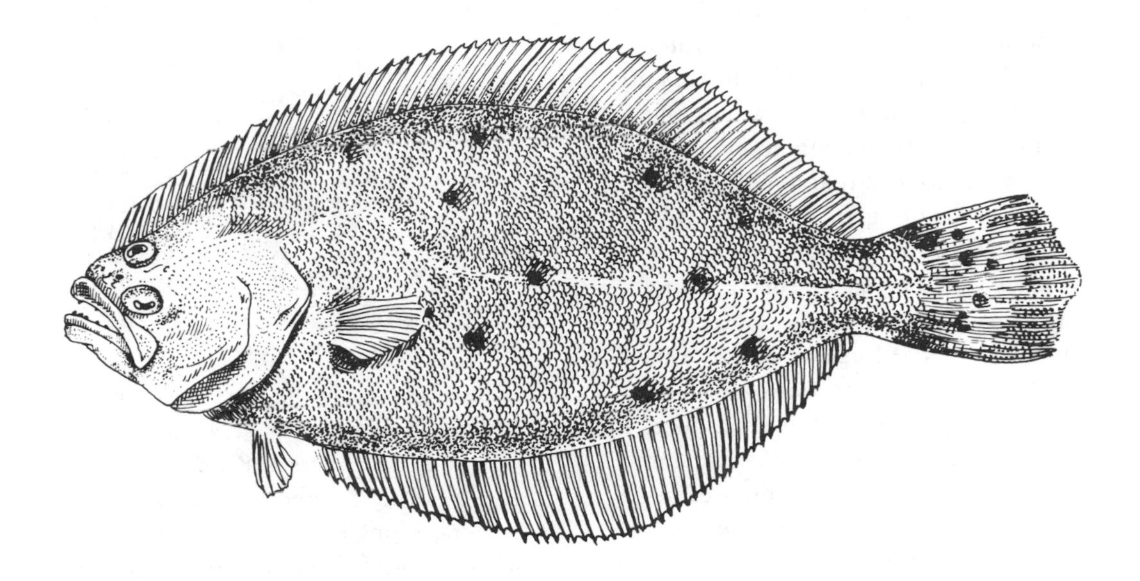

The summer flounder, or fluke, a flatfish noted for its fighting ability and flavor, is found in coastal waters from the southern Gulf of Maine to Florida. There are important recreational and commercial fisheries for this species from Cape Cod to Cape Hatteras.

Like other species of flatfish, the fluke has both eyes on one side of its head and rests on the ocean floor on its side. The fluke is a "left-handed" flatfish. This species has a very large mouth that extends below and beyond its eyes. The left-handed windowpane flounder differs from the summer flounder in having a much rounder body; in addition, the anterior rays in the dorsal fin of the windowpane are branched, forming a conspicuous fringe, and the posterior outline of the tail is rounded. The physically similar fourspot flounder differs from the fluke in having four large dark spots encircled by faint pinkish rings on its body.

Summer flounder are called the chameleons of the sea because of their ability to change color to match the bottom on which they are found. Generally, they are white below and darker above, but they can turn various shades of gray, blue, green orange, and almost black. The upper part of the fluke's body is marked with scattered spots that are darker than the general body color.

Biology

The summer flounder is distributed offshore in the winter and early spring, usually inhabiting water depths of 70 to 155 m (230 to 509 ft). This species moves inshore during the warmer seasons, first appearing in inshore waters in April and reach peak numbers by August. While inshore, fluke prefer eelgrass beds and wharf pilings because of the protection they offer. When threatened, fluke quickly bury all but their eyes in the sand or escape at surprisingly high speeds. In the summer, small and medium-sized adults are found on the sandy and muddy bottoms of bays and harbors and along the open coastline. Most of the larger fish tend to stay in somewhat deeper water (15.2 to 18.3 m; 50 to 60 ft). With the approach of fall, summer flounders migrate to more offshore waters. Fluke found in the coastal waters of New England probably overwinter east of the Hudson Canyon (off the New Jersey–Long Island coast). Young summer flounder up to age 2 are virtually absent from New England waters. This absence is due to autumn water-circulation patterns that distribute larvae produced in southern New England southward and to the absence of spawning in the area between Nantucket Shoals and Georges Bank. The adult fishery that occurs in New England is due to the general northward movement of fish from their natal and juvenile grounds to the south.

Both males and females become sexually mature at age 2 to 3. A 35.6-cm (14-in) female produces about 460,000, a 40-cm (16-in) female about 500,000, and a 65-cm (26-in) female about 3.5 million eggs in a season. Females are serial spawners, producing and shedding consecutive batches of eggs during the protracted spawning season.

Reproduction takes place in the fall, as soon as the fish begin migrating to wintering grounds. The center of spawning activity occurs off the coasts of New York and New Jersey, with less concentrated activity occurring in southern New England waters. In southern New England, peak spawning activity occurs in mid September in water temperatures of 11.7 to 18.9 degrees C (53 to 66 F) and at depths of 18.3 to 48.8 m (60 to 160 ft). As autumn proceeds, spawning activity is initiated progressively southward.

Table 39. Mean lengths at age in cm (inches in parentheses) for summer flounder

Age	*Female*	*Male*
1	12.8 (5.0)	14.8 (5.8)
2	24.3 (9.6)	24.3 (9.6)
3	34.1 (13.4)	32.3 (12.7)
4	42.4 (16.7)	38.9 (15.3)
5	49.6 (19.5)	44.5 (17.5)
6	55.6 (21.9)	49.1 (19.3)
7	60.8 (23.9)	53.0 (20.9)
8	65.2 (25.7)	56.3 (22.2)
9	69.0 (27.2)	59.0 (23.2)
10	72.2 (28.4)	61.2 (24.1)
11	74.9 (29.5)	63.1 (24.8)
12	77.2 (30.4)	64.7 (25.5)

Source: Data from Penttila et al. 1989.

The eggs float in the water column, hatching in 56 hours at 22.9 degrees C (73.2 F) to 142 hours at 9.1 degrees C (48.4 F). After they become bottom dwellers, young fluke move into bays and estuaries where they spend the early portion of their lives. Throughout much of this species' range, juveniles remain in estuaries year-round, although they tend to move offshore for their first winter in the northern reaches of their geographic distribution.

Although the largest fluke may weigh up to 12 kg (26 lb), the average adult weighs 1 to 2.3 kg (2 to 5 lb) and measures 43 to 64 cm (17 to 25 in) long. Females may live up to 20 years and weigh more than 9 kg (20 lb), but males rarely exceed 7 years of age and 1.4 to 2.3 kg (3 to 5 lb) in weight. Table 39 shows mean lengths at age for female and male fluke from Delaware Bay.

The summer flounder, which depends upon sight to capture its food, feeds most actively during daylight hours. Juveniles and smaller adults feed mostly upon small shrimp and other crustaceans; adults eat a variety of fishes, including small winter flounder, menhaden, sand lances, red hakes, silversides, bluefish, weakfish, and mummichogs, as well as invertebrates such as blue crabs, squid, sand shrimp, opossum shrimp, and mollusks. Weakfish and winter flounder constituted the greatest volume of food eaten by fluke in two separate studies. Sand shrimp are a major food for both juveniles and adults. Adults are very active predators, often chasing schools of small fishes to the surface and leaping out of the water in pursuit of them. This behavior clearly distinguishes the summer flounder from the other, more sluggish species of inshore flatfishes.

Status and Management

Historically, the summer flounder has been among the most important commercial and recreational flatfishes on the East Coast. Approximately 90% of the commercial landings come from otter trawlers. Commercial catches in the southern part of the fluke's range were stable from the 1950s to the early 1970s, whereas those in the northern portion of its range persistently declined over the same time period. For example, the commercial catch in Massachusetts dropped from 5,240 mt (11.5 million lb) in 1961 to 41 mt (90,200 lb) in 1970. In 1974 it was estimated that total commercial and recreational harvests exceeded a level that could be sustained for any extended period of time. Despite this warning, total harvest has exceeded the 1974 level in the 1980s.

Recreational harvest typically peaks each year in July and drops off sharply before the end of August. Recreational fishing has always been a major component of the total fluke harvest, often exceeding commercial catches in the Middle Atlantic states. The recreational catch for the Georges Bank/Middle Atlantic region ranged from 37% to 65% of the total harvest from 1979 to 1986. Certain regions have historically generated tremendous recreational catches. One such region, the Great South Bay of Long Island, reported as many as 2 million flukes landed yearly during the late 1950s and early 1960s.

Although population levels in the 1980s have been somewhat higher than they were in the 1960s and 1970s, persistently high harvest levels may once more reduce this species' abundance. The Atlantic States Marine Fisheries Commission developed a Summer Flounder Management Plan for inshore fisheries that was adopted by coastal states from Massachusetts to North Carolina in 1982. A minimum legal size limit was established under this plan. In 1988, the Mid-Atlantic Fishery Management Council implemented a management plan for U.S. territorial waters that established minimum size limits, required commercial vessels and recreational boats (party and charter) harvesting more than 100 pounds to obtain an annual permit, and excluded foreign vessels from the fishery. All New England states managed summer flounder harvest in their territorial waters by means of minimum size limits in 1989.

Angling and Handling Tips

Fluke are well known for the aggressive way they grab bait and battle when hooked. A wide array of anglers, young and old, novice and pro, enjoy the challenge of catching, and the pleasure of eating, this left-handed species of flatfish. The fluke offers a particular challenge to the angler bold enough to use light tackle. Average-sized fluke, sometimes called flatties, weigh about 2 to 4 pounds, whereas the aptly named doormats (so-called due to their similarity in size to a welcome mat) weigh 8 or more pounds and provide memorable battles for anglers lucky enough to hook them.

Summer flounder start to move inshore in July, provide peak action in August, and continue to be available until waters begin cooling near the end of September. They can be found on sandy or muddy bottoms of protected inshore habitats, in the beach surf, near jetties and other structures, and in river channels; fluke are particularly abundant in fast-moving rips that gather debris and bait fishes.

Shoreline anglers use medium-weight spinning gear spooled with 12-pound test line; boat anglers fishing deeper water with strong currents need 15–20-pound test line on light to medium conventional gear to match the larger fish found there. Commercial rigs with spinners are used by many anglers. One favorite is a ⅜–1½-ounce weighted bucktail that can be baited with strips of fresh or frozen squid or bait fishes such as the sand lance. If these baits are not available, 4–5-inch strips cut from the tail of a fish such as the searobin can be used. Some anglers prefer strips from the belly area of a fluke or bluefish, or half of a snapper bluefish.

A few anglers make their own rigs by tying a 1–2-ounce sinker to the end of the line and a "dropper loop" or three-way swivel 4 inches above the sinker. A 3-foot leader with a 1/0 to 3/0 hook is attached to the loop or swivel. A 4–5-inch strip of squid split along half its length is attached to the hook along with a bait fish hooked through the lips. This rig is bounced along the bottom as the angler drifts or casts.

Anglers troll, chum, still-fish, and cast for fluke, but drifting bait along the bottom is the most popular method. When drifting, the bail of the reel should be open and the line held by the fingers. Once the line stops drifting and is tugged, allow it to run freely for a moment to let the fish get the bait in its mouth before the hook is set. Casting a baited red and white bucktail rig from boat or shore can also be a rewarding approach. The jig should be retrieved with a slow, pumping action. When a fluke grabs the rig, the rod tip should be lowered to slacken the line; when the line tightens again, the hook can be set.

Unlike other flounders that delicately mouth bait, the fluke literally attacks its food. It provides a better fight than the winter flounder and should be played with patience so that the hook won't tear loose from its mouth.

The white, flaky meat of the summer flounder is highly rated because of its delicate flavor and texture. This versatile fish provides delightful dining when steamed, poached, baked, broiled, sautéed, fried, or microwaved. Large doormats can be quarter-filleted for most recipes or cut into steaks and grilled over charcoal or gas.

Also try this traditional method of cooking fluke. Put flavored bread crumbs, some flour, and a little commercial seafood spice into a plastic bag, mixing well. Cut large fluke fillets along the center line to produce two triangular pieces. Drop these into the mixture and shake to coat. Next, melt several tablespoons of butter in a pan over medium heat and add the fillets. Grind some black pepper onto each fillet and sauté until the fillets are light golden brown, gently turning each one once. Don't overcook, as the fillets will lose their delicate texture and dry out. Place on warmed plates and spoon some of the pan drippings over each. Serve with a wedge of lemon. This method brings out the best flavor and texture of the summer sole.

References

Morse, W. W. 1981. Reproduction of the summer flounder, *Paralichthys dentatus* (L.). *J. Fish Biol.* 19:189–203.

Penttila, J. A., G. A. Nelson, and J. M. Burnett, III. 1989. *Guidelines for estimating lengths at age for 18 Northwest Atlantic finfish and shellfish species.* NOAA Tech. Mem. NMFS-F/NEC-66. 39 pp.

Smith, R. W., and F. C. Daiber. 1977. Biology of the summer flounder, *Paralichthys dentatus,* in Delaware Bay. *Fish. Bull.* 75:823–830.

Windowpane Flounder
(*Scophthalmus aquosus*)

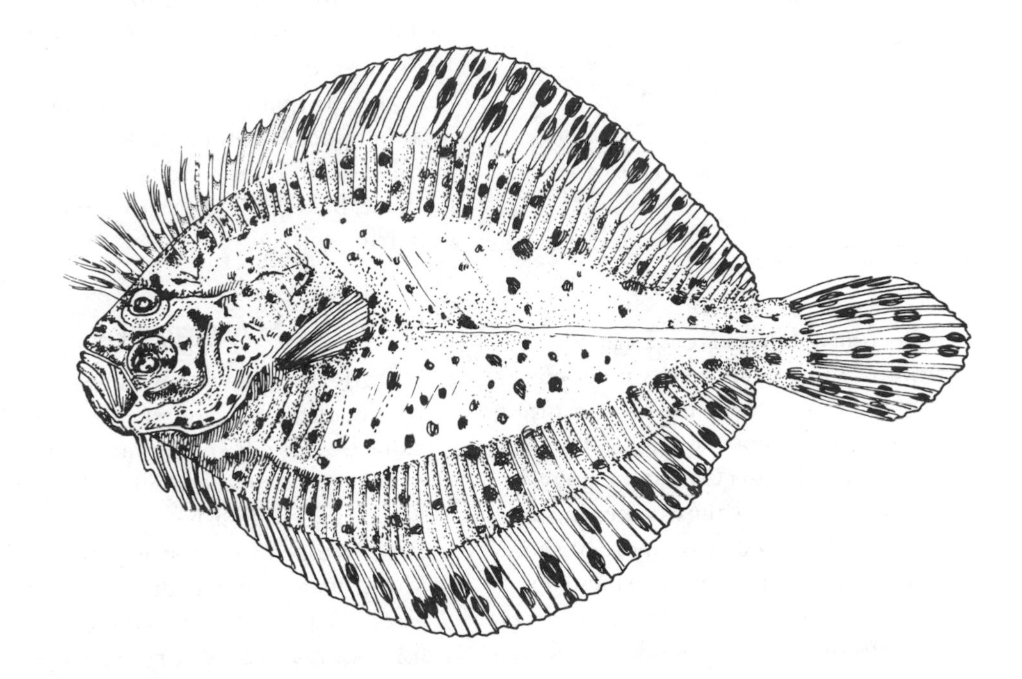

The windowpane flounder (also called the sand flounder) derives its name from the translucent appearance of its thin body when held up to a light. This small-bodied flatfish species occurs from the Gulf of St. Lawrence to South Carolina and is most abundant between Georges Bank and the Chesapeake Bay. It is found only in widely scattered areas north of Cape Cod.

Like other flatfishes, windowpane flounder are laterally flattened with both eyes on the upper side of the body. The windowpane is referred to as a left-handed species because both eyes are on the left side of the body.

The tail of the windowpane is rounded at its outer margin. The pelvic fins are broad-based and appear nearly continuous with the anal fin. The first several rays of the windowpane's dorsal fin are branched and unconnected by membrane, giving this fin a fringed, or frayed, appearance near the head. This characteristic separates the windowpane from all other Northwest Atlantic flatfishes.

The windowpane varies less in color than do most shallow-water flounders. Generally, it is pale brown or olive green and mottled with irregularly shaped small brown spots. The anal fin, dorsal fin, and tail match the body color. The blind side of the fish is white.

Biology

Windowpane flounder live in shallow water on the sandy bottom; hence, the name "sand flounder." They occur from the intertidal zone to depths of 46 m (151 ft) in the northern portion of their range, and up to 73 m (240 ft) in more southern areas. The windowpane is typically a year-round

Table 40. Length at age in cm (inches in parentheses) of the windowpane flounder from the Long Island Sound/Middle Atlantic Bight region

Age	*Length*
1	6.0 (2.4)
2	11.4 (4.5)
3	19.0–20.0 (7.5–7.9)
4	24.0–25.0 (9.4–9.8)
5	28.0 (11.0)
6	29.0–30.0 (11.4–11.8)
7	30.5–31.0 (12.0–12.2)

Source: Data from Scott and Scott 1988.

resident from southern New England northward. Individuals may wander considerable distances along the coast, causing frequent intermingling among local populations. South of Cape Cod, these flounder tend to aggregate in shallow water in summer and early fall and move more offshore in winter and spring. Windowpanes can tolerate temperatures from 2 to 21 degrees C (36 to 70 F).

Some males become sexually mature at age 2, and all individuals of both sexes are mature by 3 to 4 years of age. Reproduction peaks from spring to late summer north of Cape Cod and in October in southern New England; it occurs during much of the year in Middle Atlantic coastal areas. Spawning takes place in water less than 40 m (131 ft). Estuaries serve as critical spawning habitats north of the New York Bight but seem to be less important to the south.

The eggs are buoyant, floating near the surface for about 8 days before hatching. Larvae metamorphose into bottom-dwelling juveniles 1 to 2 months after hatching, a relatively rapid rate of growth compared to other flatfish species.

The maximum size of windowpane flounder is about 45.7 cm (18 in) in length and 0.9 kg (2 lb) in weight. Fish of 35.6 cm (14 in) weigh about 0.5 kg (1 lb). The young grow rapidly during their first several years of life, after which growth slows markedly. Females are usually slightly larger than males after the age of 3 years. Table 40 lists average lengths for 1- to 7-year-old windowpanes.

Windowpanes eat sand shrimp, mysids, amphipods, small mollusks, and small fishes such as the sand lance, tomcod, smelt, hake, pollock, striped bass, and herring.

Status and Management

Historically, the windowpane had little commercial, and modest recreational, value because its small, thin body does not produce as desirable a fillet as other, larger-bodied flatfishes. A commercial market developed for windowpanes during World War II, in part due to the dangers associated with fishing for more offshore fisheries resources during wartime. Commercial landings peaked at about 1,600 mt (3.5 million lb) in 1945. The fishery declined shortly after the end of the war.

The decline in abundance of other flatfish species, particularly the yellowtail flounder, prompted commercial vessels to land greater numbers of the windowpane beginning in the mid 1980s. New Bedford, Massachusetts, became a center of this activity. Landings increased to 2,000 mt (4.4 million lb) by 1978, much of this catch coming from Georges Bank. Although commercial and recreational harvests remained relatively modest throughout the 1980s, they did display annual increases. By the late 1980s, harvest levels had become substantial enough to cause concern about the ability of this species to withstand continued growth in fishing pressure. Although fisheries managers remained alert to problems that future growth in the fishery might cause, harvest of this species remained unregulated in U.S. and New England state territorial waters through 1989.

Angling and Handling Tips

The windowpane has traditionally held only modest interest for anglers due to the difficulty of removing fillets from its thin body. In recent years, anglers have become increasingly interested in this flavorful species because it continues to be available while other, larger inshore flatfish species become scarce.

Anglers catch windowpanes using gear and methods similar to those used for other flatfishes, such as the summer flounder. Many anglers prefer light tackle, such as a medium-action spinning rod spooled with 10–15-pound test line. Heavy line is necessary because of the likelihood of occasionally hooking other, somewhat larger species when fishing for the windowpane. No. 1 hooks can be tied to the line 12 to 18 inches below a sinker. A variety of baits can be used, such as sandworms, bloodworms, pieces of clam, or even small strips of squid. However, fish strips or sand eels, retrieved along the bottom, may provide the greatest angling success. Hooks should be checked frequently in order to keep bait fresh.

Some anglers fillet this species; others do not, because they don't want to waste the meat near the frame bones. They simply scale, gut, and remove the head from the fish. Whole, cleaned windowpanes are typically floured and pan-fried.

Filleting the windowpane may be a greater challenge than catching it. Use a thin-bladed knife to fillet this species. Cut behind the head and gill covers and slice down the lateral line to the tail. A raised ridge of bones running from the head to the tail along the middorsal portion of the body naturally divides the fillet on either side into upper and lower halves. To remove these fillets, position the fish with its tail toward you. Cut along the frame, removing the fillet from the midline bones to the dorsal fin; use the same method to remove the lower fillet from the same side. Turn the fish over and repeat. Removing these fillets so that little meat is wasted requires experience. With practice, each fish will provide four thin but tasty pieces of meat.

Windowpane fillets have a mild, sweet flavor and are excellent when prepared using any flatfish recipe. For an easily and quickly prepared meal, fry or broil this species. To fry, dredge windowpane fillets in flour. Sauté them in butter or margarine until lightly browned, turning each fillet once during cooking. Season to taste and sprinkle with parsley and lemon juice. Broiled windowpane fillets also make a delightful meal. Quick-broil under high heat for several minutes, basting with butter. Add parsley and serve with a quarter wedge of lime.

Reference

Scott, W. B., and M. G. Scott. 1988. Atlantic fishes of Canada. *Can. Bull. Fish. Aquat. Sci.* 219.

Northern Puffer

(*Sphoeroides maculatus*)

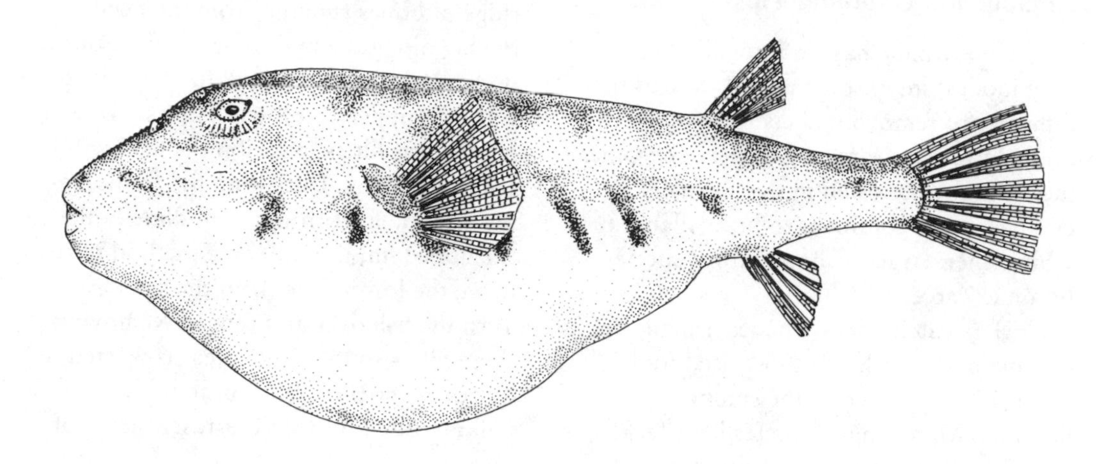

The northern puffer, also called the blowfish or balloonfish, occurs along the Atlantic coast of North America from Newfoundland to Florida, although it is uncommon north of Cape Cod.

The most unique characteristic of the puffer is its ability to inflate with air or water so extremely when disturbed that its shape becomes nearly spherical. When not inflated, it is a somewhat slender fish, tapering from the broadest point of the body at the pectoral fins to a slender snout and caudal peduncle. It has a very small, toothless mouth with extremely sharp-edged upper and lower jaws, which form a continuous cutting surface. The small gill openings are positioned directly in front of the pectoral fins. The dorsal and anal fins are very narrow at the base; this species lacks pelvic fins. The body is scaleless, but there are numerous small prickles on the head and much of the body.

The northern puffer is dark olive dorsally, fading to a greenish yellow or sometimes orange laterally, and white on the belly. It has six to eight dark bands or blotches on its sides.

Biology

The northern puffer is an inshore species most commonly occurring from the tide line to depths of perhaps 6.1 m (about 20 ft), and occasionally in brackish waters of estuaries. It is seldom found more than 1 or 2 miles from shore. This species is largely absent from the southern New England coast during the winter, when it probably moves offshore to deeper waters.

Puffers spawn from June to nearly the end of summer in southern New England waters. Both sexes reach sexual maturity at age 1. The eggs are adhesive, sticking to the substrate over which they are spawned. Fecundity varies from about 40,000 eggs for smaller females to nearly 450,000 for larger ones. Hatching occurs from 3 to 5 days after fertilization at 20

Table 41. Mean length at age in cm (inches in parentheses) for northern puffers from the Chesapeake Bay region

Age	*Male*	*Female*
1	14.6 (5.7)	13.7 (5.4)
2	18.5 (7.3)	19.4 (7.6)
3	21.3 (8.4)	23.8 (9.4)
4	22.7 (8.9)	25.9 (10.2)

Source: Data from Laroche and Davis 1973.

degrees C (68 F). Larvae are about 2.5 mm (0.1 in) long at hatching.

Although maximum length is about 36 cm (14 in), most puffers are less than 25 cm (10 in) long. Mean lengths at age are presented in Table 41. Most puffers do not live beyond 4 years of age.

Young-of-the-year puffers eat copepods and crustacean and molluscan larvae. Adults feed upon a variety of invertebrates, including crabs, shrimp, isopods, amphipods, mollusks, worms, barnacles, and sea urchins.

Status and Management

The northern puffer historically supported a commercial fishery that was most active in the Middle Atlantic region. Harvests from the Chesapeake Bay region varied from 455 to 5,455 mt (1 million to 12 million lb) per year between 1962 and 1970. After this time, landings declined precipitously. Southern New England has produced much smaller commercial and recreational landings than has the Middle Atlantic region. Due to this species' modest fishery, no New England state regulated its harvest through the 1980s.

Angling and Handling Tips

Blowfish is an excellent name for this little species, as many an angler will attest. When hooked, many puffers will inflate to a balloon shape while being retrieved. If you catch one that doesn't put on such a show, try rubbing its belly to stimulate this fright response.

Puffers can be caught with small pieces of a variety of baits, including worms, crabs, or cut fishes. They will also occasionally strike at small jigs or spinners. Equipment from a handline or cane pole to a light/medium-action spinning outfit spooled with light line can be used; the rod needs to be light enough to allow the angler to feel the puffer pecking away on the baited hook with its beaklike jaws.

Although the blowfish is not a game species, its flavor makes it a desirable catch for those willing to try it. Its historical market names of sea squab and chicken-of-the-sea suggest the comparison that is made between its white, firm, moist, and slightly sweet meat and poultry.

Unlike some of the Asiatic blowfishes such as the Japanese *fugu,* the northern puffer is not considered to be toxic; however, you shouldn't eat the skin or roe of this species. When handled properly, puffers are one of the easiest species to clean. When cleaning, hold the puffer by the head and press it down on a cutting board. Cut through the backbone behind the head, leaving the head attached. Next, grasp the head with one hand and the exposed end of the backbone with a pair of pliers. Pull the head and skin backward while holding the backbone firmly in the pliers; this is not unlike pulling a sock from your foot. In one sharp pull, the fish will be skinned and gutted as well. You will end up with a piece of meat that looks something like a chicken leg; hence the name "sea squab." The bones will not interfere with eating this delicate fish. With a little experience, you can clean sea squabs in less than half the time it takes to clean most other fishes.

Traditionally, sea squabs were breaded and deep-fried. Sautéing is another easy cooking method. To sauté, mix ¾ cup of flour, ¾ cup of flavored bread crumbs, salt, pepper, and a prepared spice mix such as Bell's Poultry Seasoning in a plastic bag. Put enough olive oil or butter into a hot heavy skillet or wok to cover the bottom. Large woks are ideal for cooking sea squabs, because the thicker parts of the "drumsticks" can be placed facing into the middle, deeper part of the pan. This allows the thicker end to cook faster, thus assuring an evenly cooked piece of meat. Sauté the fish pieces until lightly browned, about 5 minutes per side. Remove and keep warm in the oven until serving time. Serve heaped high on a large platter. Because sea squabs are ideal finger food for dipping, surround the platter with small bowls of different sauces (barbecue, soy, hot and spicy, and tartar).

Oven-frying is another excellent cooking method. Mix ¾ cup of flour, ¾ cup of flavored bread crumbs, and ¼ cup of grated hard cheese in a plastic bag to serve as a coating. Dip drumsticks into a mixture of one beaten egg and 2 tablespoons of milk, drain, and place in the coating mixture, shaking the bag to coat the drumsticks. Cover a broiling pan, large cookie sheet, or pizza pan with foil and brush the foil with a heavy layer of olive oil. Roll each drumstick on the oiled foil to coat lightly. Grind some black pepper on top and place in an oven preheated to 450 degrees F. Cook 10 minutes and serve hot on the cooking pan.

References

Laroche, J. L., and W. J. Davis. 1973. Age, growth, and reproduction of the northern puffer, *Sphoeroides maculatus*. *Fish. Bull.* 71:955–963.

Merriner, J. V., and J. L. Laroche. 1977. Fecundity of the northern puffer, *Sphoeroides maculatus*, from Chesapeake Bay. *Chesapeake Sci.* 18:81–83.

References

Acheson, J. M. 1980. Attitudes towards limited entry among finfishermen in northern New England. *Fisheries* 5(6):20–25.

Alm, G. 1959. *The connection between maturity, size and age in fishes*. Inst. Freshw. Res., Drottingholm Rep. 40:145 pp.

Almeida, F. P. 1987. Stock definition of silver hake in the New England–Middle Atlantic area. *N. Am. J. Fish. Manage.* 7:169–186.

Alperin, I. M. 1987. Management of migratory Atlantic coast striped bass: a historical perspective, 1938–1986. *Fisheries* 12(6):2–3.

Anthony, V. C., and M. J. Fogarty. 1985. Environmental effects on recruitment, growth and vulnerability of Atlantic herring (*Clupea harengus harengus*) in the Gulf of Maine region. *Can. J. Fish. Aquat. Sci.* 42:158–173.

Anthony, V. C., and D. J. Garrod. 1986. *Attempts at effort regulation in the North Atlantic—a review of problems*. Coop. Res. Report No. 139:26–41. Fifth Dialogue Meeting, ICES, Copenhagen.

Anthony, V. C., and S. A. Murawski. 1986. *Managing multi-species fisheries with catch quota regulations—the ICNAF experience*. Coop. Res. Report No. 139:42–57. Fifth Dialogue Meeting, ICES, Copenhagen.

Arnett, G. R. 1983. Introduction to the session—user group demands on the environment, ppg. 11–13. In J. W. Reintjes (ed.), *Improving multiple use of coastal and marine resources*. Am. Fish. Soc., Bethesda MD.

Backiel, T., and E. D. LeCren. 1967. Some density relationships for fish population parameters, ppg. 261–293. In S. Gerking (ed.), *The biological basis of freshwater fish production*. Wiley, New York.

Bailey, K. M. 1981. Larval transport and recruitment of Pacific hake *Merluccius productus*. *Mar. Ecol. Prog. Ser.* 6:1–9.

Baird, S. F. 1873. *Report on the condition of the sea fisheries of the south coast of New England in 1871 and 1872*. Part 1. U.S. Government Printing Office, Washington DC.

Baird, S. F. 1874. *Report of the U.S. Commissioner of Fisheries for 1872 and 1873*. U.S. Government Printing Office, Washington DC.

Balon, E. K. 1975. Reproductive guilds of fishes: a proposal and definition. *J. Fish. Res. Board Can.* 32:821–864.

Barsukov, V. V. 1972. *The wolffish (Anarhichadidae)*. Indian National Scientific Documentation Centre, New Delhi.

Baylis, J. R. 1981. The evolution of parental care in fishes, with reference to Darwin's rule of male sexual selection. *Environ. Biol. Fishes* 6:223–251.

Bell, F. T. 1940. *Report of the U.S. Commissioner of Fisheries for the fiscal year 1938*. U.S. Government Printing Office, Washington DC.

Berst, A. H., and G. R. Spangler. 1972. Lake Huron: effects of exploitation, introduction, and eutrophication on the salmonid community. *J. Fish. Res. Board Can.* 29:877–887.

Beverton, R. J. H., and S. J. Holt. 1957. *On the dynamics of exploited fish populations*. Fisheries Investigations Ministry of Agriculture,

Fisheries and Food (Great Britain) Series II 19.

Bigelow, H. B., and W. C. Schroeder. 1953. *Fishes of the Gulf of Maine*. U.S. Fish Wildl. Serv. Fish. Bull. 74, Vol. 53. U.S. Printing Office.

Bowen, J. T. 1970. A history of fish culture as related to the development of fishery programs, ppg. 71–93. In N. G. Benson (ed.), *A century of fisheries in North America*. Am. Fish. Soc. Spec. Publ. 7.

Bowers, G. M. 1901. *Report of the U.S. Commissioner of Fisheries for the year ending June 30, 1900*. U.S. Government Printing Office, Washington DC.

Bowers, G. M. 1911. *Report of the U.S. Commissioner of Fisheries for the Fiscal Year 1910*. U.S. Government Printing Office, Washington DC.

Bowman, R., J. Warzocha, and T. Morris. 1984. *Trophic relationships between Atlantic mackerel and American sand lance*. MARMAP Contribution MED/NEFC 84–19. 18 pp.

Buckley, J. 1989. *Species profiles: life histories and environmental requirements of coastal fishes and invertebrates (North Atlantic)—winter flounder*. U.S. Fish Wildl. Serv. Biol. Rep. 82(11.87). U.S. Army Corps of Engineers, TR EL-82-4. 12 pp.

Budgen, G. L., B. T. Hargrave, M. M. Sinclair, C. L. Tang, J. C. Therriault, and P. A. Yeats. 1982. *Freshwater runoff effects in the marine environment: the Gulf of St. Lawrence example*. Can. Tech. Rep. Fish. Aquat. Sci. No. 1078:1–71.

Bull, J. J., and R. Shine. 1979. Iteroparous animals that skip opportunities for reproduction. *Am. Nat.* 114:296–303.

Butman, B., and R. C. Beardsley. 1987. Physical oceanography, ppg. 88–98. In R. H. Backus and D. W. Bourne (eds.), *Georges Bank*. MIT Press, Cambridge MA.

Campbell, W. J. 1989. Too long at sea for too few fish. Reprint of 3-day series, published July 9–11, in the *Hartford Courant*.

Carlander, K. D., and P. M. Payne. 1977. Year-class abundance, population, and production of walleye (*Stizostedion vitreum vitreum*) in Clear Lake, Iowa, 1948–1974, with varied fry stocking rates. *J. Fish. Res. Board Can.* 34:1792–1799.

Carlander, K. D., and R. R. Whitney. 1961. Age and growth of walleye in Clear Lake, Iowa, 1935–57. *Trans. Am. Fish. Soc.* 90:130–138.

Chapman, D. W. 1988. Critical review of variables used to define effects of fines of redds in large salmonids. *Trans. Am. Fish. Soc.* 117:1–22.

Charnov, E. L., and W. M. Schaffer. 1973. Life history consequences of natural selection: Cole's result revisited. *Am. Nat.* 107:791–793.

Clarke, T. A. 1970. Territorial behavior and population dynamics of a pomacentrid fish, the garibaldi, *Hypsypops rubicunda*. *Ecol. Monogr.* 40:180–212.

Cohen, E. B., and M. D. Grosslein. 1987. Production on Georges Bank compared with other shelf ecosystems, ppg. 383–391. In R. H. Backus and D. W. Bourne (eds.), *Georges Bank*. MIT Press, Cambridge MA.

Conover, D. O. 1985. Field and laboratory assessment of patterns in fecundity of a multiple spawning fish: the Atlantic silverside *Menidia menidia*. *Fish. Bull.* 83:331–341.

Conover, D. O., and B. E. Kynard. 1984. Field and laboratory observations of spawning periodicity and behavior of a northern population of the Atlantic silverside, *Menidia menidia* (Pisces Atherinidae). *Environ. Biol. Fishes* 11:161–171.

Crecco, V. A., and T. F. Savoy. 1984. Effects of fluctuations in hydrographic conditions on year-class strength of American shad (*Alosa sapidissima*) in the Connecticut River. *Can. J. Fish. Aquat. Sci.* 41:1216–1223.

Crecco, V. A., T. Savoy, and L. Gunn. 1983. Daily mortality rates of larval and juvenile American shad (*Alosa sapidissima*) in the Connecticut River with changes in year-class strength. *Can. J. Fish. Aquat. Sci.* 40:1719–1728.

Dadswell, M. J. 1979. Biology and population characteristics of the shortnose sturgeon, *Acipenser brevirostrum* LeSueur 1818 (Osteichthyes: Acipenseridae), in the Saint John River estuary, New Brunswick, Canada. *Can. J. Zool.* 57:2186–2210.

Dawson, C. P., and B. T. Wilkins. 1981. Motivations of New York and Virginia marine boat anglers and their preferences for potential fishing constraints. *N. Am. J. Fish. Manage.* 1:151–158.

Deegan, L. A., and J. W. Day, Jr. 1984. Estuarine fishery habitat requirements, ppg. 315–336. In B. J. Copeland (ed.), *Research for managing the nation's estuaries: Proceedings of the Conference*. UNC Sea Grant College Publ. UNC-SG-84-08.

Dey, W. P. 1981. Mortality and growth of young-of-the-year striped bass in the Hudson River estuary. *Trans. Am. Fish. Soc.* 110:151–157.

Everhart, W. H., A. E. Eipper, and W. D. Youngs. 1975. *Principles of fisheries science*. Cornell Univ. Press, Ithaca NY.

Finkelstein, S. L. 1969. Age at maturity of scup from New York waters. *N.Y. Fish Game J.* 16:224–237.

Finkelstein, S. L. 1971. Migration, rate of exploitation and mortality of scup from inshore waters of eastern Long Island. *N.Y. Fish Game J.* 18:97–111.

Fishelson, L. 1970. Protogynous sex reversal in the fish *Anthias squamipinnis* (Teleostei, Anthiidae) regulated by presence or absence of male fish. *Nature* 227:90–91.

Forney, J. L. 1977. Evidence of inter- and intraspecific competition as factors regulating walleye (*Stizostedion vitreum vitreum*) biomass in Oneida Lake, New York. *J. Fish. Res. Board Can.* 34:1812–1820.

Fortier, L., and W. C. Leggett. 1982. Fickian transport and the dispersal of fish larvae in estuaries. *Can. J. Fish. Aquat. Sci.* 39:1150–1163.

Frank, K. T., and W. C. Leggett. 1981. Wind regulation and emergence times and early larval survival in capelin (*Mallotus villosus*). *Can. J. Fish. Aquat. Sci.* 38:215–223.

Gadgil, M., and W. Bossert. 1970. Life history consequences of natural selection. *Am. Nat.* 104:1–24.

Glebe, B. D., and W. C. Leggett. 1981. Latitudinal differences in energy allocation and use during the freshwater migrations of American shad (*Alosa sapidissima*) and their life history consequences. *Can. J. Fish. Aquat. Sci.* 38:806–820.

Graham, H. W. 1970. Management of the groundfish fisheries of the northwest Atlantic, ppg. 249–261. In N. G. Benson (ed.), *A century of fisheries in North America*. Am. Fish. Soc. Spec. Publ. 7.

Gross, M. R., and E. L. Charnov. 1980. Alternative male life histories in bluegill sunfish. *Proc. Nat. Acad. Sci. USA* 77:6937–6940.

Gross, M. R., and R. Shine. 1981. Parental care and mode of fertilization in ectothermic vertebrates. *Evolution* 35:775–793.

Grosslein, M. D., R. W. Langton, and M. P. Sissewine. 1980. Recent fluctuations in pelagic fish stocks of the Northwest Atlantic, Georges Bank region, in relation to species interactions. *Rapp. P.-v. Reun. Cons. Int. Explor. Mer.* 177:374–404.

Hachey, H. B., F. Hermann, and W. B. Bailey. 1954. *The waters of the ICNAF Convention area*. Sci. Pap. ICNAF Ann. Mtg., ppg. 68–102.

Hart, J. L. 1973. *Pacific fishes of Canada*. Bull. Fish. Res. Board Can. 180:740 pp.

Hartley, P. H. T. 1948. Food and feeding relationships in a community of fresh-water fishes. *J. Anim. Ecol.* 17:1–14.

Harville, J. P. 1983. Cooperation among users-fisheries, ppg. 45–57. In J. W. Reintjes (ed.), *Improving multiple use of coastal and marine resources*. Am. Fish. Soc., Bethesda MD.

Hennemuth, R. C., and S. Rockwell. 1987. History of fisheries conservation and management, ppg. 430–446. In R. H. Backus and D. W. Bourne (eds.), *Georges Bank*. MIT Press, Cambridge MA.

Howell, P., D. Simpson, and G. Matlezos. 1984. Effects of a length limit on the recreational catch of scup (*Stenotomus chrysops*) in Connecticut. *Trans. Northeast Fish. Wildl. Conf.* 41:239.

Hunt, R. L. 1975. Angling regulations in relation to wild trout management, ppg. 66–74. In W. King (ed.), *Wild trout management*. Trout Unlimited, Denver CO.

Hunter, J. R. 1972. Swimming and feeding behaviour of larval anchovy, *Engraulis mordax*. *Fish. Bull.* 70:821–838.

Iselin, C. 1955. *Coastal currents and the fisheries*. Pap. Mar. Biol. Ocean, Deep-Sea Res. Suppl. to Vol. 3.

Jackson, H. W., and R. E. Tiller. 1952. Preliminary observations on spawning potential in the striped bass (*Roccus saxatilis* Walbaum). *Ches. Biol. Lab. Publ.* 93:1–16.

Jenkins, R. M. 1970. Reservoir fish management, ppg. 173–182. In N. G. Benson (ed.), *A century of fisheries in North America* Am. Fish. Soc. Spec. Publ. 7.

Keast, A. 1965. *Resource subdivision amongst cohabiting fish species in a bay, Lake Opini-*

con, Ontario. Proc. 8th Conf. Gt. Lakes Res., Univ. Michigan. Ppg. 106–132.

Keast, A. 1966. *Trophic interrelationships in the fish fauna of a small stream.* Proc. 9th Conf. Gt. Lakes Res., Univ. Michigan. Ppg. 51–79.

Keast, A., and D. Webb. 1966. Mouth and body form relative to feeding ecology in the fish fauna of a small lake, Lake Opinicon, Ontario. *J. Fish. Res. Board Can.* 23:1845–1874.

Keats, D. W., G. R. South, and D. H. Steele. 1985. Reproduction and egg guarding by Atlantic wolffish (*Anarhichas lupus:* Anarhichidae) and ocean pout (*Macrozoarces americanus:* Zoarcidae) in Newfoundland waters. *Can. J. Zool.* 63:2565–2568.

Lagler, K. F., J. E. Bardach, R. R. Miller, and D. R. May Passino. 1977. *Ichthyology.* Wiley, New York.

Lambert, T. C. 1984. Larval cohort succession in herring (*Clupea harengus*) and capelin (*Mallotus villosus*). *Can. J. Fish. Aquat. Sci.* 41:1552–1564.

Larkin, P. A. 1956. Interspecific competition and population control in fresh-water fish. *J. Fish. Res. Board Can.* 13:327–342.

Lasker, R., H. M. Feder, G. H. Theilacker, and R. C. May. 1970. Feeding, growth and survival of *Engraulis mordax* larvae reared in the laboratory. *Mar. Biol.* 5:345–353.

Laurence, G. C. 1974. Growth and survival of haddock, *Melanogrammus aeglefinus,* larvae in relation to plankton prey concentration. *J. Fish. Res. Board Can.* 31:1415–1419.

Leggett, W. C., and J. E. Carscadden. 1978. Latitudinal variation in reproductive characteristics of American shad (*Alosa sapidissima*): evidence for population specific life history strategies in fish. *J. Fish. Res. Board Can.* 35:1469–1478.

Leggett, W. C., K. T. Frank, and J. E. Carscadden. 1984. Meteorological and hydrographic regulation of year-class strength in capelin (*Mallotus villosus*). *Can. J. Fish. Aquat. Sci.* 41:1193–1201.

Loiselle, P. V. 1978. Prevalence of male brood care in teleosts. *Nature* 276:98.

Magnusson, K. G., and O. K. Palsson. 1989. On the trophic ecological relationships of Iceland cod. *Rapp. P.-v. Reun. Cons. Int. Explor. Mer.* 188:206–224.

May, R. C. 1974. Larval mortality in marine fishes and the critical period concept, ppg. 3–19. In J. H. S. Blaxter (ed.), *The early life history of fishes.* Springer-Verlag, Berlin.

McHugh, J. L. 1970. Trends in fisheries research, ppg. 25–56. In N. G. Benson (ed.), *A century of fisheries in North America.* Am. Fish. Soc. Spec. Publ. 7.

Meador, K. L., and A. W. Green. 1986. Effects of a minimum size limit on spotted seatrout recreational harvest. *N. Am. J. Fish. Manage.* 6:509–518.

Miller, A. W. [Sparse Grey Hackle]. 1971. Fishless days, angling nights. Crown, New York.

Mockek, A. D. 1973. Spawning behavior of the lumpsucker *Cyclopterus lumpus* (L.). *J. Ichthyol.* 13:615–619.

Moller, H. 1984. Reduction of a larval herring population by jellyfish predator. *Science* 224:621–622.

Moody, J., and B. Butman. 1980. *Semidiurnal bottom pressure and tidal currents on Georges Bank and in the Mid-Atlantic Bight.* U.S. Geol. Survey Open-File Rep. 80-1137. 22 pp.

Murawski, S. A., and J. S. Idoine. In press. *Multispecies size composition: a conservative property of exploited fish systems?* NAFO SCR Doc 89.

Murphy, G. J. 1968. Patterns in life history and the environment. *Am. Nat.* 102:390–404.

Murphy, G. I. 1977. Clupeiods, ppg. 283–308. In J. A. Gulland (ed.), *Fish population dynamics.* Wiley, London.

Nelson, W. R., M. C. Ingham, and W. E. Schaaf. 1977. Larval transport and year class strength of Atlantic menhaden, *Brevoortia tyrannus. Fish. Bull.* 75:23–42.

Nikolsky, G. V. 1963. *The ecology of fishes.* Academic Press, New York.

Nixon, S. W. 1982. *The ecology of New England high salt marshes: a community profile.* U.S. Fish Wildl. Serv., Office Biol. Serv., Washington DC. FWS/OBS-81/55. 70 pp.

Norcross, J. J., S. L. Richardson, W. H. Massman, and E. B. Joseph. 1977. Development of young bluefish (*Pomatomus saltatrix*) and distribution of eggs and young in Virginian coastal waters. *Trans. Am. Fish. Soc.* 103:477–497.

Odum, E. P. 1980. The status of three ecosystem-level hypotheses regarding salt marsh estuaries: tidal subsidy, outwelling, and detritus-based food chains, ppg. 485–495. In V. S. Kennedy (ed.), *Estuarine perspectives.* Academic Press, New York.

O'Reilly, J. E., C. Evans-Zetlin, and D. A. Busch. 1987. Primary production, ppg. 220–233. In R. H. Backus and D. W. Bourne (eds.), *Georges Bank*. MIT Press, Cambridge MA.

Payne, P. M., J. R. Nicolas, L. O'Brien, and K. D. Powers. 1986. The distribution of the humpback whale, *Megaptera novaeangliae,* on Georges Bank and the Gulf of Maine in relation to densities of the sand eel, *Ammodytes americanus. Fish. Bull.* 84:271–277.

Pearcy, W. 1962. *Ecology of an estuarine population of winter flounder,* Pseudopleuronectes americanus (*Walbaum*). Bull. Bingham Oceanogr. Collect. Yale Univ. 18:1–78.

Peden, A. E., and C. A. Corbett. 1973. Commensalism between a liparid fish, *Careproctus* sp., and the lithodid box crab, *Lopholithodes foraminatus. Can. J. Zool.* 51:555–556.

Perrone, M., Jr., and T. M. Zaret. 1979. Parental care patterns of fishes. *Am. Nat.* 113:351–361.

Pitcher, T. J., and P. J. B. Hart. 1982. *Fisheries ecology.* Avi, Westport CT.

Pratt, H. J., Jr. 1979. Reproduction in the blue shark, *Prionace glauca. Fish. Bull.* 77:445–470.

Quinn, S. P., and M. R. Ross. 1985. Nonannual spawning in the white sucker, *Catostomus commersoni. Copeia* 1985:613–618.

Reid, G. K., and R. D. Wood. 1976. *Ecology of inland waters and estuaries.* Van Nostrand, New York.

Robertson, D. R. 1972. Social control of sex reversal in a coral reef fish. *Science* 1977: 1007–1009.

Roff, D. A. 1981. Reproductive uncertainty and the evolution of iteroparity: why don't flatfish put all their eggs in one basket? *Can. J. Fish. Aquat. Sci.* 38:968–977.

Roff, D. A. 1982. Reproductive strategies in flatfish: a first synthesis. *Can. J. Fish. Aquat. Sci.* 39:1686–1698.

Ross, M. R. 1977. Aggression as a social mechanism in the creek chub (*Semotilus atromaculatus*). *Copeia* 1977:393–397.

Ross, M. R. 1983. The frequency of nest construction and satellite male behavior in the fallfish minnow. *Environ. Biol. Fishes* 9:65–70.

Ross, M. R., and F. P. Almeida. 1986. Density-dependent growth of silver hakes. *Trans. Am. Fish. Soc.* 115:548–554.

Ross, M. R., and R. J. Reed. 1978. The reproductive behavior of the fallfish *Semotilus corporalis. Copeia* 1978:215–221.

Roussou, G. 1957. Some considerations concerning sturgeon spawning periodicity. *J. Fish. Res. Board Can.* 14:553–572.

Royce, W. F. 1987. *Fishery development.* Academic Press, New York.

Royce, W. F. 1989. A history of marine fishery management. *Rev. Aquatic Sci.* 1:27–44.

Schaffer, W. M. 1974. Optimal reproductive effort in fluctuating environments. *Am. Nat.* 108:783–790.

Schaffer, W. M., and P. F. Elson. 1975. The adaptive significance of variations in life history among local populations of Atlantic salmon in North America. *Ecology* 56:577–590.

Schopf, T. J. M., and J. B. Colton, Jr. 1966. Bottom temperature and faunal provinces: continental shelf from Hudson canyon to Nova Scotia. *Biol. Bull.* 131:406.

Scogin, W. M., Jr. 1983. The licensing of marine recreational fishermen. *N. Am. J. Fish. Manage.* 3:276–282.

Scott, W. B., and M. G. Scott. 1988. Atlantic fishes of Canada. *Can. Bull. Fish. Aquat. Sci.* 219.

Sherman, K. 1981. Zooplankton of Georges Bank and adjacent waters in relation to fisheries ecosystems studies. In UNH 173: *3rd informal workshop on oceanography of the Gulf of Maine and adjacent seas.* Univ. of New Hampshire, Durham.

Sherman, K., C. Jones, L. Sullivan, W. Smith, P. Berrien, and L. Ejsymont. 1981. Congruent shifts in sand eel abundance in western and eastern North Atlantic ecosystems. *Nature* 291:486–489.

Skud, B. E. 1982. Dominance in fishes: the relation between environment and abundance. *Science* 216:144–149.

Smith, C. L. 1967. Contribution to a theory of hermaphroditism. *J. Theoret. Biol.* 17:76–90.

Smith, H. M. 1921. *Report of the U.S. Commissioner of Fisheries for the fiscal year ending June 30, 1920.* U.S. Government Printing Office, Washington DC.

Spencer, H. 1872. *The principles of psychology* (2nd ed., 2 vols.). Williams and Norgate, London.

Stearns, S. C. 1976. Life-history tactics: a review of the ideas. *Quart. Rev. Biol.* 51:3–47.
Stolte, L. 1981. *The forgotten salmon of the Merrimack.* U.S. Government Printing Office, Washington DC.
Sutter, F. C. 1980. *Reproductive biology of anadromous rainbow smelt,* Osmerus mordax, *in the Ipswich Bay area, Massachusetts.* Unpublished master's thesis, Univ. of Massachusetts, Amherst.
Swingle, H. S., and E. V. Smith. 1942. *Management of farm fish ponds.* Alabama Agric. Exper. Station, Auburn Univ. Bull. 254.
Taubert, B. D. 1980. Reproduction of the shortnose sturgeon (*Acipenser brevirostrum*) in Holyoke pool, Connecticut River, Massachusetts. *Copeia* 1980:114–117.
Teal, J. M. 1986. *The ecology of regularly flooded salt marshes of New England: a community profile.* U.S. Fish Wildl. Serv. Biol. Rep. 85(7.4). 61 pp.
Thayer, G. W., W. J. Kenworthy, and M. S. Fonseca. 1984. *The ecology of eelgrass meadows of the Atlantic coast: a community profile.* U.S. Fish Wildl. Serv., Biol. Serv. Program, Washington DC. FWS/OBS-81/01. 125 pp.
Uchapi, E., and J. A. Austin, Jr. 1987. Morphology, ppg. 25–30. In R. H. Backus and D. W. Bourne (eds.), *Georges Bank.* MIT Press, Cambridge MA.
Walters, C. J., M. Stocker, A. V. Tyler, and S. J. Westrheim. 1986. Interactions between Pacific cod (*Gadus macrocephalus*) and herring (*Clupea harengus pallasi*) in the Hecate Strait, British Columbia. *Can. J. Fish. Aquat. Sci.* 43:830–837.
Ware, D. M. 1985. Life history characteristics, reproductive value, and resilience of Pacific herring (*Clupea harengus pallasi*). *Can. J. Fish. Aquat. Sci.* 42 (Suppl. 1):127–137.
Warner, R. R. 1975. The adaptive significance of sequential hermaphroditism in animals. *Am. Nat.* 109:61–82.
Warner, R. R. 1984. Mating behavior and hermaphroditism in coral reef fishes. *Am. Sci.* 72:128–136.
Warner, W. W. 1983. *Distant water: The fate of the North Atlantic fisherman.* Little Brown, Boston MA.
Weatherly, A. H. 1972. *Growth and ecology of fish populations.* Academic Press, New York.
Welsh, B. L. 1980. Comparative nutrient dynamics of a marsh-mudflat ecosystem. *Estuarine Coastal Mar. Sci.* 10:143–164.
Werner, E. E., and D. J. Hall. 1976. Niche shifts in sunfishes: experimental evidence and significance. *Science* 191:404–406.
Werner, R. G., and J. H. S. Blaxter. 1980. Growth and survival of larval herring, *Clupea harengus,* in relation to prey density. *Can. J. Fish. Aquat. Sci.* 37:1063–1069.
Whitlatch, R. B. 1982. *The ecology of New England tidal flats: a community profile.* U.S. Fish Wildl. Serv., Biol. Serv. Program, Washington DC. FWS/OBS-81/01. 125 pp.
Williams, G. C. 1966. Natural selection, the cost of reproduction, and a refinement of Lack's principle. *Am. Nat.* 100:687–690.
Woodhead, A. D. 1979. Senescence in fishes, ppg. 179–205. In P. J. Miller (ed.), *Fish phenology: anabolic adaptiveness in teleosts.* Symp. Zool. Soc. Lond. 44.
Yentsch, C. S., and N. Garfield. 1981. Principal areas of vertical mixing in the waters of the Gulf of Maine, with reference to the total productivity of the area. In UNH 173: *3rd informal workshop on oceanography of the Gulf of Maine and adjacent seas.* Univ. of New Hampshire, Durham.
Zaret, T. M., and A. S. Rand. 1971. Competition in tropical stream fishes: support for the competitive exclusion principle. *Ecology* 52:336–342.

Glossary of Technical Terms

Anadromous Fishes that migrate into fresh water to spawn but spend much of the rest of their lives in salt water.

Brooding Carrying young after fertilization, through embryological development.

Competition The sharing of a resource, such as food, that is necessary for survival, the use of which by one individual, or species, potentially limits its availability to others.

Conservation Planned management of a natural resource to prevent its decline or destruction and to insure its availability in the future.

Continental shelf Underwater expanses of the coastlines of continents, with water depths less than 200 m.

Continental slope Area immediately seaward from the continental shelf, where water depths rapidly increase to those characteristic of the open ocean.

Deferred maturity Condition whereby one sex within a species typically reaches sexual maturity at a significantly older age than the other sex.

Demersal Occurring on the substrate rather than in the open water column well above the substrate.

Detritus Fine material suspended in flowing water that will settle to the substrate when flow ceases. Organic detritus comes from the decomposition of plant and animal matter.

Fecundity Number of viable eggs produced by a female. Fecundity is usually referenced in terms of either a specific reproductive season or the lifetime of the female.

Hermaphroditic Reproductive condition in which an individual will produce both viable eggs and sperm at some time during its life.

Intertidal Area of the shoreline that is inundated at high tide but above the water line at low tide.

Iteroparity Life cycle in which individuals of a population have some level of probability of living through more than one reproductive season.

Juvenile fish Stage of the life cycle after a larval fish metamorphoses into the body type and form characteristic of the adult stage but previous to sexual maturity.

Larval fish Stage of the life cycle after hatching but before the fish metamorphoses fully into the juvenile and adult body type and form.

Metabolism All of the internal chemical processes in plants or animals that cause the growth or breakdown of living tissues.

Metamorphosis Process whereby a larval fish changes to a juvenile/adult body form. In many fishes this process may be subtle; in others, such as flounders and eels, it may include major changes in basic body form.

Nursery grounds Habitats that provide all requirements for survival and growth of newly hatched larvae and young juvenile fishes.

Ovoviviparous Reproductive condition in

which a female incubates its young within the reproductive system but nourishes the young wholly through yolk that is deposited in the egg before fertilization occurs.

Pelagic Occurring in the open water column well above the substrate and away from shoreline or other shelter areas.

Plankton Small animals and plants that are suspended in the water column and are passively moved about by currents.

Primary production Production of plant cells and tissues through metabolic processes and growth.

Protandry Hermaphroditic condition in which an individual produces sperm and reproduces as a male when it first matures and produces eggs and reproduces as a female later in its life.

Protogyny Hermaphroditic condition in which an individual produces eggs and reproduces as a female when it first matures and produces sperm and reproduces as a male later in its life.

Recruitment Addition of new members to an aggregation. *Population recruitment* refers to all individuals that survive early life stages to become part of a population; *fishery recruitment* refers to all individuals that have grown to a size that makes them vulnerable to fishing gear being used to harvest the population.

Redd Gravel nest constructed by the female salmon and trout.

Salinity Weight in grams of all salts dissolved in 1 kg of seawater.

Semelparity Life cycle in which all individuals of a population reproduce during one season in their lifetime and then die.

Stock Group of fish that form a reproductive unit and are reproductively separated from other individuals of the same species, frequently due to geographic separation. Stock is essentially synonymous with the term "population."

Subtidal Below the water line along the shore at low tide.

Year class All fish within a population that were hatched, or born, in a specific year. When identifying year classes, reference is usually made to the calendar year in which the fish were hatched (e.g., the 1985 year class of Gulf of Maine silver hakes).

Zooplankton Portion of the plankton made up of various animal groups.

Index

Page numbers in bold face indicate species profiles